THE BIOLOGY OF BATS

THE BIOLOGY OF
BATS

GERHARD NEUWEILER

Translated by Ellen Covey

NEW YORK OXFORD
OXFORD UNIVERSITY PRESS
2000

Oxford University Press

Oxford New York
Athens Auckland Bangkok Bogotá Buenos Aires Calcutta
Cape Town Chennai Dar es Salaam Delhi Florence Hong Kong Istanbul
Karachi Kuala Lumpur Madrid Melbourne Mexico City Mumbai
Nairobi Paris São Paulo Singapore Taipei Tokyo Toronto Warsaw

and associated companies in
Berlin Ibadan

Originally published in 1993 by Georg Thieme Verlag, Rüdigerstraße 14,
D-7000 Stuttgart 30, Germany.
Original German title: *Biologie der Fledermäuse*

Published by Oxford University Press, Inc.
198 Madison Avenue, New York, New York 10016

Library of Congress Cataloging-in-Publication Data
Neuweiler, Gerhard, 1935–
[Biologie der Fledermäuse. English]
The biology of bats / by Gerhard Neuweiler;
translated by Ellen Covey.
p. cm.
Includes bibliographical references and index.
ISBN 978-0-19-509951-5
1. Bats. I. Title.
QL737.C5N4813 1998
599.4–dc21 97-26735

Printed in the United States of America

PREFACE

MY OWN INTEREST in bats was originally an interest in echolocation. I wanted to discover how the auditory brain of a mammal can represent the acoustic properties of the outside world so perfectly that a bat can pursue and capture tiny, erratically flying insects in total darkness, while at the same time avoiding bumping into bushes or hitting wires and lamp posts. As I continued to study the ears and auditory brain of these little creatures, and especially after I had the opportunity to closely observe their nightly acrobatic flight activities and their daytime life in sultry dungeons and caves, I realized that in order for these flying mammals to carry on their daily life, successfully raise offspring, and survive droughts, harsh winters, and the impact of human civilization, many different bodily functions have to be integrated and intricately coordinated. Therefore, I started to study not only audition, but also all of the other functions and aspects of a bat's life. All of the information that I collected was eventually integrated into lectures on the biology of bats. These lectures form the basis of this book. I freely admit that I have written this book not only for students, but also for my own education.

Over the last three decades, biology as a science has become highly differentiated into many disciplines, with new ones still springing up. Molecular methods and computational tools expand the realm of scientific analysis into the dynamics of molecules on the one hand, and population dynamics and multifactorial ecological interactions on the other. All of these highly specialized disciplines have their own methods and their own language. There is, therefore, not much crosstalk among disciplines. Knowledge about different aspects of life is left as unconnected as the scattered pieces of a puzzle and is only rarely integrated into a comprehensive view of organismic life. This book attempts to offer an antidote to reductionistic specialization and serves to remind the student of modern laboratory biology that the dynamics of ion channels, the affinities of receptors for molecules, and the effects of a specific gene only make sense when considered as a tiny contribution to the fully integrated functioning of a single, unique organism. If biologists lose sight of the integrity and individuality of each organism, they will, indeed, become "blind watchmakers."

Biology has two faces. The deep insight we have obtained into how living matter functions has put an understanding of the animate world at our fingertips. On the other hand, the more we learn about how living organisms function, the more we must admire and respect the variability and perfection of organismic forms which the history of life on earth has handed down to us.

Bats provide a good demonstration of the way in which knowledge promotes awareness of the necessity to preserve animals and their biotopes. Decades ago bats were feared and were considered to be repulsive nocturnal spirits who had to be kept out of houses by any means possible. They were even fumigated as if they were pests, and were "cleaned" from barns by the use of wood-preserving chemicals and other means. Since bats have become subjects of biological science, we have gained much insight into their magnificent auditory world, and we have learned about their extraordinary ability to precisely control body temperature. We now know that bats pollinate and disperse the seeds of many important tropical plants and therefore constitute indispensable links in both tropical and desert ecosystems. Knowledge has changed our attitude toward nature.

In ecology, bats have become a major focus of study. In many countries, bat protection clubs and associations have sprung up to provide housing for bats, to protect bat caves from tourism, and to promote the conservation of bats. In the United States, an international bat conservation association has been successfully established. This organization has effectively promoted the knowledge of bats and hence their protection on an international scale. In some industrial countries, the disastrous decrease of bat populations has been halted and even reversed.

Thus, bats will remain an important subject of study. An aim of this book is to inform the student of biology about what science has revealed about bats. The reader will soon realize that surprisingly small changes in the general mammalian structure have sufficed to create such bat-specific capabilities as flight, echolocation, and hibernation. Therefore, this book may also serve as an introduction to the general structure and function of mammals.

Biology of Bats was first published in 1993 in German. The translation into English gave me the opportunity to update the text and the literature citations and to eliminate a number of mistakes.

ACKNOWLEDGMENTS

I AM VERY grateful to Dr. Ellen Covey for translating this book. Ellen is not only a leading expert in bat audition but is also fluent in German. O. V. Helversen from Erlangen University was one of the colleagues who read the German version from cover to cover and helped me eliminate a number of errors. Finally, I thank Kirk Jensen from Oxford University Press for his cooperation and patience with the translation process and with me.

CONTENTS

Abbreviations ix
Introduction 3

1 | FUNCTIONAL ANATOMY AND LOCOMOTION 9
1.1 The Spine and Pelvis 10
1.2 The Wings 11
1.3 The Legs 21
1.4 Aerodynamics 24
1.5 Energetics of Flight 29
1.6 Flight Techniques 32
References 40

2 | THE CIRCULATORY AND RESPIRATORY SYSTEMS 43
2.1 The Heart 43
2.2 The Blood Vessels 47
2.3 Circulation in the Wing Membrane 49
2.4 The Blood 53
2.5 Respiration and Gas Exchange 55
References 60

3 | HEAT AND WATER BALANCE 63
3.1 Homeothermy 64
3.2 Torpor 68
3.3 Hibernation 70
3.4 Water Turnover 82
References 94

4 | DIET, DIGESTION, AND ENERGY BALANCE 98
4.1 The Teeth 98
4.2 Insectivorous and Carnivorous Bats 103
4.3 Vampire Bats 109
4.4 Flower and Fruit Feeders 112
References 115

5 | CENTRAL NERVOUS SYSTEM 117
5.1 Encephalization 117
5.2 Anatomy of the Central Nervous System 118
5.3 Aging and the Central Nervous System 137
References 137

6 | ECHOLOCATION 140
6.1 The Discovery and Basic Principles of Echolocation 140
6.2 Echolocation Calls and Their Production 144
6.3 The Auditory System 156
6.4 Echolocation Performance 178
6.5 Adaptation to Different Biotopes 192
References 206

7 | VISION, OLFACTION, AND TASTE 210
7.1 Vision 210
7.2 Olfaction 224
7.3 Taste 233
References 233

8 | REPRODUCTION AND DEVELOPMENT 236
8.1 Male Genital Organs 236
8.2 Female Genital Organs 237
8.3 Reproductive Cycles 241
8.4 Control of Rhythms 248
8.5 Reproductive Behavior and Rearing of the Young 250
References 259

9 | ECOLOGY 262
9.1 Geographical Distribution 262
9.2 Bat Migrations 264
9.3 Bats and Plants 267
9.4 Living Quarters 277
9.5 Bats and Human Activities 282
References 285

10 | PHYLOGENY AND SYSTEMATICS 287
10.1 Phylogeny 287
10.2 Systematics 288
References 299

Index 301

ABBREVIATIONS

The following abbreviations are used throughout the book:

ca., circa

bw, body weight

M., muscle

N., nerve

Nc., nucleus (a group of neurons in the central nervous system whose boundaries can be defined on the basis of histological criteria)

CF, constant frequency component of an echolocation signal

FM, frequency modulated component of an echolocation signal

For physical weights and measurements, the standard international units and abbreviations are used. For a short description of the different families of bats, see pp. 5–8.

THE BIOLOGY OF BATS

INTRODUCTION

THE PURPOSE OF this book is twofold: As the title implies, its primary aim is to summarize the large body of information about bats that the scientific community has amassed over the years, and thus to document what is known about this highly successful and fascinating order of mammals. In addition to this obvious purpose, however, a secondary aim is to inform the reader about the biology of mammals in general. In order to accomplish this, it is necessary to include some topics that are no longer areas of active research by scientists who study bats. These sections (e.g., the skeleton) are necessarily more descriptive in their character than are certain other topics (e.g., flight and echolocation) that have been the focus of intensive research over the past few decades. These latter topics are discussed in a more dynamic and less formal way.

Some chapters include boxes listing concepts and formulas that recur frequently in the text and that are important for understanding the topic under consideration. Abbreviations used in the text and figure legends are listed in the front matter.

Because this book was conceived as a textbook, references to the literature have been deliberately omitted from the text. Nevertheless, at the end of every chapter there is a list of references citing the most important work that served as the basis for the text. Publications that provide a particularly good introduction to a topic are marked with an asterisk (*). Books about bats that either deal with a broad range of topics or that represent collected papers published as symposium proceedings are listed at the end of this section.

Although there are close to a thousand species of bats, only a few of these have been the subject of experiments, and the names of these species will appear repeatedly throughout the book. To avoid any ambiguity, all species are designated by their scientific (Latin) names. In cases where the family to which a species belongs is not obvious from the genus and species designation, it is included in parentheses. For readers who do not work with bats, these names will have little meaning. For that reason, a summary of the taxonomy of bats is included below, in which the salient characteristics of each family are described.

For readers who prefer not to have to refer to the taxonomic listing, the following information should prove helpful.

WHAT ARE BATS?

All bats belong to the order Chiroptera. This order includes two major suborders, the Microchiroptera, or true bats, currently comprising 782 species, and the Megachiroptera, or flying foxes, comprising 175 species.

Flying foxes are found only in the tropical regions of Asia, Australia, and Africa. Flying foxes are relatively large, with body weights ranging from 100 g to 1000 g. They are vegetarians, feeding on fruit, nectar, and pollen. During the day, flying foxes live in colonies that hang in trees. During their nocturnal flights to feed on fruit- or flower-bearing trees, they depend exclusively on their large eyes for visual orientation. The only flying foxes that echolocate are those of the genus *Rousettus*, a species that lives in caves. *Rousettus* produce clicks with their tongues and use these echolocation signals to orient in complete darkness.

The true bats, or Microchiroptera, without exception possess a highly developed echolocation system; their echolocation signals are produced by the larynx. Thanks to the combination of flight and echolocation, most Microchiroptera have become skillful nocturnal hunters of insects. The group with the largest number of species and the widest distribution worldwide is the vespertilionids. In the chapter on echolocation, the Old World horseshoe bats (Rhinolophidae) and the Central and South American species *Pteronotus parnellii* (Mormoopidae) play a prominent role. These species possess a highly specialized echolocation system that allows them to recognize flying insects based on the pattern of their wingbeats. Some tropical species, especially those belonging to the large family Phyllostomidae, have taken advantage of food sources other than insects. Phyllostomids are found only in South and Central America. Many phyllostomid species are facultative or obligate frugivores and flower feeders, playing the same role in the neotropics that flying foxes do in the Old World. Therefore, many New World plants are dependent on bats for pollination. Some larger tropical Microchiroptera, the megadermatids for example, mainly forage close to the ground, feeding on small vertebrates such as frogs, lizards, birds, mice, and even other bats. The most unusual diet is that of vampire bats, which feed by licking blood. There are three species of true vampires, two of which are rare; all are restricted to Central and South America. As implied by their name, Microchiroptera are quite small, with body weights between 5 g and 20 g. A minority of Microchiroptera, including mainly carnivorous species and a few frugivorous and omnivorous phyllostomids, may weigh up to 100 g.

TAXONOMY OF BATS

The classification of bats used today is based on the system developed by Miller in 1907 (see figure 10.1).

Order Chiroptera—Bats
Characteristics: Front limbs adapted to function as wings; elongated fingers; keeled sternum; legs rotated so that the knees are oriented toward the back.

Suborder Megachiroptera—Flying Foxes
Characteristics: second finger clawed, humerus small.
Family **Pteropodidae** with 42 genera and 175 species.
Suborder Microchiroptera—True Bats
Characteristics: second finger clawless and tightly connected to third finger; large humerus: 144 genera and currently 782 species.
Superfamily Emballonuroidea
Family **Rhinopomatidae** (mouse-tailed bats): 1 genus and 3 species.
Family **Emballonuridae** (sheath-tailed bats): 13 genera and 51 species.
Family **Craseonycteridae** (bumblebee bat): 1 genus and 1 species.
Superfamily Rhinolophoidea
Family **Nycteridae** (slit-faced bats): 1 genus and 12 species.
Family **Megadermatidae** (false vampires): 4 genera and 5 species.
Family **Rhinolophidae** (horseshoe bats): 1 genus and 69 species.
Family **Hipposideridae** (Old World leaf-nosed bats): 9 genera and 60 species.
Superfamily Phyllostomoidea
Family **Mystacinidae** (short-tailed bat): 1 genus and 1 species.
Family **Noctilionidae** (bulldog bat and fisherman bat): 1 genus and 2 species.
Family **Mormoopidae** (mustached bats): 2 genera and 8 species.
Family **Phyllostomidae** (New World leaf-nosed bats): 51 genera and 147 species.
Noctilionidae, Mormoopidae, and Phyllostomidae are regarded by many taxonomists as the oldest families of bats.
Superfamily Vespertilionoidea
Family **Natalidae** (funnel-eared bats): 1 genus and 4 species.
Family **Furipteridae** (smoky bats): 2 genera and 2 species.
Family **Thyropteridae** (disc-winged bats): 1 genus and 2 species.
Family **Myzpodidae** (sucker-footed bats): 1 genus and 1 species.
Family **Vespertilionidae** (evening bats): 42 genera and 326 species.
Family **Molossidae** (free-tailed bats): 12 genera and 88 species.
Altogether there are 18 recent families (Mega- and Microchiroptera) with 186 genera and 957 species. However, these figures change continuously as new species are described.

BRIEF DESCRIPTIONS OF FAMILIES

PTEROPODIDAE—FLYING FOXES

Flying foxes of the Old World tropics; with the exception of the genus *Rousettus*, and possibly *Stenonycteris*, these animals do not echolocate. Large eyes and good nocturnal vision; diet consists of fruit and flowers; nonfunctional claw on second finger; genus *Pteropus* includes large species (> 1 kg body weight, wingspan up to 1.7 m).

RHINOPOMATIDAE—MOUSE-TAILED BATS

Insectivorous bats distributed in arid regions of North Africa and India; average sized; long tail is used to scan rocks for cracks into which the bat can retreat; rudimentary nose leaf; often enters summer torpor with fat reserves.

EMBALLONURIDAE—SHEATH-TAILED BATS

Average-sized insectivorous bats found worldwide throughout the tropics. Tail protrudes from interfemoral membrane to lie freely on the upper side of the membrane; frequently fast-flying, with many different foraging niches.

CRASEONYCTERIDAE—BUMBLEBEE BAT

One species only, first discovered in Thailand in 1973; very small; forearm length only 2.5 cm and body weight only 3–4 g; insectivorous; begins foraging early, at dusk.

NYCTERIDAE—SLIT-FACED BATS

Two long skin folds begin behind the nose and end in a deep, closeable pouch on the side of the forehead; large tail membrane; end of tail T-shaped; mainly feed on arthropods, even foraging close to the ground; large pinnae, which they use to detect rustling prey; found in Africa, two species in Southeast Asia.

MEGADERMATIDAE—FALSE VAMPIRES

Medium-sized predatory bats of tropical Africa, Asia, and Australia; nose leaf; large pinnae joined together at the midline; feed on arthropods and small vertebrates captured from the ground and water; detect prey based on rustling noises; broadband echolocation signals, very low intensity and short duration.

RHINOLOPHIDAE—HORSESHOE BATS

Nose leaf forms a complex horseshoe shape; feed on flying insects which they detect based on the pattern of their wingbeats, using a pure-tone echolocation signal that is well adapted for foraging in thick vegetation; pinnae lack tragus; distributed in tropical and temperate zones of the Old World.

HIPPOSIDERIDAE—OLD WORLD LEAF-NOSED BATS

Nose leaf similar to that of the Rhinolophidae but without a median fold (sella); pinnae lack tragus; echolocation system similar to that of Rhinolophidae, except that duration of pure-tone component is less than 10 ms and therefore less well adapted to foraging in dense vegetation; feed primarily on flying insects; distributed in tropical and subtropical regions of the Old World.

MYSTACINIDAE—SHORT-TAILED BAT

A species endemic to New Zealand; thick and leathery portions of the wing and tail membrane near the body are used to cover the folded wings; good runners; forage for arthropods on the ground and in flight; diet also includes fruit, pollen, nectar, small vertebrates, and carrion.

NOCTILIONIDAE—FISHERMAN BAT AND BULLDOG BAT

New World tropical bats; large claws on feet are used for fishing; mainly feed on arthropods and small fish from the surface of the water.

MORMOOPIDAE—MUSTACHED BATS

Tubelike oral cavity, presumably for directional beaming of the echolocation signal; wings high on body; good flyers that forage for insects at relatively high speed, close to obstacles; found in tropics of New World.

PHYLLOSTOMIDAE—NEW WORLD LEAF-NOSED BATS

Large family of New World tropical bats with leaf-shaped nose leaf; occupy a variety of foraging niches; feed on insects, small vertebrates, fruit, pollen, nectar, and, in the case of vampire bats, blood; echolocation signals generally short, broadband, and low intensity.

NATALIDAE—FUNNEL-EARED BATS

Insectivorous; tube-shaped pinnae; no nose leaf; forage for insects during slow, butterfly-like flight; found in tropics of New World.

FURIPTERIDAE—SMOKY BATS

Small neotropical bats, presumably insectivorous; very rare; thumb rudimentary and nonfunctional.

THYROPTERIDAE—DISC-WINGED BATS

Suction pad at base of thumb and foot allows these bats to cling head-up inside rolled leaves; insectivorous; neotropical.

MYZPODIDAE—SUCKER-FOOTED BATS

Also possess suction pads at base of thumb and foot; endemic to Madagascar; very rare; little known about lifestyle.

VESPERTILIONIDAE—EVENING BATS

Family with largest number of species; insectivorous; cosmopolitan, distributed worldwide, even to timberline; echolocate, but do not possess nose leaf.

MOLOSSIDAE—FREE-TAILED BATS

Tail extends well beyond small interfemoral membrane; powerful insectivores; most species fly high and fast, but can also run fast on the ground; most advanced family of bats based on wing morphology; distributed worldwide in the tropics and subtropics.

1

FUNCTIONAL ANATOMY AND LOCOMOTION

DURING THE COURSE of evolution, land animals have developed progressively more sophisticated means of locomotion. A primitive vertebrate such as a lizard has short, stiff legs that stick out on either side of its body. With every step, the whole animal sways from side to side so that its gait appears heavy and awkward. This form of locomotion is inefficient, and it is useful only for short distances. In contrast, in a highly evolved vertebrate such as a gazelle, the legs are flexible and in line with the body. The gazelle is able to move quickly and gracefully over even the roughest terrain so that it appears almost weightless.

This enormous improvement in speed and efficiency of locomotion among four-legged animals is due to three specialized features of the mammalian body (fig. 1.1):

1. The only link between the shoulder and the skeleton is through muscles. The advantage of this arrangement is that it allows the forelimb to move in practically any direction and provides a high degree of mobility that allows the animal to compensate for unevenness in the terrain even when running at high speeds.
2. During movement of the forelimb, the shoulder blade is moved along with the limb. Whenever the animal steps forward, the shoulder blade turns with the front leg. This means that in mammals, the pivot point of the forelimb lies high on the body, somewhere between the shoulder blade and the ribs rather than in the upper joint of the limb as it does in amphibians and reptiles. As a result of this high pivot point, leverage and stride length are greatly increased.
3. The limbs are placed directly beneath the body, and movement of the limbs is mainly in the vertical dimension. As a consequence, the entire length of the limb from shoulder to fingers acts as one arm of a lever, allowing the whole limb to be used for movement.

Bats have exploited these specialized features of the mammalian body to achieve an entirely different means of locomotion than that used by the gazelle and other land mammals. Through flight, bats have gained access to a niche—the air—that is normally inaccessible to mammals. Along with flight, bats have developed an unusual upside-down resting posture. Through a few surprisingly minor changes in the basic structure of the mammalian body, bats have achieved a status comparable to birds as flying vertebrates.

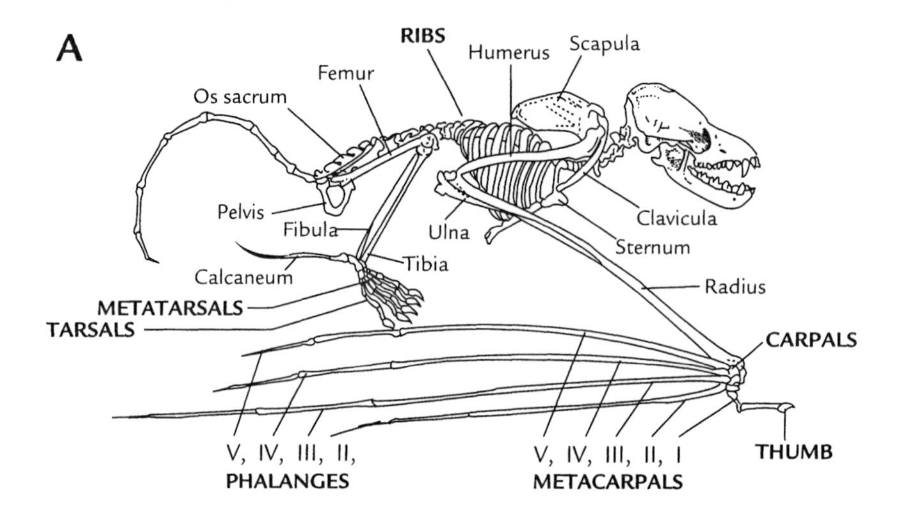

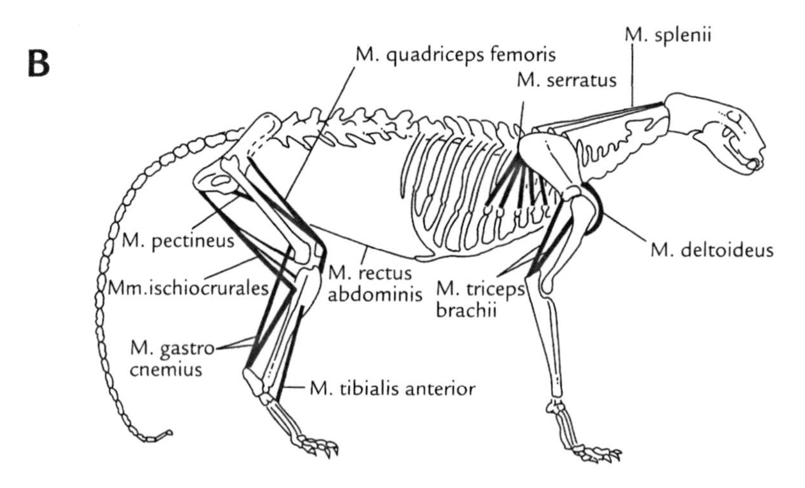

Figure 1.1 Comparison of a bat skeleton with that of a typical land mammal. (A) *Myotis myotis* (Vespertilionidae); (B) *Panthera tigris*, the tiger. The principal muscles that attach the head and legs to the trunk are drawn in black. (A) Adapted from Wimsatt (1970); (B) From Starck (1982).

1.1 THE SPINE AND PELVIS

SPINE

The axial skeleton of bats is similar to that of other mammals (fig. 1.1). In bats, the spine is made up of 7 cervical vertebrae, 11 thoracic vertebrae, and up to 10 caudal vertebrae (in humans, only 4 caudal vertebrae are present; these are fused together to form the coccyx). In the Megachiroptera, the caudal vertebrae are entirely absent. To support the large flight muscles, the thoracic vertebrae are tightly connected to one another to form a stiff column. In some families of bats, the lowest cervical vertebra is fused to the uppermost thoracic vertebra.

PELVIS

The pelvis forms the connection between the spine and the legs. In mammals, the pelvis consists of two symmetrical halves, which are joined in the back by the sacrum. Each half of the pelvis is made up of three separate bones: the ischium, the ilium, and the pubis. These three bones are fused together to form a strong supporting structure. In most mammals the two pubic bones are joined at the front by the pubic symphysis, a flexible plate of cartilage and connective tissue. The ilium inserts tightly into a groove in the sacrum so that only small bending movements are possible. Thus, the mammalian pelvis forms a relatively immobile support structure. In bats, the pelvis has undergone several adaptations that make it especially well suited for flight. In the Microchiroptera, the ilium and the sacrum are fused together up to the level of the acetabulum (the socket into which the head of the femur fits; see section 1.3). Thus, there is no mobility at the iliosacral joint. In flying foxes the two bones are entirely fused. In these animals, the pelvis and thoracic column form a sturdy support for the upper and lower limbs, which, in turn, provide a framework over which the wing membrane is stretched.

1.2 THE WINGS

SKELETAL STRUCTURE OF THE WING

From a mammalian forelimb that is constructed for locomotion on the ground, bats have evolved a highly efficient wing (box 1.1). The wing consists of all the same elements that make up the mammalian forelimb (fig. 1.1). The large, strong upper arm bone, or humerus, is suspended from the shoulder blade, or scapula. The elongated lower arm consists of two bones: the radius, which is thick and sturdy, and the ulna, which in bats is thin. The elements corresponding to the mammalian hand or forefoot include six carpals or wrist bones, five metacarpals or hand bones, and five sets of phalanges or finger bones. The second to fifth metacarpals are greatly elongated so that they form a supporting structure of long but rigid spokes like those of an umbrella. The thumb is the only digit that retains its function as a freely moving grasping tool and is the only digit with a functional claw.

WING MEMBRANE

Stretched between the side of the body and the collapsible support system formed by the arm, hand, and finger bones is a broad, two-layered membrane (fig. 1.2). The part of the membrane that lies between the rump and the fifth finger is called the plagiopatagium; the small portion of the membrane that stretches between the shoulder and wrist is the propatagium. Together, these two sections of the wing membrane support the weight of the body during flight. The part of

Box 1.1 The Evolution of the Wing

Flight in vertebrates developed three times during evolution: once in the pterosaurs, animals that became extinct more than 65 million years ago, a second time in birds, and a third time in bats. Each of the three groups developed a very different flight technique. It is thought that the pterosaurs flew slowly, with a flapping motion of the wings. Birds with their "featherweight" wings were able to achieve extremely high wingbeat rates and high flight speeds. Bats with their cutaneous wing membranes fly rather slowly, but are remarkably agile. To obtain sufficient lift, all three groups had to undergo a lengthening of the forelimbs. At the same time, the muscles used in flight had to remain close to the body's center of gravity to minimize the weight of the wings (see p. XX). To prevent the wing from being deformed by the upward forces created during the wingbeat, the framework of the wing and the surface that acts as an airfoil had to be rigid (see fig. 1.4). In the pterosaurs, the necessary rigidity was achieved through the large and sturdy arm bones, the humerus, radius, and ulna, as well as the metacarpals and the greatly elongated phalanx of the fourth finger. The wing of the pterosaurs was protected against deformation by fine tubelike rods in tightly packed layers that extended from the leading to the trailing edge of the wing.

Because the long feathers of birds are resistant to bending, there was no need for birds to develop elongated fingers. The stiff skeletal structure of the arm transmits lift to the shoulder, which in turn is tightly anchored to the body via the triangle formed by the coracoid bone, the scapula, and the furcula (the "wishbone," formed through the fusion of the two clavicles). In bats, the thin membrane that forms the wings cannot be deformed during flight because it is tightly stretched between the elongated, spokelike fingers, the edge of the body, and the legs. In contrast to birds, bats can alter the camber of their wings, thereby optimizing the wing shape for any given set of flight conditions. The variable camber of the wings and movable shoulder of bats permit them to fly with an agility and maneuverability that could never be possible in birds.

the wing membrane that lies between the fifth finger and the wing tip is called the chiropatagium and functions mainly to propel the bat forward during flight. The part of the membrane that stretches between the legs and the tail is the uropatagium.

EXTENSION OF THE WINGS

When the bat is at rest, the wings are folded together like an accordion. The second to fifth fingers lie parallel to one another and are drawn in close to the side of the body. When the bat takes flight, a rapid extension of the spokelike structure formed by the fingers occurs, the result being a full extension of the wing. Basically, this wing extension is accomplished through the contraction of a single muscle, the supraspinatus (fig. 1.3). The contraction of this muscle sets in motion a chain reaction that automatically spreads the wing with no additional muscle contractions. The basis for this chain reaction is the fact that one section of the triceps, the muscle that extends the lower arm, is not attached to the upper arm at

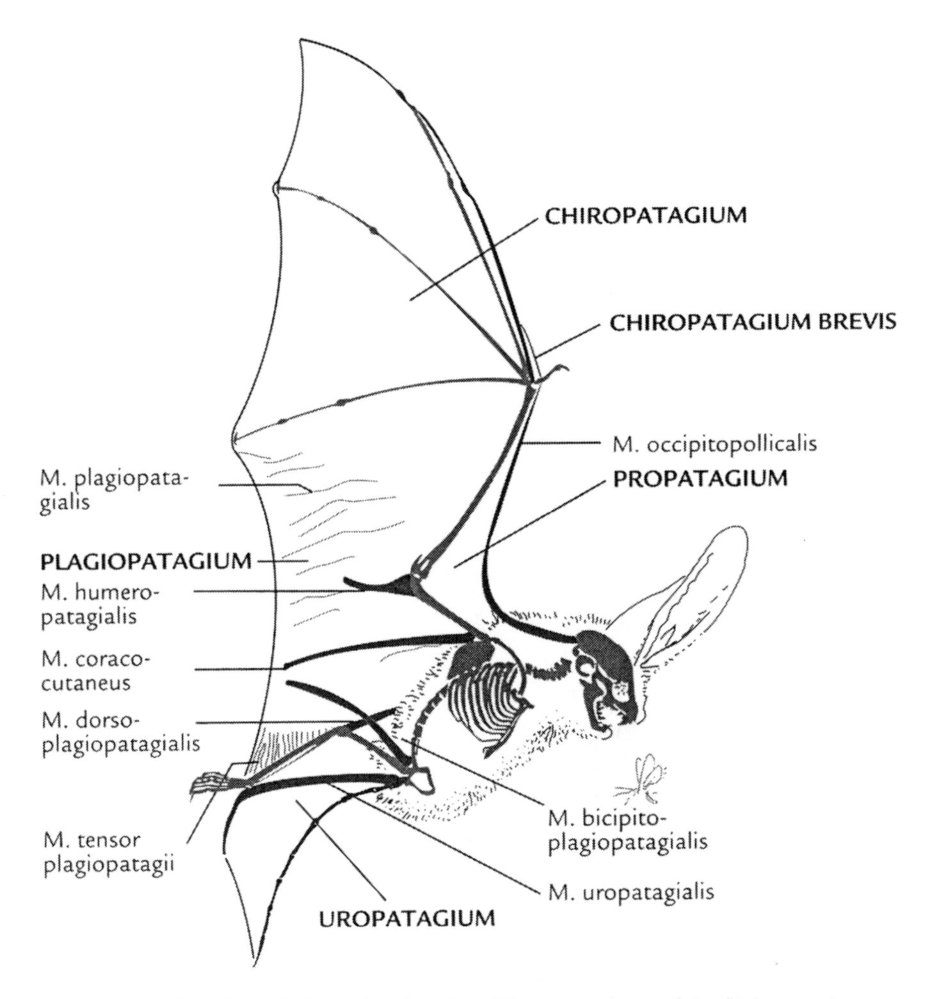

Figure 1.2 The wing of a bat, showing the different sections of the flight membranes and the cutaneous muscles unique to bats that help stretch the flight membranes. From J. Gebhard (1982).

any point as it is in other mammals, but instead is attached completely to the shoulder blade. As a consequence, the extension of the upper arm through contraction of the supraspinatus causes a shortening of the distance between the origin of the triceps on the shoulder blade and its insertion on the lower arm. This shortening causes an automatic extension of the elbow.

The fingers are extended through a similar mechanism (fig. 1.3). The extensor carpi radialis muscles are attached to the upper arm and, through their strong links to the first three metacarpals, extend the fingers. When the upper arm and elbow are extended, this increases the distance between the origin of the muscles on the upper arm and their insertion on the fingers, tensing the muscles and thereby extending the fingers. Because the fourth and fifth fingers are connected to the third

WING EXTENDED

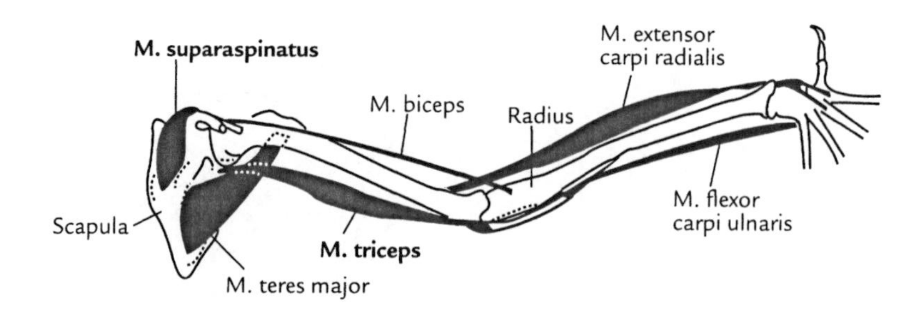

WING FLEXED

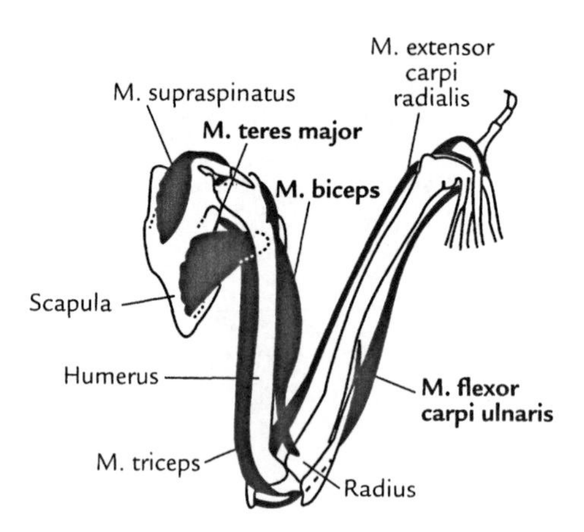

Figure 1.3 Extension and folding of the wing. The muscles that extend the lower arm (triceps) and the fingers (extensor carpi radialis) originate on the shoulder blade and the upper arm, respectively. Consequently, when the upper arm is extended through contraction of the supraspinatus, the lower arm and the fingers are automatically extended. The names of the muscles involved in extension and flexion are boldfaced. From Vaughan (1970).

finger through the wing membrane (fig. 1.4), they are passively extended. Thus, the passive extension of the elbow causes the entire chiropatagium to be outstretched.

Folding of the wings is accomplished through a similar mechanism (fig. 1.3). The biceps muscle, which flexes the lower arm, is attached to the shoulder blade, and the flexor carpi ulnaris, which flexes the fingers, is attached to the upper arm. Thus, when the upper arm is drawn in toward the body through contraction of the teres major, there is a passive tensing of the biceps and flexor carpi ulnaris, which causes the entire arm and hand portion of the wing to fold inward.

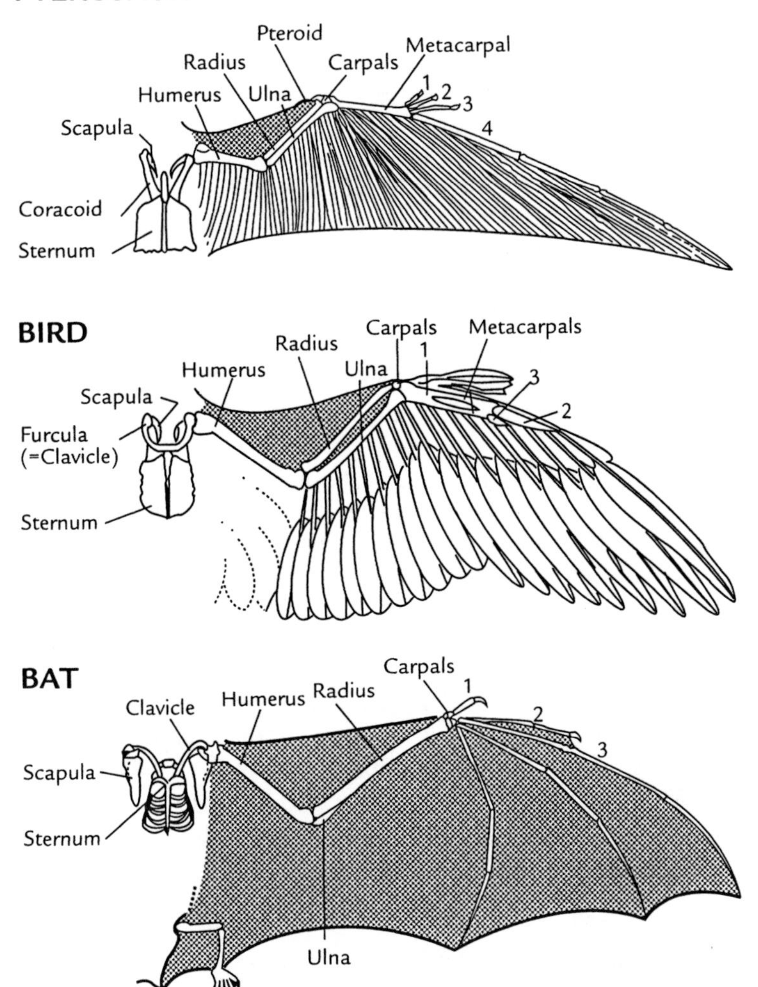

Figure 1.4 Comparison of wing structure in pterosaurs, birds, and bats. In pterosaurs, the supporting framework for the wing consisted mainly of the metacarpals and the fourth finger. In birds, the wing is supported mainly by the lower arm and metacarpals, and in bats by the elongated lower arm and second to fifth fingers. Thus, among the vertebrates, flight evolved independently three times. From Padian (1985).

Thanks to these cascaded movement systems, the only muscles that need to actively contract to extend or fold the wing are the shoulder muscles, which lie close to the trunk. This allows the arm and hand musculature to be very small, thereby reducing the weight of the wing. Of the 20 muscles normally present in the mammalian hand, only 7 are found in the wing of most bat species, and these are greatly reduced in size. The most important are the interossei muscles, which function to hold the outstretched fingers apart.

A few muscles peculiar to bats are responsible for keeping the elastic wing membrane tightly stretched across its supporting framework (fig. 1.2). The most important muscle of this group is the occipito-pollicalis, a thin muscle that extends from the back of the skull to the shoulder, where it fuses into a strong tendon. This tendon forms the anterior border of the wing. It extends from the shoulder across the thumb to terminate on the second finger. The occipito-pollicalis muscle keeps the front edge of the wing tightly stretched, thereby improving the wing's aerodynamic properties. Other important muscles are the coraco-cutaneus, which extends from the shoulder to the posterior border of the plagiopatagium, and the bicipito-plagiopatagialis, which extends between the hip and the plagiopatagium (fig. 1.2). Both of these muscles help keep the wing membrane taut by pulling it in toward the body, thereby counteracting the distal pull of the outstretched fingers. A bony spur, the calcar (fig. 1.1), extends from the heel bone along the edge of the tail membrane and keeps it from flapping during flight.

The bat's arm and wrist joints are also adapted to the special requirements of flight. Both the elbow and the wrist joints articulate through grooved, interdigitating joints so that movement can only occur in the horizontal plane of the outstretched wing. During spreading of the wing, the heads of the long bones hit against bony processes that extend from the socket, thereby preventing an overextension of the wing and maintaining the spread wing in an optimal position. The second to fifth metacarpals overhang the wrist in such a way as to prevent all movement in the dorsoventral plane. The fifth finger, which forms a crosspiece for the wing (fig 1.4), is especially well protected against upward movement by the pisiform bone, which overlies the joint formed by the wrist and the fifth metacarpal. This additional source of rigidity protects the supporting structure of the wing against the air pressure generated during each downward stroke of the wing and prevents the wing from being flipped over.

As already mentioned, the shoulder girdle is connected to the trunk exclusively through muscles. The resulting mobility of the shoulder blade facilitates flight in bats. In birds, the point of rotation for the wing lies in the joint of the upper arm. In birds, the scapula, coracoid, and clavicle together form a tripod that keeps the shoulder joint rigidly pulled in toward the body. In bats the shoulder blade is free to follow every wing movement. As in humans, the shoulder blade of bats is rotated 90° so that its long axis is parallel to the spine (fig. 1.4). On the head of the humerus is a protrusion that hits against the socket on the shoulder blade whenever the wing is lifted upward (fig. 1.5a). This locking mechanism increases the amount of leverage by the width of the shoulder blade, about 40%, and the wing then pivots at its joint with the collarbone (fig. 1.5a). In addition, with every wingbeat the scapula moves about 20° across the chest. This movement shifts the base of the wing toward the middle of the body near the sternum, thereby improving stability during flight. The joint between the collarbone and sternum then becomes an additional pivot point for the wing.

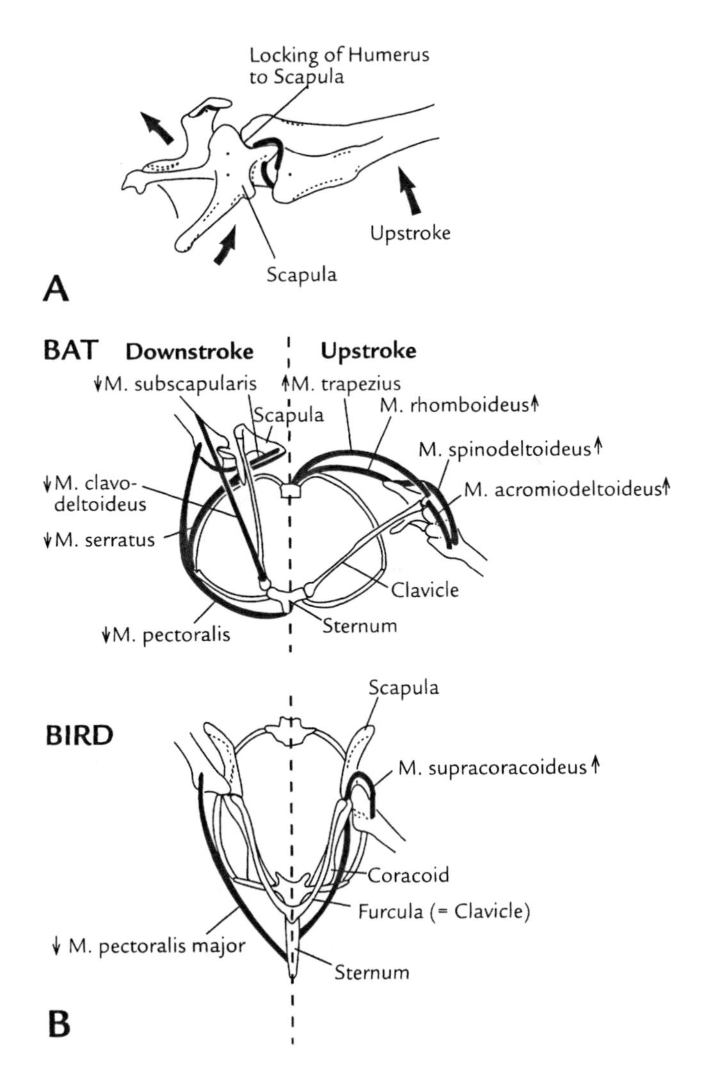

Figure 1.5 Comparison of the flight musculature and pivot point of the wing in bats and birds. (A) In bats, during the upstroke of the wing the upper arm is mechanically locked to the shoulder blade, causing it to participate in the movement of the arm (arrow). (B) In birds, the point of rotation of the wing is located at the upper arm joint; in bats, it is located at the joint between the sternum and the clavicle. In each diagram, the left side shows muscles used during the downstroke, the right side shows muscles used during the upstroke. The arrows next to the names of the muscles indicate whether they are active during the upstroke or the downstroke. From Hill and Smith (1984).

FLIGHT MUSCLES

All of the power for the wingbeat is provided by the chest and shoulder muscles. The same set of muscles that in four-legged animals provides support for the shoulder and foreleg when standing and brings the leg forward when walking or running, serves as the flight muscles in bats (fig. 1.6). In birds, just two muscles are sufficient to control the wingbeat (fig. 1.5b), but in bats, 17 different muscles are involved in wing movements. The function of most of these muscles is to adjust the position of the wing; only a few extremely powerful muscles are responsible for the wingbeat itself (fig. 1.5b).

During flight, most of the work is done by the muscles that control the *downstroke* of the wing. These are the pectoralis, serratus, and clavodeltoideus muscles (figs. 1.5a and 1.7). The serratus muscles connect the second through eleventh ribs to the medial border of the scapula and form a kind of sling, which suspends the shoulder blade from the chest. When the shoulder blade is raised during the upward stroke of a wingbeat, these supporting muscles are stretched. The serratus muscles contract in response to the stretch, thus initiating the downward phase of the wingbeat. The main power for the downstroke is supplied by the pectoralis muscles, which originate along the entire length of the sternum and insert on the upper arm. The pectoralis muscles weigh four times as much as all of the other flight muscles put together. Because the chest muscles are attached to the front of the body, each contraction not only pulls the wing downward, but also moves it forward. Depending on which part of the pectoral muscle contracts more strongly, the upper arm may assume different rotational positions, thereby adjusting the angle of the wing according to flight conditions. The clavodeltoideus muscle, which extends from the collarbone to the upper arm and scapula, also contributes to the downstroke of the wing by pulling it downward and forward. The subscapularis muscle originates along the entire concave side of the shoulder blade and inserts on the humerus. The powerful contraction of this muscle pulls the upper arm downward relative to the shoulder blade and controls the position of the wing mainly through rotational movement. In terrestrial mammals, the subscapularis muscle maintains the front leg in a position beneath the body's center of gravity. In humans, the subscapularis muscle rotates the arm inward. Physiologically, the downstroke of the wing corresponds to stepping and pushing off with the outstretched foreleg.

The *upstroke* of the wing is accomplished mainly through the action of three muscle groups. The rhomboideus muscle (fig. 1.5; not illustrated in fig. 1.7) extends between the medial border of the scapula and the cervical vertebrae. Its function is to pull the shoulder blade across the torso in the direction of the spine. The trapezius muscle performs a similar function. It has its origin in the region that extends from the seventh cervical vertebra to the third lumbar vertebra; it inserts on the outer side of the scapula and the cranial part of the collarbone. The deltoideus muscle extends from the shoulder blade to the upper arm. It lifts and rotates the upper arm. The serratus muscle, which functions mainly to initiate the downstroke, also provides some fine control at the beginning of the upstroke.

The main muscles involved in *reversing* the direction of the wingbeat are the latissimus dorsi, which originates in the region from the eleventh thoracic verte-

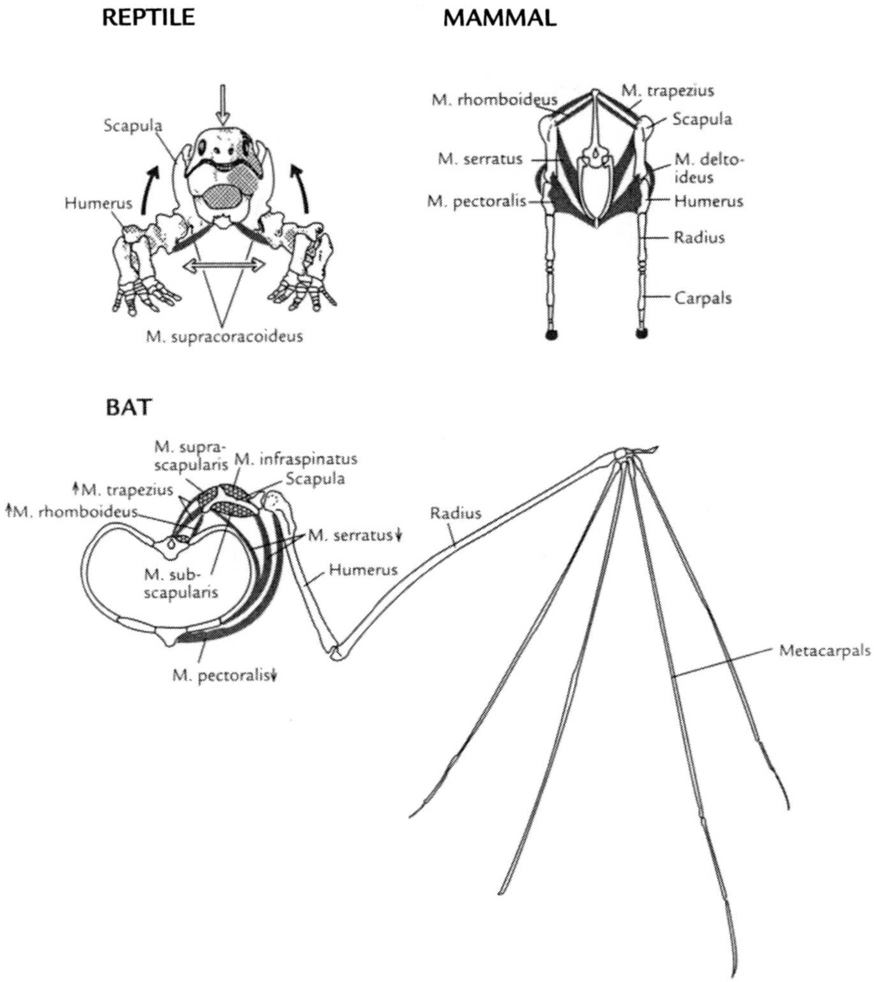

Figure 1.6 Comparison of supporting musculature in reptiles and mammals and its transformation into flight musculature in bats. In reptiles, the legs protrude on either side, so that the body weight (white arrow) must be supported through the pulling action of the supracoracoideus muscles. In mammals, the line of the legs is directly beneath the body, and the legs are connected to the trunk through many different muscles. In bats, the muscles that normally support the foreleg are transformed into flight muscles (compare muscles in generalized mammal and bat). The arrows next to the names of the muscles indicate the phase of flight during which they are active. From Starck (1979, 1982) and Kluge (1977).

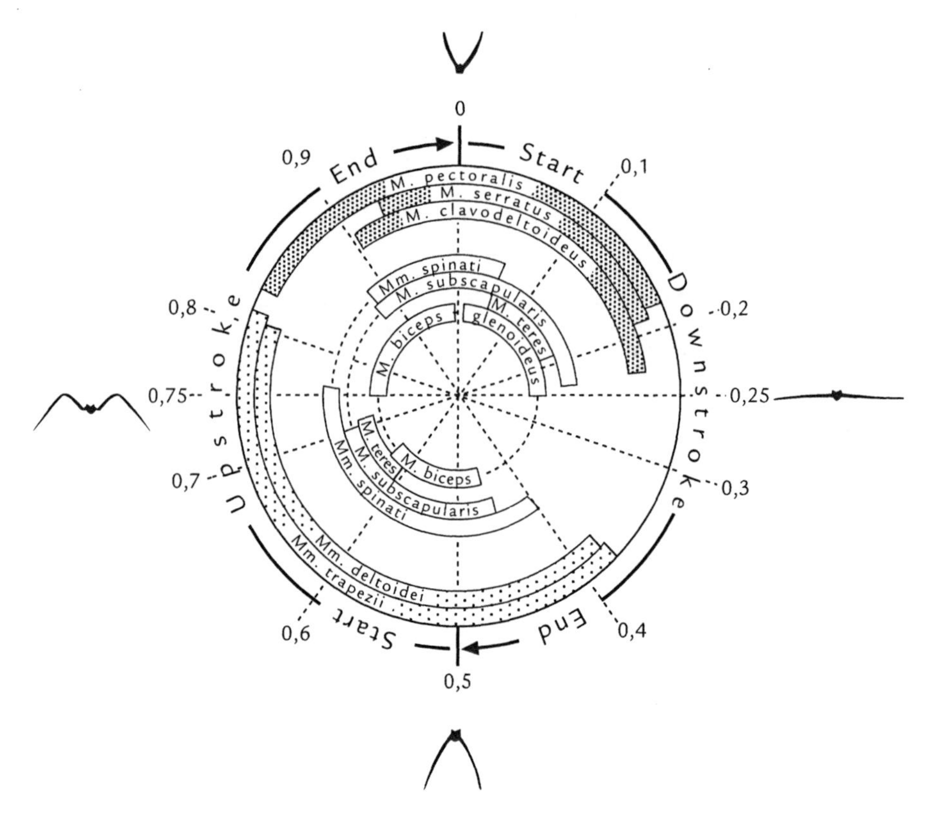

Figure 1.7 Cyclic activity of flight muscles during slow flight in *Antrozous pallidus* (Vespertilionidae), indicated by the gray and white bars. Gray bars indicate the muscles that are most important in powering the bat's flight. Muscles that power the downstroke: pectoralis, serratus, clavodeltoideus. Muscles that power the upstroke: deltoideus, trapezius. Muscles that power both phases and also help control wing position: spinatus, subscapularis, teres major, biceps glenoideus. Drawn from data of Hermanson and Altenbach (1983).

bra to the fourth lumbar vertebra, and the infraspinatus, which extends from the outer side of the shoulder blade to the humerus, and stabilizes the shoulder joint by counteracting the enormous forces generated by contraction of the chest muscles. The supraspinatus, which takes a similar course, is active during the initial phase of both the upstroke and the downstroke.

The reversal of wing direction uses more energy than any other phase of flight. At the beginning of the downstroke, 14 of the 17 flight muscles are active, whereas during the middle of the downstroke only the subscapularis muscle remains active. To provide a smooth reversal of wing direction while avoiding abrupt and uneconomical movements, the set of muscles that will initiate one phase have already begun to contract during the end of the preceding phase (fig. 1.7). For example, during the last third of the upstroke, the pectoral muscles have already started to contract so that the point at which the forces responsible for the upstroke and downstroke are balanced (position 0 in fig. 1.7) marks the reversal point and the beginning of the downstroke. Thus, the muscles responsible for the upstroke and the downstroke are active only at the beginning or the end of each phase and are relaxed during the actual movement. The momentum generated at the beginning of the upstroke or downstroke is sufficient for the wing movement to be completed without any further muscle contraction. In many cases little force is required for the upstroke because the air pressure under the surface of the wing passively forces it upward. The flight muscles of the bat are adapted for the metabolism of fatty acids rather than glycogen [respiratory quotient (RQ) = 0.79]. This adaptation probably contributes to a reduction in weight during flight because fat reserves contain eight times as much energy as a glycogen reserve of equal weight.

As demonstrated in flying foxes, the wing bones are most stressed by rotational and shear forces. Such stress mainly occurs at the lowest point of the downstroke and the top of the upstroke. During the upstroke, mechanical stress on the wing bones is negligible. The bones of the forelimb in bats are adapted to sustain high levels of torsion and shear by being relatively wider and thinner walled than those of land-dwelling mammals.

1.3 THE LEGS

During flight, the legs of a bat serve merely as a framework to support the wing membranes, so they are small, thin, and not very muscular. The small size of the legs may be one reason bats do not normally stand on their legs, but rather hang upside-down from twigs or the ceilings of caves. In addition, it is much easier for the bat to take off and fly from an elevated hanging position than it would be if it were standing on the ground. The legs of a bat are different from those of a walking animal in that they are adapted for pulling rather than pushing.

In most mammals, the hip socket, or acetabulum, into which the head of the femur fits, is located at the junction of the ilium, ischium, and pubis. Because the hind legs of most mammals lie under the rump, the hip sockets open in the ventrocaudal direction. In bats, the position of the hip socket is adapted for hanging. In contrast to other mammals, the opening of the acetabulum is shifted upward and opens toward the back. This configuration is ideal for hanging, but it does not allow the legs to be positioned in a straight line beneath the body. For this reason the hind legs of a bat crawling on the ground stick out on either side like the legs of a reptile (fig. 1.8).

The large head of the femur (thigh bone) provides a wide range of movement for the leg. It is necessary for the femur to have this freedom of rotation because, when the bat is hanging, the legs are oriented parallel to the long axis of the body and point directly backward. In flight, the legs are angled toward the side and rear; when the bat is crawling along the ground, the legs are angled forward. Some species such as horseshoe bats can rotate their body nearly 360° around their thin legs. When the horseshoe bat hangs from a twig, it can rotate back and forth and echolocate in all directions.

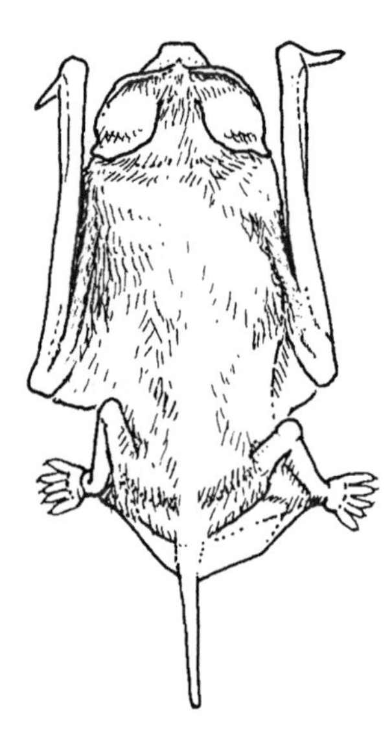

Figure 1.8 *Tadarida brasiliensis*, a Molossid bat, crawling on the ground. When a bat crawls, it supports itself on its wrists and ankles, with the lower legs angled toward the rear.

In addition, the legs of bats are rotated 180° around their long axis so that their knees face the rear instead of the front as they do in other mammals. For this reason, bats are unable to kneel down. When a bat crawls along the ground, its body does not rest directly over its legs. Instead, the body is suspended between the wings and legs, which stick out sideways and are in turn supported by the wrists and ankles, which contact the ground (fig. 1.8).

The thighs of bats are moved by the same muscles as in other mammals. The dorsal hip muscle, the psoas minor, is especially powerful in bats. This muscle extends from the lumbar vertebrae to the tip of the pelvis so that when it contracts it lifts the pelvis toward the spinal column. When a bat lands, the psoas minor acts as a spring that damps the sudden force exerted on the pelvis. This muscle also contributes to forming the tail membrane into a pouch, which the bat uses to scoop insects out of the air during flight.

Unlike other mammals, bats have only one bone in the lower leg—the tibia. The fibula in bats is vestigial. The quadriceps femoris muscle acts to extend the lower leg, and the ischiocruralis acts to flex the lower leg (compare with fig. 1.1b).

GRASPING

When a bat is at rest, its entire body weight hangs suspended from its toes. Like birds of prey, most bats have evolved a locking mechanism that keeps their claws bent without any muscular contractions (fig. 1.9). Exceptions are the phyllostomids and a few other species that have a far less sophisticated tendon-locking mechanism. An important element of this mechanism is the special structure of

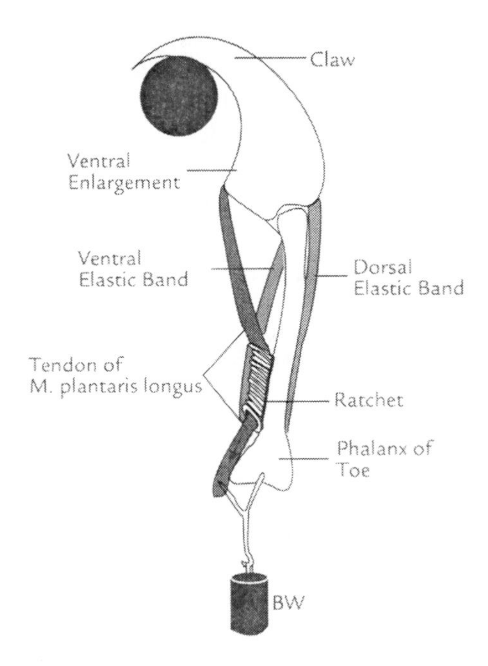

Figure 1.9 Diagram of the tendon-locking mechanism, maintained by the weight of the bat's body, through which the toes flex to grip a support. The suspended weight (BW) symbolizes the bat's body mass. From Grasse (1955).

the claws of the hindlimbs, which are compressed into a bladelike structure. The claw protrudes ventrally relative to the joint so that the joint and tendon of the flexor and plantaris muscles, which insert on the claw and act to flex it, are located at a considerable distance from the toe bone. Near the proximal end of the phalanx, the tendon passes through a tough sheath attached to the bone. This sheath consists of 19–50 rings, oriented at an angle so that the inside surface is ridged. When the muscles pull on the tendon, this presses the rough part of the tendon against the ridged surface of the sheath and secures it in place. Even after the muscle relaxes, the tendon is held in place through the tension exerted by the body weight. The tendon cannot slip through the sheath because the sharp ridges inside the sheath dig into the rough surface of the tendon and hold it fast. Due to this ratchet mechanism, with no muscle contraction whatsoever, the claws continue to grip the surface on which the bat is hanging as long as the weight of the body is suspended from the tendon. When the bat takes off in flight, the tension on the tendon is released, and the claw releases its hold. Thus, the function of the plantaris muscle is to initially bend the claws so that they grasp the surface when the bat lands. The locking mechanism is further aided by a ventral elastic band of connective tissue that pulls the sheath proximally and narrows its inner diameter. This tendon-locking mechanism works so well that bats remain hanging on the ceilings of caves even after they have died. A similar locking mechanism has been described in the thumb claws of flying foxes.

Some tropical species that spend the day in rolled-up leaves or among bamboo stems have evolved a different mechanism for effortless hanging. For example, *Thyroptera tricolor* (Thyropteridae), a small neotropical bat that weighs only 4–5 g, attaches itself to the slick surfaces of rolled-up leaves through suction cups at the base of its thumbs and on the soles of its feet. The two species of *Tylonycteris* (Vespertilionidae) have sticky pads on their thumbs and feet.

OTHER TYPES OF LOCOMOTION

Bats use their legs primarily as a framework for stretching their wings and for hanging. Nevertheless, bats can move quite well on the ground. Species that are good at walking, such as the mouse-eared bat, do not point their thighs backward, but rather hold them straight out to the side. Bats that are especially good at walking on the ground, as many molossid species are, have powerful legs with fully developed calf muscles. Many species of bats hop along on the ground, propelled by a sort of rowing motion of the arms. Vampire bats (Desmodontidae) have long legs and are good walkers and jumpers. Usually they land close to their intended victim. They then approach cautiously and must be constantly on the lookout for unexpected defensive movements of their prey so that they can make a rapid getaway if necessary. The vampire's high degree of mobility on the ground is manifested in the strong bones of the thighs and lower legs, which are deeply notched to provide good attachment sites for the powerful leg muscles. In fact, vampire bats are the only ones capable of launching themselves from the ground into flight by a powerful jump into the air.

Fishing bats use their hind claws to catch their prey. When fishing, they extend their long legs and spread their long sharp claws so that they can grasp the widest possible area on the fish's body.

Bats can even swim. If a bat falls in the water, it immediately tries to get its wings up above the surface. Once this is accomplished, the bat can use rowing movements of its wings to propel itself across the surface of the water. The mouse-eared bat can move in this way at speeds of 0.5 m/s. Many species are able to take off from the water surface and become airborne within as few as two to six wingbeats. This ability seems to be more than just an emergency measure in case of accidents, as bats are able to light briefly on the surface of the water to drink.

1.4 AERODYNAMICS

"Aerodynamics is not an exact science," according to the Swedish flight researcher U. Norberg. "It is full of assumptions, oversimplifications and anecdotal evidence." This statement characterizes all of the attempts that have been made so far to describe the aerodynamics of bat flight. Until the late 1970s, the only theoretical framework that had been applied to the analysis of animal flight was the same as that used to analyze airfoils and propellers, a system that had been developed to deal strictly with rigid, motionless surfaces. The vortex theory, first proposed by Ellington in 1978 and further developed by Rayner in 1979, does a somewhat better job of describing flight with beating wings, which must provide not only lift, but also propulsion.

For an animal to fly at a constant speed and elevation, two forces must be overcome. First, the animal's own mass must be counterbalanced by an opposing force, lift (*L*) (fig. 1.10), which is generated through the animal's own energy. Second, the resistance of the body and wings, or drag (*D*), must be counterbalanced by an opposing force—thrust.

LIFT AND THRUST

Lift and thrust (fig. 1.10) are produced by the wingbeat in the following way. The downstroke of the wing gathers in masses of air and presses them downward, creating a region of high pressure under the wing and a region of low pressure above it. This pressure difference is expressed as an aerodynamic force, *R*. This force *R* can be broken down into two vectors. The vertical vector is lift (*L*). The horizontal vector is thrust, which acts to overcome drag (*D*) and propel the animal forward. Thus, for a given aerodynamic force *R*, the smaller the drag *D*, the greater the thrust; *R* depends on the volume and speed of the air mass moved by the wing as it traverses a cross-section of air space in one unit of time. The diameter of the cross-section of air space traversed by the wings, termed the wing disk, is equivalent to the diameter of the wingspan. In terms of defining air pressure, it does not matter whether movement of air through the wing disk is due to wind or to the animal's own wingbeats.

According to the vortex theory, forces are created in complicated ways. Air pressure gradients at different points around the wing give rise to air currents that flow upward over the front edge of the wing toward the rear, around the rear edge of the wing, and back toward the front. This circular air current dissipates at the wingtips and the tail in the form of a vortex. The rotational movement of these vortices provides the force *R*. Because the animal is positioned at the upper, leading front of the vortices where the direction of air movement is downward and to the rear, the resulting rotational movement carries the animal's body forward and upward (fig. 1.10c). There have been no investigations of how the air flow and air circulation around the wing interact with different surface areas, shapes, and positions of the wing.

The amount of lift and thrust produced by each wingbeat depends not only on the wingbeat itself, but also on the wing profile and the position of the wing relative to the air stream. Aerodynamically, the wing of a bat takes the form of a thin plate of asymmetrical cross-section, curved at the front. When such a curved, or cambered, plate moves horizontally through the air, the air beneath the curved surface moves more slowly than the air above, resulting in an increased pressure under the plate. The unequal pressures on the two sides result in a net upward pressure or lift (fig. 1.10b). The induced lift is proportional to the speed of the air stream, the camber of the wing profile, and the angle of attack, α. The angle of attack is defined as the angle between the wing chord (the straight line connecting the leading and trailing edges of the wing) and the direction of air flow (fig. 1.10b).

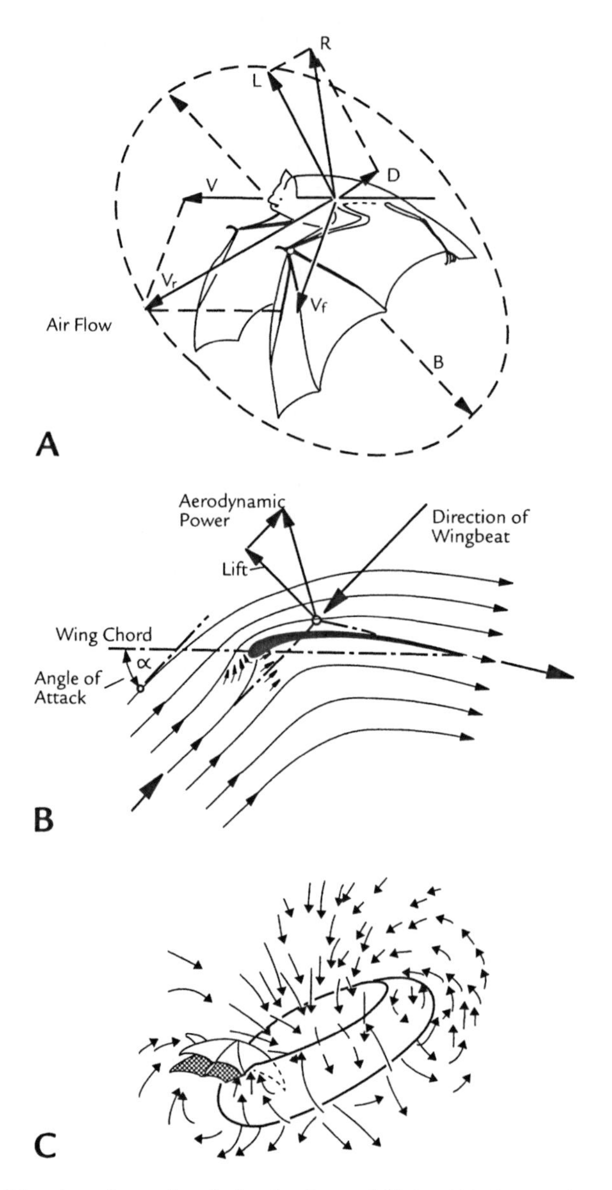

Figure 1.10 Aerodynamics during horizontal flight. (A) Generation of lift and thrust by the wingbeat. *D*, drag; *L*, lift; *R*, aerodynamic power; *V*, flight velocity, V_f, wingbeat velocity, V_r, resultant velocity. The diameter of the wing disk (dashed oval) is equal to the wingspan, *B*. (B) Generation of aerodynamic power beneath the wing as air streams across the cambered airfoil. (C) Vortex created in air by the downstroke in a flying bat (*Plecotous auritus*, Vespertilionidae) traveling at 1.5 m/s. The animal "rides" on the front edge of a vortex in which air flows backward and downward at an angle. (A) From Norberg (1986); (B) From Hertel (1963); (C) From Rayner et al. (1986).

For reasons having to do with the mechanics of air flow, the angle of attack must be relatively flat. If the angle becomes too steep, the laminar air flow moving over the surface of the wing separates from the wing surface due to turbulence (fig. 1.11, box 1.2).

REYNOLDS NUMBER

In bats, the danger of laminar flow separation and consequent stalling is especially acute. This is related to a factor called the "Reynolds number," a value that expresses the relationship between the viscosity and inertia of the air flowing around the wing. The Reynolds number is calculated using the formula:

$$\text{Reynolds number} = \frac{\text{speed of air flow} \times \text{chord length}}{\text{kinetic viscosity of air}}$$

The maximal speed of air flow in bats is 8–10 m/s, and the maximal chord length of the wing (width of the wing) is 15–20 cm. At 20°C, the kinetic viscosity of the air at the bat's flight elevation is about 15×10^{-6} m^2/s. The Reynolds number of 10^4–10^5 calculated on the basis of these values is technically poor—airplanes have Reynolds numbers of 10^6–10^8. A value of 10^4 lies within the range in which lift is subject to sudden large decreases due to separation of laminar flow caused by increased pressure at the trailing edge of the wing.

IMPROVEMENT OF AERODYNAMICS

Nature, like technology, has found ways to prevent sudden large decreases in lift. One strategy is through optimization of the wing profile. For a Reynolds number of 10^4, a cambered wing profile is optimal. The wings of both bats and insects are constructed in this form.

Aeronautical engineers have developed the "leading edge flap," an adjustable surface on the leading edge of the airfoil that optimizes air flow over the wing surface, thereby permitting a steeper angle of attack and improving lift. Bats have a similar adjustable leading edge flap in the propatagium, the angle of which can be made flatter or steeper through movements of the thumb (fig. 1.11b). When the bat is flying slowly or executing difficult maneuvers, the angle of the propatagium is increased, thereby improving lift while avoiding separation of the laminar flow from the wing surface.

The camber of the wing can also be adjusted through flexion of the fifth finger. This finger is flexed by the adductor digiti quinti muscle, which originates at the wrist and inserts on the underside of the joint between the metacarpal and proximal phalanx. This muscle can stretch the thin fifth finger like a bow, causing the entire wing membrane to curve upward (fig. 1.11b). The pisiform bone fixes the finger in place and prevents it from pulling out of the joint due to the force exerted by the stretched wing membrane.

The finely adjustable camber of the wing is not the only means of minimizing the danger of separation of laminar air flow from the wing. Such a separation can also be prevented through the creation of a weakly turbulent boundary layer at the

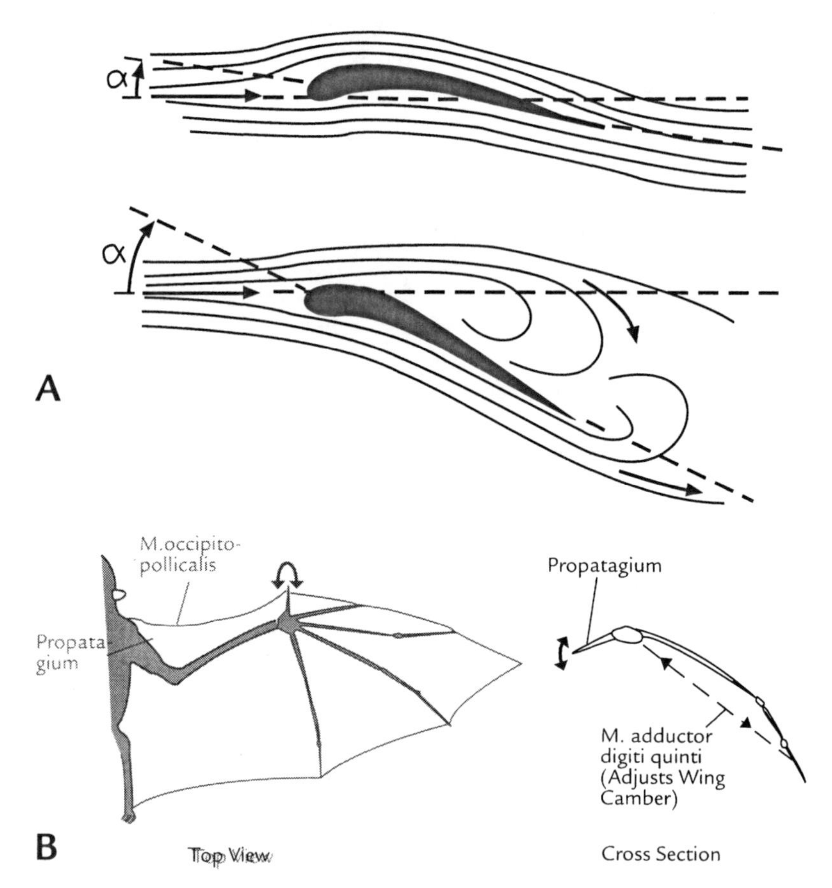

Figure 1.11 Laminar air flow. (A) Laminar air flow at a shallow angle of attack (upper drawing) and breaking away of laminar flow due to turbulence that originates at the rear edge of the wing when the angle of attack is steep (lower drawing). (B) The adjustable leading edge flap can prevent disruption of laminar flow at a steep angle of attack. The propatatium of bats forms such a leading edge flap, the angle of which can be adjusted by movement of the thumb (double-headed arrows). The occipitopollicalis muscle provides a rigid front edge for the leading edge flap. The camber of the wing can also be increased through contraction of the adductor digiti quinti muscle. From Rayner (1981) and Pennycuick (1971).

wing surface. In aeronautics, such a boundary layer can be created by roughening the surface of the wing. In bats, the bones of the arm are slightly raised with respect to the wing membranes, and parts of the wing membrane are often covered with hair. This naturally occurring unevenness in the wing structure serves the same purpose in bats.

Thus, through their relatively steep angle of attack and their high camber, the wings of bats create a powerful lift. This enables the bat to remain airborne even

Box 1.2 Separation of Laminar Flow

The friction of the air against the wing causes the formation of a thin layer of air that moves at reduced velocity, the boundary layer. To maintain lift, this boundary layer must remain adjacent to the surface of the wing. Under certain conditions, there is a danger that the boundary layer may separate completely from the surface of the wing. This is likely to happen if the angle of attack of the wing is too high or if excess air pressure builds up at the rear edge of the wing. During horizontal flight, the highest air pressure is at the leading edge. At the peak of the wing curvature, where the air flow velocity reaches a maximum, air pressure reaches a minimum. At the trailing edge of the wing, the air current that rises from the underside of the wing creates a higher pressure. If this becomes too great, it can block the laminar flow of the boundary layer and cause it to separate from the surface of the wing.

during slow flight. Wings of this type are not so well suited for fast flight, mainly because they have a high resistance to the air, creating drag. Consequently, wing structure and style of flight represent a compromise between maximal lift and minimal air resistance.

1.5 ENERGETICS OF FLIGHT

The total power (P) required to produce the aerodynamic force for flight includes the aerodynamic power output needed to remain airborne as well as the inertial power output (P_{in}) needed to move the wings up and down. At medium and high flight speeds, P_{in} for wing acceleration is negligible. However, during slow horizontal flight and hovering in the air (see section 1.6), P_{in} is a significant part of the bat's energy expenditure.

The aerodynamic power output consists of three components (fig. 1.12, box 1.3):

- P_i: The power *induced* by the wingbeat that creates lift and propulsion;
- P_{par}: The *parasite* power to overcome the resistance caused by friction between the body and the air;
- P_{pro}: The *profile* power to overcome the resistance caused by friction between the wings and the air.

The drag of the beating wings (D_{pro}) is a function of both the wing profile and the resistance created by the motion of the wing. The latter value is difficult to calculate because every portion of the wing has its own characteristic resistance and because these values change in a periodic manner. Measurements in birds indicate that the P_{pro} necessary to overcome D_{pro} is about twice the minimal power, P_{mp}, necessary to fly without loss of altitude. Thus, the total flight power is the sum of all of the individual components:

$$P_{sum} = P_{in} + P_i + P_{par} + P_{pro}.$$

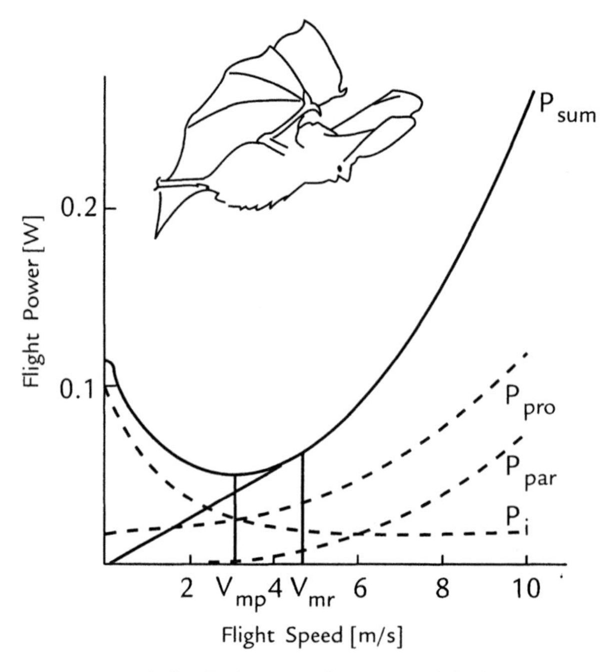

Figure 1.12 Power used for flight as a function of flight speed (*Plecotus auritus*, Vespertilionidae). The solid curve shows the total energy, P_{sum}, that must be provided by the animal for flight. The dashed curves show the different components that make up the total: the power to overcome the drag of the wing (P_{pro}) and the body (P_{par}) as well as the power (P_i) to drive the wingbeat to provide lift and thrust. At high speeds, most energy is expended in overcoming drag. During hovering flight ($V = 0$), the greatest amount of energy is spent in overcoming the inertia and induced drag of the wing and in creating lift. V_{mp} is the flight speed that requires the smallest energy output, and V_{mr} is the speed at which the longest distance can be covered with the minimum energy output. V_{mr} is the point of intersection of the total energy curve with a tangent originating at the zero point. From Norberg (1976).

HOVERING FLIGHT

When a bat hovers (flight speed = 0), there is no horizontal propulsion to provide air flow through the wing disk. Consequently, the total flow of air through the wing disk must be generated through beating the wings. Stationary flight is, therefore, energy intensive and can only be maintained for short periods of time. In a nectar-feeding bat, *Glossophaga soricina*, weighing 105 g, the mechanical power needed for hovering was calculated to be 0.34 W, a value that is considerably higher than the one calculated for *Plecotus auritus* in fig. 1.12. The formula for the induced speed of air flow during hovering (V_i, see box 1.3) shows that short, broad wings are best suited for hovering flight.

Box 1.3 Definitions and Formulas

V = Flight velocity; $V \approx (m \cdot g/S)^{1/2}$
D_{par} = Drag of the body
D_{pro} = Drag of the wings
P_{sum} = Total power for flight
P_{in} = Power for movement of the wings
P_i = Power to produce lift and thrust
P_{par} = Power to overcome the drag of the body
P_{pro} = Power to overcome the drag of the wings

For horizontal flight: $D_{par} = V^3$
$P_i = (m \cdot g)^2/B^2 \cdot V$

For hovering flight: $P_i \approx [(m \cdot g)^3/B]^{1/2}$

m = mass
g = gravitational force
B = wingspan
S = wing surface area

HORIZONTAL FLIGHT

As horizontal flight speed increases, ever greater masses of air are moved through the wing disk per unit time, resulting in increased lift. The induced power, P_i, that is required thus decreases with increasing flight speed (fig. 1.12). However, the drag of the body (D_{par}) quickly limits these energy savings because it is proportional to the third power of the flight speed (see formula in box 1.3). Thus the energy savings gained at increased speed must be paid for by additional energy expenditure to overcome air resistance (P_{par} and P_{pro}). Figure 1.12 illustrates the relationship between flight speed and energy expenditure. The intersection of the curves for P_i and P_{pro} represents the optimal flight speed, V_{mp}, at which minimal muscle power is required to fly without loss of altitude. A bat flies at this speed when it wants to remain airborne as long as possible with limited energy reserves. The point of intersection between the flight power curve, P_{sum}, and a tangent to this curve originating from the zero point on the axes (fig. 1.12) represents the speed at which a given weight can be transported one unit of distance with minimal energy output (V_{mr}). This is the speed at which a bat flies in order to travel the longest possible distance using a limited energy supply, without regard to time. For example, this is the speed at which a bat would fly during migration.

The most economical travel speed, V_{mr}, is 1.3–2.0 times greater than the flight speed at which minimal energy is used, V_{mp}. Norberg calculated V_{mp} for the long-eared bat, *Plecotus auritus* (*Vespertilionidae*) to be 3.1 m/s. At this speed the bat could remain in flight for the longest possible time without stopping to "refuel." The optimal speed for long-distance flights was between 4 and 6 m/s. These values correspond to the average flight speeds for most species of bats.

A flower-visiting bat, *Leptonycteris curasoae*, for example, commutes about 25 km from its roost to its foraging grounds. These bats fly at speeds of 7–10 m/s, which closely correspond to the calculated V_{mr} for this species.

EFFICIENCY

The mechanical efficiency of flight, expressed as the relationship between flight power and metabolic energy, is relatively low—0.10–0.28. The experimentally determined value of metabolic energy required to power flight is

$$P_{fm} \approx 4(P_i + P_{par} + P_{pro}) + P_b.$$

Where P_{fm} is flight metabolism and P_b is the basal metabolic rate of the animal and represents the minimal metabolic energy necessary to maintain bodily function when the animal is at rest. The basal metabolic rate could be thought of as equivalent to the cost of preparedness plus the cost of body maintenance.

1.6 FLIGHT TECHNIQUES

MANEUVERABILITY

The wings of bats are flexible and adjustable through the action of many different muscles. This gives them an agility and maneuverability that could never be achieved by birds. Bats are capable of rapid acceleration and braking. This capability is especially useful when the bat is hunting insects. Insectivorous bats are unbelievably nimble when they are flying in pursuit of prey. They can make right angle turns, dive straight down, climb nearly vertically, fly on their side, turn flips, and even occasionally fly upside down. Small species of bats are so skillful at flying in a confined space that they can capture insects from among thick underbrush. The broad wings of these small species of bats are aided in these complicated maneuvers by the uropatagium, which can be spread during flight by spreading the legs. When this happens, the uropatagium opens like an umbrella and provides a braking action, thus initiating a change in flight direction. It is only through their extreme maneuverability that bats are able to live on the walls and ceilings of caves. In order to land, their legs must be pointed upward. To achieve this position, the bat makes a 180° turn around the long axis of the body shortly before it reaches its roosting site.

The extraordinary agility of bats is coupled with a relatively slow speed of flight. Unlike birds, bats cannot increase their flight speed by shortening the breadth of their wings. The fastest flight speed that has been measured in a bat is 27 m/s (*Tadarida brasiliensis*, Molossidae) while flying out of a cave on the way to its foraging territory.

PATTERNS OF WING MOVEMENT DURING HORIZONTAL FLIGHT

To stay airborne, a bat requires 8–15 wingbeats/s. When flying straight ahead in a horizontal course both wings move from about 60° above the long axis of the body to about 30° below. Both wings move forward at the same time, to a point below the level of the head. During the upstroke, the chiropatagium remains fully outstretched, while other wing surfaces are slightly bent. Thus the surface of the wing is reduced to about 65% of its full size.

During one wingbeat cycle, the tip of the wing passes through an ellipse oriented at an angle to the long axis of the body, extending from above the body in the rear to below the body in the front (fig. 1.13a). The downstroke, which takes up 50–60% of the wingbeat cycle, is primarily responsible for generating lift (fig. 1.13b). At fast flight speeds, the amplitude of the wingbeat is reduced and the wings are buoyed up almost passively by the air pressure under the wing membranes (fig. 1.14). Although the upstroke costs virtually nothing in terms of energy, it contributes little to lift or propulsion, and may even consume aerodynamic force due to the drag of the wing. At low flight speeds, the upstroke can either consume power or generate power (active upstroke), depending on the flight style.

A characteristic feature of flight in bats is the propellerlike action of the wingtips, which consists of the following movements. During the rapid phase of the upstroke, the tips of the wings are held in an almost vertical position relative to the direction of air flow, with the underside of the wing pointing forward. At the end of the upstroke, the wingtip is suddenly flipped upside down. This "flick of the wingtip" provides a powerful thrust that propels the body in the direction of flight (fig. 1.14, upper illustration). On the downstroke, the chiropatagium is turned so that its upper side faces forward; at the end of the downward phase, it is again in a vertical position with the upper side facing the direction of flight. Thus, during each wingbeat cycle, the tip of the wing describes a path shaped like a figure eight. During parts of this cycle, the wingtips move at extremely high speeds. It has been shown that in the long-eared bat, a slow-flying species, the propellerlike movement of the wingtips produces mainly thrust (fig. 1.13b). In the long-eared bat, up to 80% of the power used to propel the bat forward comes from the upward phase of the wingbeat cycle.

HOVERING FLIGHT

In order for a bat to hover and remain stationary in the air (fig. 1.13), the wings are raised high relative to the body so that they beat in a forward direction rather than a downward one. The wing chord is at approximately a 50° angle relative to the body, and the tips of the wings move at speeds up to 10 m/s. When the wings are in such a near-vertical position, they mainly generate lift rather than thrust, thus keeping the body airborne. Whatever thrust is generated is canceled out by the air resistance of the beating wings. Because all forces act together to provide lift only, the bat hangs beneath its rapidly beating "hand propellers" like a hovering helicopter beneath its rotating blades.

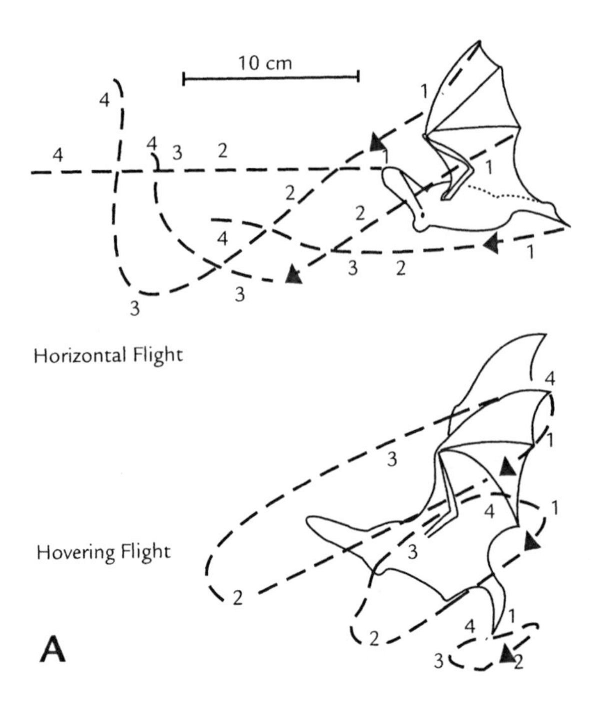

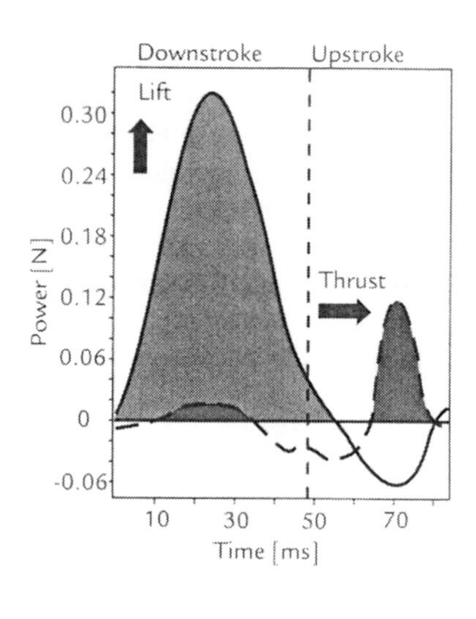

Figure 1.13 Flight of the long-eared bat, *Plecotus auritus*, Vespertilionidae. The position of the wings at each of four 10-ms intervals is indicated by the numbers 1–4 on the dashed lines. (A) Path taken by the wings during horizontal flight (top, flight speed = 2.35 m/s) and hovering flight (bottom, flight speed = 0). (B) Production of lift and thrust during one wingbeat cycle of horizontal flight. At a flight speed of 2.35 m/s, *Plecotus* generates thrust mainly during the upstroke. From Norberg (1976).

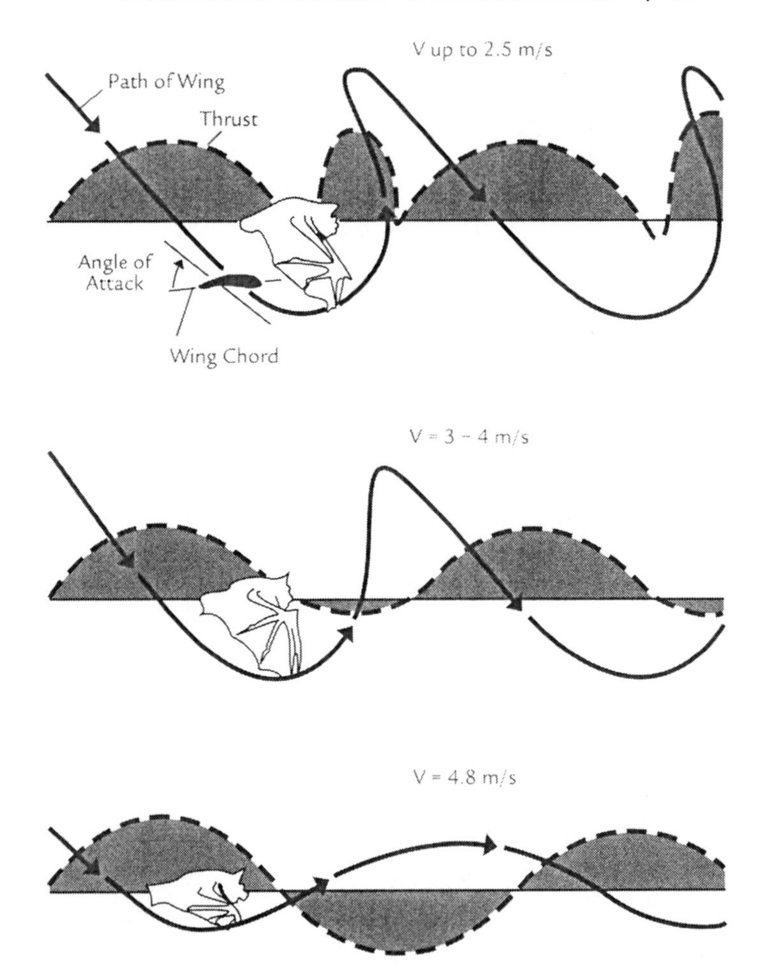

Figure 1.14 Horizontal flight by *Rhinolophus ferrumequinum* at different velocities (*v*). At speeds ≤ 2.5 m/s, thrust is generated not only on the downstroke, but also on the caudally directed upstroke. Such a wing movement is therefore called an "active upstroke." From Aldridge (1987).

Nectar-feeding bats of the genus *Glossophaga* are also thought to hover by using the upstroke of the wingbeat cycle to generate lift. The different parts of the wing perform different functions in this type of flight (fig. 1.15). The proximal, arm portion, of the wing, or plagiopatagium, remains folded to reduce air resistance, while the distal, hand portion, of the wing, or chiropatagium, is extended and supinated so that the surface that is anatomically the underside of the wing faces upward. In this configuration, both lift and thrust are generated on the upstroke.

Hovering flight consumes a great deal of energy. In *Glossophaga soricina* (*Phyllostomidae*), it has been calculated that the power required for hovering is about 2.5 times greater than that for horizontal flight at a speed of 4.2 m/s. This is

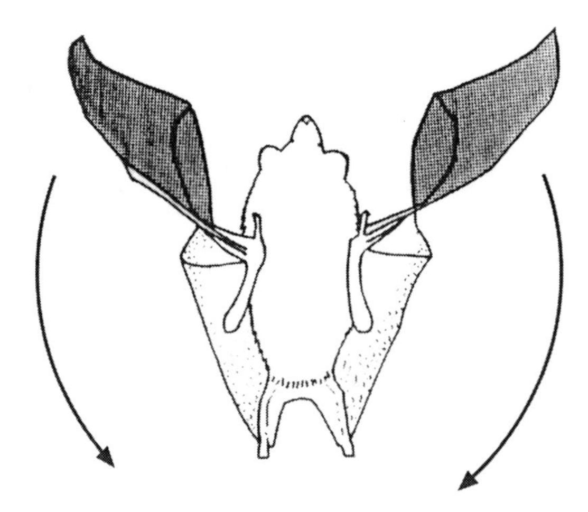

Figure 1.15 Hovering flight of the nectar-feeding bat *Glossophaga soricina* (Phyllostomidae), viewed from above. The arm portion of the wing is folded to produce a small surface with low drag, but the chiropatagium is spread and supinated so that the underside, shown in gray, points upward. When the wings flap backward (arrows), lift and thrust are generated. From Helversen (1986).

mainly due to the large inertial power (55% of total power) needed to accelerate the wings during hovering.

RELATIONSHIP BETWEEN FLIGHT STYLE AND FORM OF THE WINGS

Lift and propulsion depend not only on pattern of movement, but also on the size and length of the wings. For this reason, the form of the wings in any given bat species provides a clue about its preferred flight style and foraging strategy (Box 1.4). The broader the wings and the larger the area of the plagiopatagium and uropatagium relative to the total flight membrane area, the more agile the bat, and the more slowly it flies. Species with ellipsoid wings fly at speeds of 2.5–6 m/s. Most bats that can hover and remain airborne in small spaces have broad wings, relatively long wingtips, and a low aspect ratio (B^2S, see box 1.4). In species that hunt in and around foliage (*Hipposideros* and *Rhinolophus* spp.), the hand portion of the wing is relatively short. The advantage of having a large wing surface area is that it reduces wing loading (see box). Reduced wing loading permits hovering flight by gleaning bats that search for insects among foliage, carnivorous bats that search the ground for prey, and nectar-feeding bats such as the South American *Glossophaga*.

Bats that hunt in open spaces above the vegetation pursue their prey with rapid flight, just as swallows do. The average flight speed of these bats if 9–15 m/s. It is easy to recognize this type of bat because of its long, narrow, pointed wings (fig. 1.16). The hand portion of the wing in fast-flying bats is relatively short. This type

Box 1.4 Parameters That Characterize Wing Shape

1. Aspect ratio: ar = B^2/S
   B = Wingspan from tip to tip
   S = Wing surface area

2. Tip length ratio: tl = l_{hw}/l_{aw}
   l_{hw} = length of the third finger (hand length)
   l_{aw} = length of the arm from shoulder to wrist

3. Wing loading: wl = $m \cdot g/S$ (N/m^2)
   Wing loading is directly proportional to body weight and inversely proportional to wing surface area. Thus, for a given body weight, the greater the wing surface area, the lower the wing loading. Table 1.1 lists the most important wing parameters for several different species of bats. Figure 1.16 shows the relationship between the form of the wings and foraging style.

of wing structure results in a high degree of wing loading, which optimizes fast flight. However, the ability to fly at high speed is gained at the expense of maneuverability. Even so, hunting bats are astonishingly agile. They can roll around the long axis of their body, dive straight down, and fly straight up. The members of the Molossid family, considered advanced on the basis of their wing structure, include many rapidly flying hunters. The wings of Molossid bats are tough and leathery in texture, with rigid edges. In addition, their wings are flat, with almost no camber. Although this low degree of curvature reduces air resistance, it produces little lift. It is clear from the case of the Molossids that species with narrow wings have no choice—they must fly fast to produce enough air flow to stay airborne.

Some bat species, for example, *Noctilio leporinus, Myotis daubentoni*, and *Pizonyx vivesi* (fig. 1.16) prefer to fly over bodies of water where they use their claws to snare arthropods and occasionally small fish from the surface of the water. This style of hunting requires an economical flight technique and a great deal of endurance, combined with a high degree of maneuverability. All of these properties are provided through lengthening the hand portion of the wing, which ensures low wing loading even with a medium to large aspect ratio.

Frugivorous and flower-visiting bats and flying foxes perform commuting flights to reach the trees on which they feed. Thus, their flight style does not require high speed or agility, but rather strength, endurance, and economy. For example, bats of the neotropical nectar-feeding species *Leptonycteris curasoae* commute an average of 27 km at a flight speed of about 8 m/s on daily one-way flights to reach their foraging areas. Similar commuting distances and flight speeds have been reported for Australian flying foxes. The powerful musculature of flying foxes increases their body weight and consequently increases wing loading. This high degree of wing loading is compensated for by the short, broad form of the wings. Despite the fact that such wings should theoretically provide a high degree of maneuverability, flying foxes, due to their body weight, are not very agile. They

Table 1.1. Wing measurements

Species	M (g)	B (cm)	S (cm²)	ar	M_g/S (N/m²)	Tl	Diet
Pteropus vampyrus	**1180**	**130**	**2000**	8.4	**57.8**	1.30	fn
Rousettus aegyptiacus	140	57	558	5.9	24.6	1.08	fn
Cynopterus sphinx	41	41	258	6.7	15.6	1.50	fn
Rhinopoma hardwickei	16	28	114	6.9	14.0	**0.89**	i
Saccopteryx bilineata	7	27	125	6.1	5.9	1.29	i
Taphozous kachhensis	50	46	219	9.5	22.4	1.41	i
Craseonycteris thongl.	**2**	**16**	**36**	7.1	5.2	1.07	i
Nycteris thebaica	1	31	171	5.5	6.3	1.57	i
Megaderma lyra	38	44	312	6.2	11.8	1.70	ci
Rhinolophus ferrumequinum	23	33	182	6.1	12.2	1.22	i
Hipposideros bicolor	6	25	100	6.2	6.1	1.00	i
Hipposideros speoris	11	28	121	6.5	8.9	0.97	i
Noctilio leporinus	59	58	380	9.0	15.2	1.55	pif
Pteronotus suapurensis	8	30	110	8.0	7.3	1.06	i
Phyllostomus hastatus	107	56	417	7.6	25.2	1.63	fic
Phyllostomus discolor	42	42	262	6.6	15.8	1.33	fin
Anoura geoffroyi	14	28	111	7.2	12.5	1.68	nfi
Carollia perspicillata	19	32	165	6.1	11.4	1.61	fin
Stenoderma rufum	22	27	120	5.9	18.2	**2.19**	f
Desmodus rotundus	28	37	200	6.7	14.0	1.12	s
Thyroptera discifera	3	21	75	5.9	**4.1**	1.46	i
Myotis bechsteini	10	26	110	6.0	9.0	1.02	i
Myotis daubentoni	7	25	98	6.3	7.0	1.22	i
Myotis emarginatus	7	24	93	5.9	7.1	1.16	i
Myotis lucifugus	7	24	93	6.0	7.5	1.05	i
Myotis myotis	26	38	233	6.3	11.2	1.22	i
Myotis mystacinus	5	21	75	6.0	7.1	1.14	i
Myotis nattereri	7	27	113	6.4	6.1	1.31	i
Pizonyx vivesi	25	45	271	7.4	9.0	1.41	ip
Pipistrellus mimus	3	19	53	6.6	6.2	1.23	i
Pipistrellus pipistrellus	5	22	63	7.5	8.1	1.18	i
Nyctalus noctula	26	34	161	7.4	16.1	1.43	i
Eptesicus fuscus	16	32	166	6.4	9.4	1.15	i
Barbastella barbastellus	10	26	111	6.0	9.1	1.20	i
Plecotus auritus	9	27	124	5.7	7.1	1.19	i
Plecotus austriacus	10	28	124	6.1	7.9	1.27	i
Miniopterus schreibersi	14	31	137	7.0	10.2	1.46	i
Antrozous pallidus	17	36	210	6.1	8.1	1.16	ifc
Mystacina tuberculata	13	27	108	7.0	12.3	1.19	ifn
Tadarida brasiliensis	12	29	106	8.2	11.5	1.38	i
Otomops martiensseni	35	47	234	9.3	14.9	1.42	i
Molossus ater	29	36	120	11.1	23.4	1.59	i

M, body mass; *B*, wingspan (tip to tip); *S*, wing area; ar, aspect ratio (B^2S); M_g/S, wing loading; Tl, tip length ratio. Diet: i, insectivorous; f, frugivorous; c, carnivorous; p, piscivorous; s, sanguivorous. Highest and lowest values of all parameters are indicated in bold type. (From Norberg and Rayner (1987) and Webb et al. (1992).

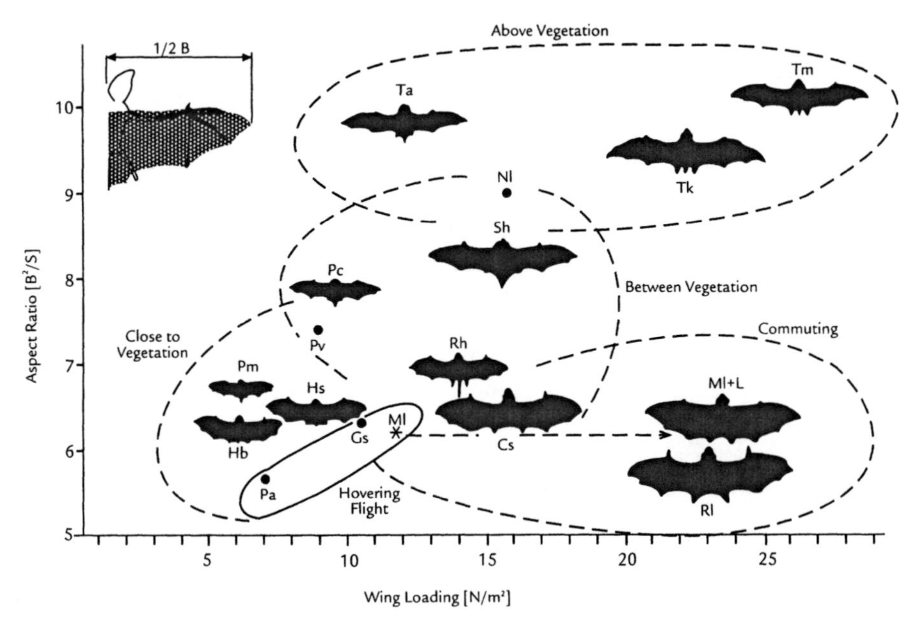

Figure 1.16 Adaptation of the form of the wing and the resulting flight style to different foraging environments. Fast fliers (e.g., *Taphozous*), the "swallows" of the bat world, hunt above the vegetation. These species have narrow, long wings, wide wingspans, and high wing loading. Species with good maneuverability (e.g., *Hipposideros*) forage near vegetation. These species have short, broad wings and low wing loading. Fruit-eating bats (flying foxes) and carnivorous bats that take prey from the ground (e.g., *Megaderma* flying with a load, Ml+L) have large wings and high wing loading. Cs, *Cynopterus sphynx* (flying fox); Gs, *Glossophaga soricina* (nectar-eating Phyllostomatid); Hb, *Hipposideros bicolor*; Hs, *Hipposideros speoris*; Ml+L, *Megaderma lyra* with load (prey); Ml, *Megaderma lyra*, no load; Nl, *Noctilio leporinus* (fishing bat); Pa, *Plecotus auritus*; Pc, *Pipistrellus ceylonicus*; Pm, *Pipistrellus mimus*; Pv, *Pizonyx vivesi* (fishing bat); Rh, *Rhinopoma hardwickei*; Rl, *Rousettus leschenaulti* (flying fox); Sh, *Scotophilus heathi*; Ta, *Tadarida aegyptiaca*; Tk, *Taphozous kachhensis*; Tm, *Taphozous melanopogon*.

cannot fly fast because their wings are too broad (fig. 1.16). Although these characteristics would theoretically make flying foxes ill-suited for exploiting their environment, in practice they provide an economical flight style that enables them to carry heavy loads. Thus, female flying foxes are able to carry their young with them for several weeks after birth, held in the fold where the wing joins the body. This increases their weight by about 50%.

Carnivorous bats that prey on frogs, mice, reptiles, and birds all have broad wings, but a surprisingly low level of wing loading (fig. 1.16). However, their wings are adapted to carry a much heavier load than their body weight. These bats can pick up and carry prey equal to their own body weight and fly with it to a secluded spot where they can devour it (fig. 1.16).

Insectivorous bats, with their medium to wide wingspans, would seem maladapted for hunting among vegetation, where insects are most numerous. However, insectivorous bats achieve good maneuverability through their low body weight, which results in a low level of wing loading. In fact, there are no large insectivorous bats. The largest insectivorous species is the fast-flying molossid *Cheiromeles torquatus*, with a body weight of 140 g.

Rhinopoma (fig. 1.16) is notable for the extreme shortness of the hand portion of its wing, which is shorter than the proximal arm segment of the wing. This wing structure does not fit the typical pattern for bats that forage in open spaces as *Rhinopoma* does. Because the elongation of the hand portion of the wing is considered to be a progressive phylogenetic characteristic, many taxonomists place *Rhinopoma*, with its short wings, at the bottom of the species radiation.

The fact that bats occupy many different ecological niches and foraging environments manifests itself in a great variety in the form and size of their wings. Through adaptive changes in wing structure, a large variety of bat species can live sympatrically within the same area and exploit all of the available sources of food. As will be shown in section 6.5, adaptation to a particular niche requires not only specific flight capabilities, but also specific echolocation abilities.

References

Skeleton

Bennett MB (1993). Structural modifications involved in the fore- and hindlimb grip of some flying foxes. J Zool 229:237–248.

Grassé PP (1955). Traité de Zoologie Vol. 17. Masson, Paris.

Jespen GL (1970). Bat origins and evolution. In W.A. Wimsatt, ed., Biology of Bats, Vol. I, pp. 1–64. Academic Press, New York.

Kwiecinski GG, Krook L, Wimsatt WA (1987). Annual skeletal changes in the little brown bat *Myotis lucifugus* with particular reference to pregnancy and lactation. Am J Anat 178:410–420.

Starck D (1982). Vergleichende Anatomie der Wirbeltiere, Vol. 3. Springer Verlag, Heidelberg.

*Vaughan TA (1970). The skeletal system. In W.A. Wimsatt, ed., Biology of Bats, Vol. I, pp. 98–138. Academic Press, New York.

*Vaughan TA (1970). The muscular system. In W.A. Wimsatt, ed., Biology of Bats, Vol. I, pp. 140–194. Academic Press, New York.

Flight

Aldridge HDJN (1987). Body accelerations during the wingbeat in six bat species: the function of the upstroke in thrust generation. J Exp Biol 130:275–293.

Aldridge HDJN (1987). Turning flight of bats. J Exp Biol 128:419–425.

Carpenter RE (1985). Flight physiology of flying foxes. *Pteropus poliocephalus*. J Exp Biol 114:619–647.

Carpenter RE (1986). Flight physiology of intermediate sized fruit bats (Pteropodidae). J Exp Biol 120:79–103.

Fenton MB, Racey P, Rayner JMV, eds. (1987). Recent Advances in the Study of Bats. Cambridge University Press, Cambridge.

Gebhard J (1982). Unsere Fledermäuse. Veröffentl. Naturhistor Mus Basel, No. 10.

Helversen von O (1986). Blütenbesuch bei Blumentledermäusen: Kinematik des Schwirrfluges und Energiebudget im Freiland. In W. Nachtigall, ed., Bat Flight—Fledermausflug Biona-Report 5, pp. 107–126. G. Fischer Verlag, Stuttgart.

Hermanson JW, Altenbach JS (1983). The functional anatomy of the shoulder of the pallid bat *Antrozous pallidus*. J Mammal 64:62–75.

*Hertel H (1963). *Struktur, Form, Bewegung*. Krauskopf Verlag, Mainz.

Hill JEH, Smith JD (1984). Bats, a Natural History. British Museum (Natural History), London.

Kluge AG (1977). Chordate structure and function. Macmillan, New York.

*Nachtigall W, ed. (1986). Bat Flight—Fledermausflug. Biona-Report 5. G. Fischer Verlag, Stuttgart.

Norberg U (1976). Aerodynamics, kinematics, and energetics of horizontal flapping flight in the longeared bat, *Plecotus auritus*. J Exp Biol 65:179–212.

Norberg U (1976). Aerodynamics of hovering flight in the longeared bat *Plecotus auritus*. J Exp Biol 65:459–470.

Norberg U (1985). Flying, gliding, and soaring. In J Hildebrand, ed., Functional Vertebrate Morphology, pp. 129–158. Harvard University Press, Cambridge, Mass.

Norberg U (1986). On the evolution of flight and wing forms in bats. In W. Nachtigall, ed., Bat Flight—Fledermausflug Biona-Report 5, pp. 13–26. G. Fischer Verlag, Stuttgart.

*Norberg U (1990). Vertebrate Flight. Zoo-physiology, Vol. 27. Springer Verlag, Heidelberg.

Norberg U, Kunz TH, Steffensen JF, Winter Y, Helversen von O (1993). The cost of hovering and forward flight in a nectar-feeding bat. *Glossophaga soricina*, estimated from aerodynamic theory. J Exp Biol 182:207–227.

*Norberg UM, Rayner JMV (1987). Ecological morphology and flight in bats: wing adaptations, flight performance, foraging strategy and echolocation. Phil Trans R Soc Lond B 316:335–427.

Padian K (1985). The origins and aerodynamics of flight in extinct vertebrates. Palaeontology 28:413–434.

Pennycuick CJ (1971). Gliding flight of the dogfaced bat *Rousettus aegyptiacus* observed in a wind tunnel. J Exp Biol 55:833–845.

Quinn TH, Baumel JJ (1993). Chiropteran tendon locking mechanism. J Morphol 216: 197–208.

Rayner JMV (1981). Flight adaptations in vertebrates. Symp Zool Soc Lond 48:137–172.

Rayner JMV (1986). Vertebrate flapping flight mechanics and aerodynamics, and the evolution of flight in bats. In W. Nachtigall, ed., Bat Flight—Fledermausflug Biona-Report 5, pp. 27–74. G. Fischer Verlag, Stuttgart.

*Rayner JMV, Jones G, Thomas A (1986). Vortex flow visualizations reveal change in upstroke function with flight speed in bats. Nature 321:162–164.

Sahley CT, Horner MA, Fleming TH (1993). Flight speeds and mechanical power outputs of the nectar-feeding bat. *Leptonycteris curasoae*. J Mammal 74:594–600.

Siefer W, Kriner E (1991). Soaring bats Naturwissenschaften 78:185.

Speakman JR, Racey PA (1991). No cost of echolocation for bats in flight. Nature 350:421–423.

Starck D (1979). Vergleichende Anatomie der Wirbeltiere, Bd 2 (1979) and Bd 3 (1982). Springer Verlag, Heidelberg.

Swartz SM, Bennett MB, Carrier DR (1992). Wing bone stresses in free flying bats and the evolution of skeletal design for flight. Nature 359:726–729.

Thomas SP, Suthers RA (1972). The physiology and energetics of bat flight. J Exp Biol 57:317–335.

Thomas ST (1975). Metabolism during flight in two species of bats, *Phyllostomus hastatus* and *Pteropus gouldii*. J Exp Biol 63:273–293.

*Vaughan TA (1970). Adaptations for flight in bats. In B.H. Slaughter, D.W. Walton, eds., About Bats, pp. 127–143. Southern Methodist University Press, Dallas, Tx.

Webb PI, Speakman JR, Racey PA (1992). Inter- and intra-individual variation in wing loading and body mass in female pipistrelle bats: Theoretical implications for flight performance. J Zool 228:669–673.

Wimsatt WA ed. (1970). Biology of Bats, Vol. 2. Academic Press, New York.

Wimsatt WA ed. (1977). Biology of Bats, Vol. 3. Academic Press, New York.

2

THE CIRCULATORY AND RESPIRATORY SYSTEMS

THE DEVELOPMENT OF wings meant that bats had to undergo many related adaptations as well, especially for the systems concerned with gas exchange and thermoregulation. For long distances, flight is the most efficient method of locomotion, but it requires three times as much energy output per unit time as does walking or running. Therefore, the circulatory and respiratory systems of bats are adapted to meet these high energy requirements and to operate under highly demanding conditions.

The blood vessels are the body's transportation system. They transport not only oxygen and carbon dioxide, but also many other types of "goods"—nutrients, waste products, ions, hormones, antibodies, and heat. The circulating blood carries its own repair system with it in the form of platelets and fibrinogen. Organs with a high rate of metabolism such as the liver, intestine, kidneys, and brain, are especially well supplied with capillaries. In fact, it is possible to estimate the importance of an organ for an animal based on capillary density. In echolocating bats, for example, the inferior colliculus, a structure that is of great importance for hearing, is especially densely supplied with capillaries. The pectoralis muscle has 3660–6400 capillaries/mm^2 as determined from a cross section through the muscle, making it more densely supplied with capillaries than any other muscle in mammals.

2.1 THE HEART

The high demand of the body for oxygen during flight is reflected in the properties of the heart (fig. 2.1). In relative terms, bats have the largest and most muscular heart of any mammal. It represents 0.6–1.3% of the bat's body weight and is two to three times as heavy as the heart of a mouse of the same size. The increase in weight is mainly due to the larger muscle mass of the right ventricle, one of the two lower chambers of the heart, and the larger size of the right atrium, one of the two upper chambers. The right side of the heart is presumably enlarged to accommodate the greatly increased return of venous blood to the heart during flight and to pump it to the lungs.

The cardiac muscle of bats is well adapted to the high demands of flight. Because the individual muscle fibers are thinner than those of other mammals, more

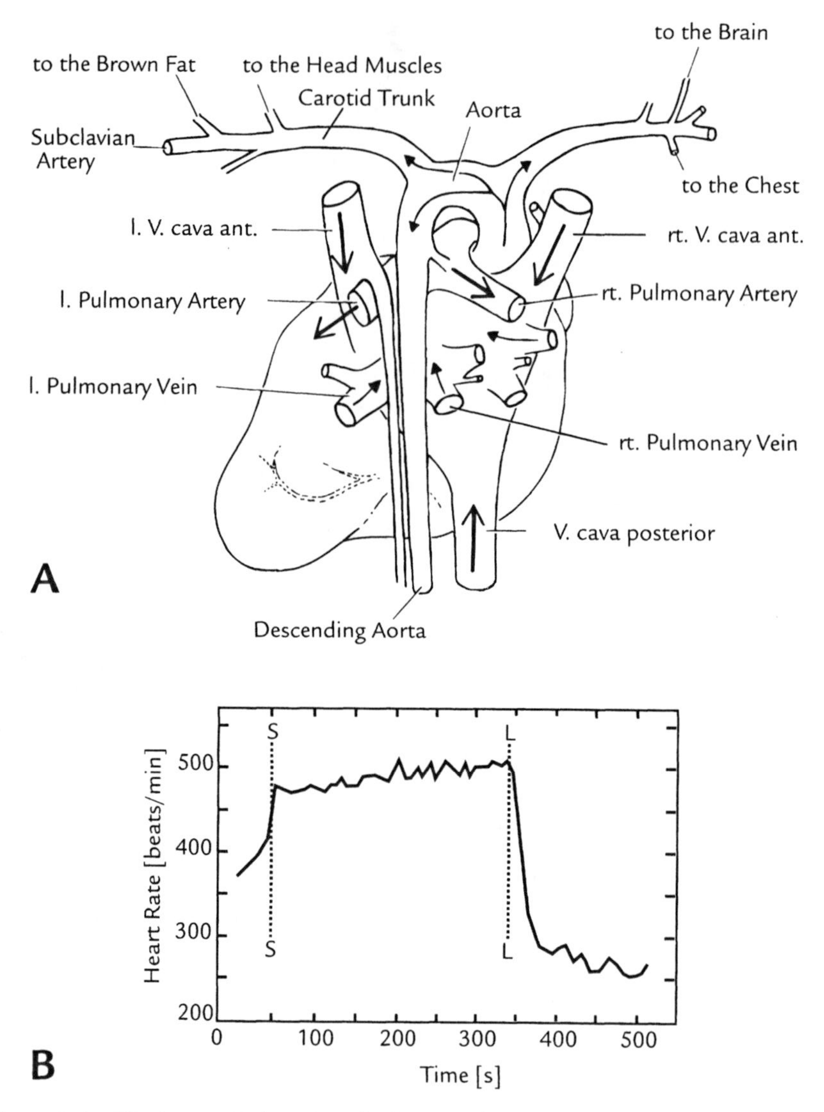

Figure 2.1 The heart of a bat. (A) The heart and large blood vessels of the flying fox *Eidolon helvum* viewed from the rear (dorsal). Arrows indicate direction of blood flow. From Kallen (1977). (B) Heart rate of the flying fox *Pteropus poliocephalus* during flight. S, start; L, landing. Flight speed 7 m/s. From Carpenter (1985).

fibers can be packed into a given volume of muscle tissue, resulting in a more powerful contraction. Appropriately, the cardiac muscle fibers of bats contain the highest level of energy reserves in the form of ATP (adenosine triphosphate) that has been measured in the heart of any animal. To meet the high oxygen requirements of bats in flight, the heart is more densely supplied with capillaries than it is in any other mammal.

Table 2.1 Heart function during exercise

Species	Body weight (g)	Heart rate (beats/min)	Cardiac output (l/h-kg)	Stroke volume (ml/kg)
Phyllostomus hastatus	87	780	89.8	1.92
White-footed mouse (*Peromyscus leucopus*)	24	700	59.5	1.42
Evening grosbeak (*Hesperiophona vesp.*)	59	840	170.0	3.38
Bird (calculated)	100	730	119	2.7

From Thomas and Suthers (1972).

STROKE VOLUME AND BLOOD PRESSURE

In bats, the stroke volume (the amount of blood pumped with each heartbeat) is about average for a mammal: 0.03 ml in an *Eptesicus fuscus* weighing 22 g, and 0.23–0.26 ml in a *Phyllostomus hastatus* weighing 120 g. In humans, the stroke volume is about 65 ml (see table 2.1 for additional comparative data). In an anesthetized *Myotis lucifugus*, the systolic blood pressure (the pressure during pumping of blood by the ventricle) measured near the heart is about 105 mm Hg. Diastolic pressure (measured during filling of the atrium) is about 40 mm Hg. The blood pressure in the small-diameter arteries of the flight membrane is maintained around 100 mm Hg, even though venous pressure in the flight membrane sinks as low as 27 mm Hg.

It would be interesting to know exactly how cardiac muscle in bats functions during hibernation (see fig. 2.2). The only thing that is known for certain is that the muscle enzymes and the contractile mechanism must be temperature tolerant because the heart muscle is capable of contracting and pumping blood even at near-freezing temperatures. However, if a hibernating bat is awakened through the action of norepinepherine, it runs the risk of fibrillation because at low body temperatures the mechanical contraction of the cardiac muscle is too slow to follow the neural volleys fired by the sinus node.

When the bat's oxygen requirements increase due to strenuous physical activity, the heart rate increases (figs. 2.1b and 2.3). There are few other mammals with such a wide range of heart rates—from 4/min during hibernation to 1100/min during flight—and few other mammals that can alter their heart rate as rapidly as bats do. As shown in fig. 2.3, the heart rate of a bat can change within a matter of seconds from 500/min to 1100/min. In humans, the heart rate varies from about 70/min at rest to a maximum of 180/min during strenuous work.

As in other mammals, there are two factors that regulate the heartbeat in bats: the volume of venous blood returning to the heart and the autonomic innervation of the heart, which acts to adjust the heart rate to different physiological conditions as they occur. (An isolated bat heart kept at body temperature beats with an intrinsic rhythm of 450 beats /min).

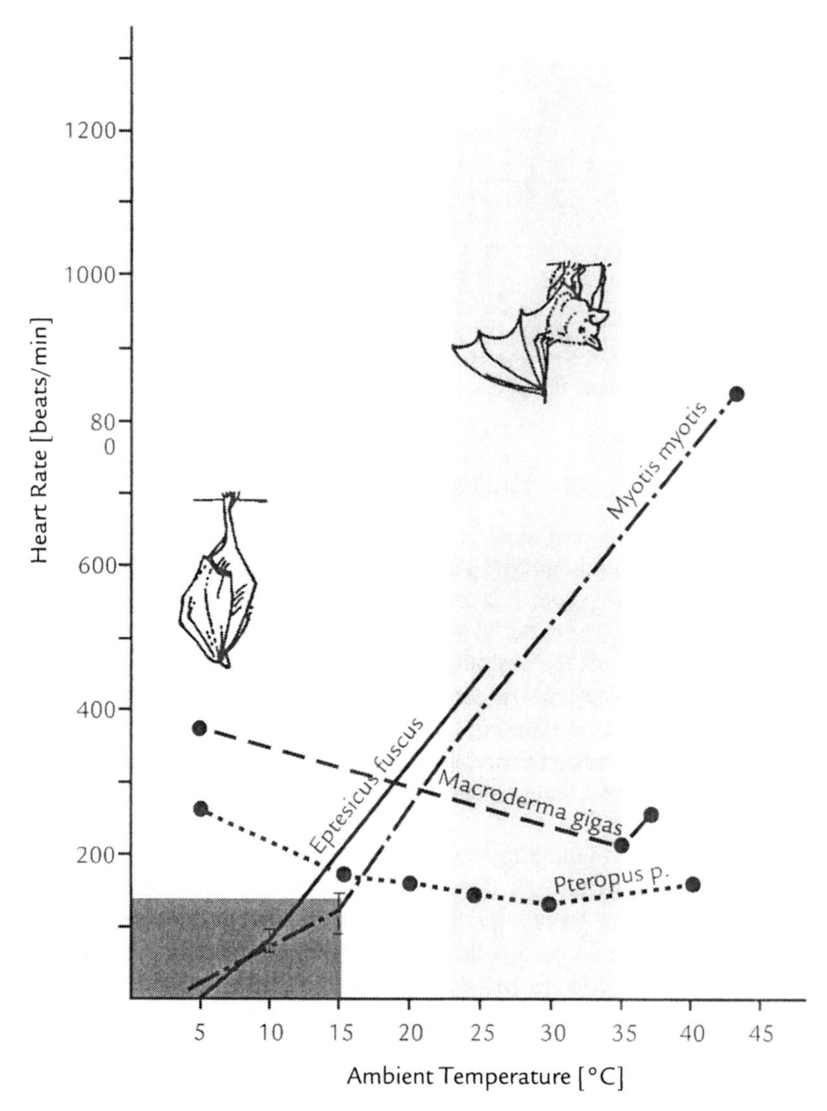

Figure 2.2 Dependence of heart rate on ambient temperature in normothermic bats (*Macroderma gigas* and *Pteropups poliocephalus*) and in hibernating Microchiroptera (*Eptesicus fuscus* and *Myotis myotis*). Left densely stippled area indicates temperature range over which hibernation occurs.

The sympathetic nervous system, which uses norepinepherine as a neurotransmitter, speeds up the heart rate. The rapid increases in heart rate at the start of flight or during arousal from hibernation are initiated through dense adrenergic innervation. Bats also have a powerful cholinergic innervation of the ventricles of the heart, which is uncommon in mammals. Acetylcholine, the neurotransmitter of the parasympathetic system, slows the heart rate.

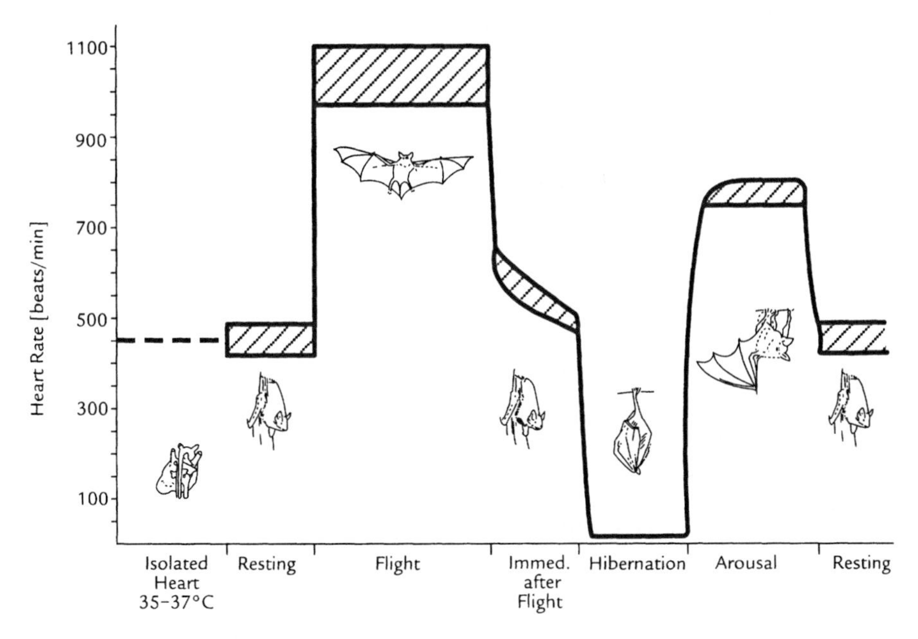

Figure 2.3 Average heart beat frequency of Microchiroptera during different activities, during hibernation, and measured from an isolated heart. The striped bars indicate the range of variability.

2.2 THE BLOOD VESSELS

The course of the blood vessels in bats follows the typical mammalian plan. Blood is pumped from the left ventricle into the descending aorta. The two pulmonary arteries, which are phylogenetically derived from the sixth branchial arch of the fish, exit from the right atrium. To meet the dynamic requirements of the circulation in different organs, the arteries are appropriately differentiated.

THE VEINS

The venous system in bats retains much of the original network characteristics of the mammalian circulatory system. The thin-walled venules collect blood from the capillary networks in the various organs and channel it into two large-diameter veins, the superior vena cava and the inferior vena cava. Both of these veins empty into the right atrium (fig. 2.1). Because the venous system of bats also functions as a storage reservoir for blood, it has several unusual characteristics.

As already mentioned, the venous right half of the heart has a comparatively thick and muscular wall and a higher capacity than that of flightless mammals. Because the large wing area must be constantly supplied with blood, the anterior vena cava in bats terminates in the right ventricle with two branches instead of one (fig. 2.1). The posterior vena cava, which ascends from the rump through the abdomen, is large in diameter and elastic until the point where it passes through the diaphragm. The short remaining portion between the diaphragm and the atrium has a narrow lumen and is extremely muscular. It has been suggested that this portion of the vena cava functions as an elastic blood reservoir of variable diameter that regulates the return of blood to the atrium. When an animal is at rest, the diameter narrows, limiting the amount of blood returned to the heart. Thus, part of the animal's blood supply remains in the storage basin formed by the abdominal section of the posterior vena cava. When the bat takes flight, the muscular thoracic portion of the vena cava relaxes, so that the stored blood is immediately made available for the increased circulatory requirements of flight. Unfortunately, no experiments have been done to test this hypothesis.

In considering the many adaptations bats have undergone to make them suited for their unique lifestyle, one of the first that comes to mind is the upside-down position in which they rest. This position aids circulation because the only part of the system in which blood must be moved against gravity is in the veins from the head, which lie close to the heart. The oxygen-poor blood from regions distant from the heart is returned with the help of gravity. The only morphological adaptation to the upside-down position is the ventral sterno-pericardial ligament that helps stabilize the position of the pericardium in the thorax.

THE BLOOD SUPPLY TO THE HEAD

In bats, blood is supplied to the head through two arteries that branch off from the two thick arteries that supply the arms. The paired vertebral arteries unite to form the basilar artery, which supplies large parts of the brain and the inner ear. In Megachiroptera, a branch of the basilar artery—the ophthalmic artery—goes to the eye. The ophthalmic artery is absent in species with very small eyes, such as horseshoe bats and vespertilionids. The other two major arteries that supply the head are the internal and external carotid arteries. These arteries originate from the common carotid artery. They primarily supply the muscles and other soft tissues of the head. In lower vertebrates, the internal carotid branches to form an artery that supplies the jaw region and the soft tissues of the mouth. During mammalian embryogenesis, this artery runs through the stapes of the middle ear and is therefore called the stapedial artery. In most adult mammals, the stapedial artery has disappeared, and its function has been taken over by the external carotid. This is also true in Megachiroptera. In many species of Microchiroptera, this artery persists into adulthood. It delivers blood to the meninges and eye sockets and, in horseshoe bats and vespertilionids, it is even larger than the internal carotid. In *Artibeus jamaicensis* (Phyllostomidae) and in vampire bats, the stapedial artery has completely supplanted the internal carotid.

The external carotid artery provides the blood supply to the tongue and jaw region. One branch of the external carotid, the laryngeal artery, is extremely large in species that echolocate. In Pteropid flying foxes, on the other hand, this artery is so thin that it is named after the thyroid, which it also supplies.

2.3 CIRCULATION IN THE WING MEMBRANE

The large surface area of the wing membrane requires good circulation. To accomplish this, the thick subclavian artery diverges at the level of the shoulder into many different branches, three of which are among the most important arteries for flight (fig. 2.4a): (1) branches of the ulnar artery supply the front part of the plagiopatagium, (2) branches of the radial artery and collaterals of the ulnar artery supply the arm and the propatagium, and (3) the median artery runs alongside the bones of the arm to the wrist, where it branches to form the arteries of the fingers, which, in turn, supply the entire hand portion of the wing membrane.

ANASTOMOSES

The arteries of the wing membrane not only branch to form a fine capillary network, but also form anastomoses—shunts formed by the direct connection of arteries with the elastic venous system. Anastomoses are especially prominent in the fingers and at the edges of the flight membrane. The shunting circuit in the arm area is formed by the median artery, the anastomoses of the finger regions, and the large cephalic vein that runs along the front edge of the propatagium. On the arterial side, the anastomoses can be closed through the action of sphincters, muscular valves that control flow (fig. 2.5a). When the sphincters are closed, all of the arterial blood flows through the capillary network, and circulation in the flight membrane is maximal. When the sphincters are open, much of the blood pumped to the flight membrane is shunted directly back to the veins and does not pass through the capillary network.

There is controversy regarding the conditions under which these shunts open and close. Some authors contend that the anastomoses open during flight because at high heart rates more blood is pumped into the flight membrane than is necessary to maintain it. When the bat is resting in its daytime roost, the anastomoses close so that blood flows through the highly branched capillary network. Only when the bat flies out in the evening do the cephalic veins fill with blood due to the opening of the anastomoses at takeoff. The regulation of circulation by anastomoses must certainly play a role in thermoregulation, but unfortunately there have been no experiments on this topic.

CAPILLARY NETWORK

Blood flow in the capillary network (fig. 2.5) is driven by the pressure gradient between the incoming and outgoing vessels of the network. The blood pressure in the arterioles of the fingers is approximately 90–95 mm Hg. The pressure gradient arises through the resistance of the blood vessels. The pressure on the arterial side is about three times that on the venous side. The pressure in the capillaries is directly proportional to the ratio of the arterial resistance to the venous resistance, while the velocity of blood flow is inversely proportional to the sum of the vascular resistance.

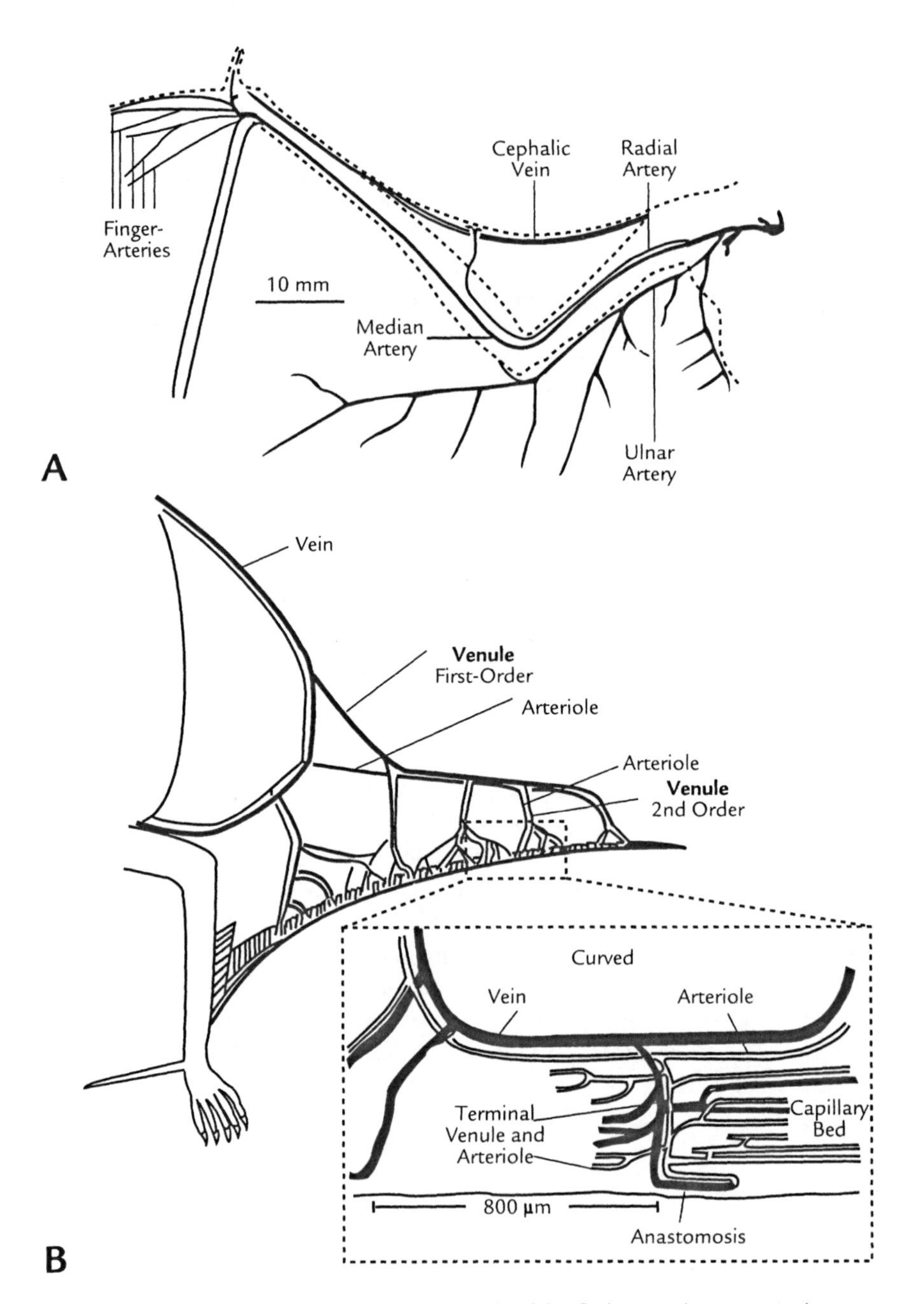

Figure 2.4 Blood vessels and capillary network of the flight membrane. (A) The most important blood vessels in the flight membrane of *Myotis lucifugus*. (B) Schematic illustration of the capillary supply and an anastomosis at the rear edge of the flight membrane. From Slaaf et al. (1987).

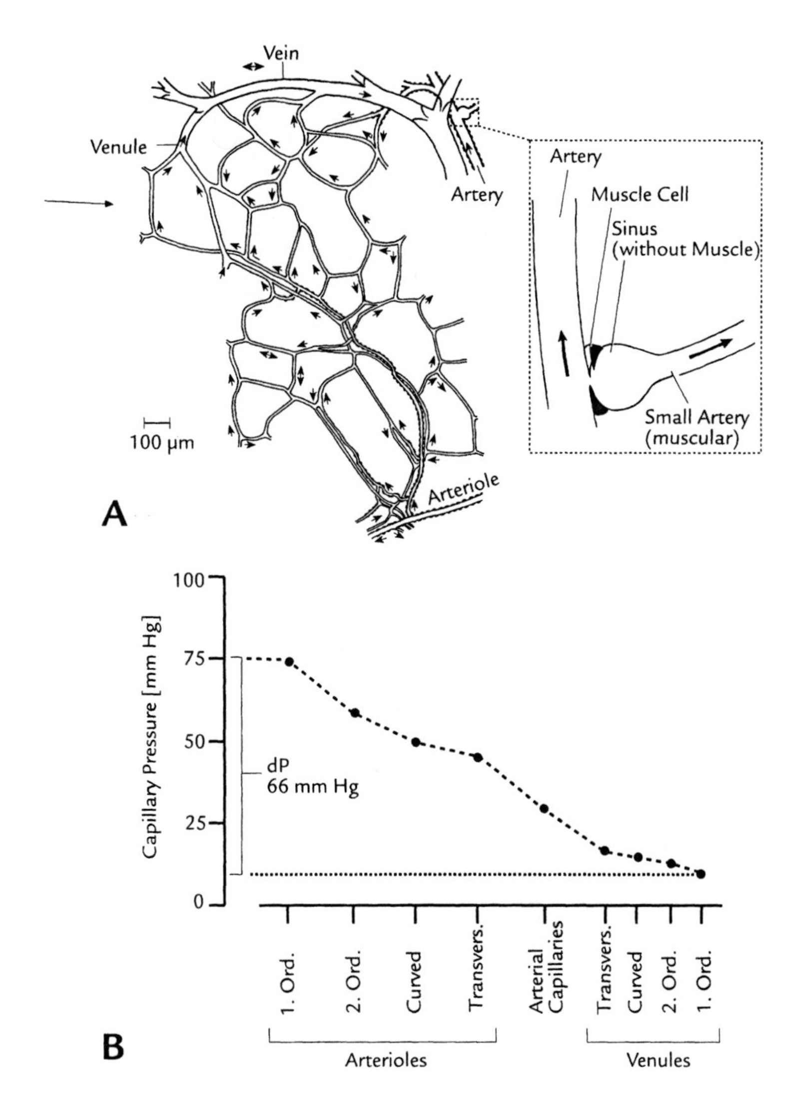

Figure 2.5 Example of capillary circulation in the flight membrane. (A) Blood flow (arrows) from an arteriole to a vein. Dotted vessels indicate those with muscular walls. The inset shows the right-angle branching of a blood vessel with a muscular sphincter and the inflation of the elastic, nonmuscular sinus which serves as a regulator. From Kallen (1977). (B) Sequential changes in blood pressure. There is a progressive drop of about 66 mm Hg in blood pressure going from the first order arterioles to the first order venules. From Slaaf et al. (1987).

The large vessels of the flight membrane give rise to first-order venules and arterioles, which in turn give rise to secondary branches (fig. 2.4). The second-order blood vessels give rise to curved venules and arterioles, which in turn give rise to the terminal (transverse) arterioles and venules, which branch off at right angles. These are the vessels that are connected with one another through the capillaries. Unlike the arterioles and venules just described, capillaries do not have a muscular wall. The fact that the terminal arterioles branch off at right angles uncouples them from the high pressures in the parent vessel, thereby increasing the pressure gradient and facilitating blood flow. At the points where the terminal arterioles branch off at right angles, the small-diameter side has a sphincter that functions specifically to regulate blood pressure and thereby control blood flow. The arterioles that flow into the capillary bed are thin-walled and elastic just behind the sphincter. This part of the arteriole acts as a regulator to transform pulsatile pressure waves caused by the pumping of the heart into a continuous pressure level (fig. 2.5a, inset).

The regulation of pressure by contraction and dilatation of the muscular walls of the curved terminal arterioles is the most important mechanism controlling circulation in the wing membrane. This mechanism is capable of maintaining a constant rate of blood flow by compensatory changes in vascular resistance and hence in blood pressure. The sphincters at the branch points of the arterioles provide a mechanism to maintain a constant, optimal blood pressure gradient overall, while varying local blood flow in accordance with specific conditions. Through such mechanisms, the local capillary beds connecting terminal arterioles and venules can quickly switch to a high rate of flow and, like the anastomoses, act as shunts within the circulatory system. This type of high flow-rate capillary bed is especially common around the borders of the wing membrane (fig. 2.5a).

"VENOUS HEARTS"

The long distance to and from the heart and the centrifugal force created by the beating of the wings appear to have led to a particular kind of adaptation in the wings of bats. The smooth muscle fibers in the walls of the blood vessels not only contract tonically as they do in other mammals, but they also contract with a rhythmic, peristaltic action. These pumping contractions occur to some extent in the arteries of the flight membrane, and they occur even more frequently and powerfully in the veins of the wing. These contractions have given rise to the rather confusing term *Venenherzen* or "venous hearts."

Each contraction begins at one of the venous valves and proceeds in the direction of the heart until it encounters the next valve, where it then dies out. In a resting bat, at 25°C, the veins contract about 12 times a minute, but the rate can increase to as much as 50 times a minute. During each contraction, the diameter of the vein is reduced by 14–56%. The contractions in the arteries are not as frequent or as strong as those in the veins. The contractions begin spontaneously, without neural input. As in the heart, they arise through the intrinsic rhythmic contractile properties of the musculature itself.

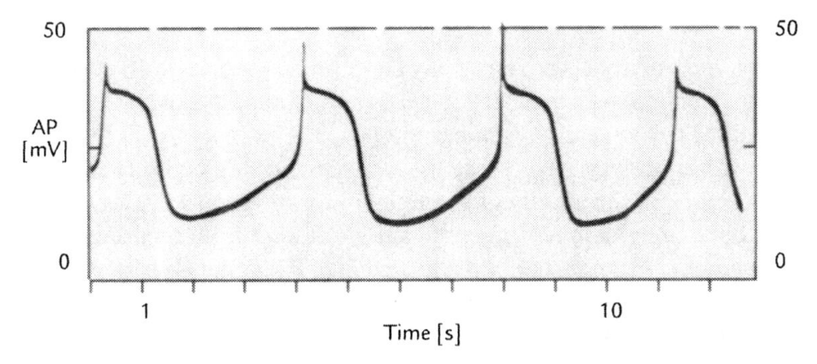

Figure 2.6 Action potentials (AP) recorded from a "venous heart" in *Myotis lucifugus*. From Kallen (1977).

Action potentials in the venous musculature begin with a steep depolarization which causes contraction of the muscle (fig. 2.6). As in the case of cardiac muscle, repolarization to the point where reexcitation can occur is slow; the fibers remain depolarized for several hundred milliseconds. It is thought that elevated blood levels of CO_2 lead to an increased rate of contraction, while elevated oxygen levels lead to a decreased rate of contraction. However, because smooth muscle reacts to stretch with depolarization and contraction, it seems more probable that venous contraction is controlled by the degree of filling of the veins. Veins that have been isolated from the wing membranes of bats and filled with bovine blood continue to contract for up to 10 days after their removal from the animal. In fact, recently it has been experimentally demonstrated that intravenous pressures modulate the contraction rates of wing veins.

The highly differentiated and specialized blood supply of the wing membrane points to the vital role of the wing for Chiroptera. The contractile properties of the veins and arteries, the anastomoses, and the locally controlled capillary beds with their right-angle branching all act together to supply blood flow to an enormous surface area compared to the rest of the body, and to provide a dynamic regulatory system that allows an instantaneous transition from rest to flight. In addition, the wing membrane acts as a blood storage reservoir when the bat is at rest and as a heat exchanger during flight and under conditions of heat stress.

2.4 THE BLOOD

The ability of the circulatory system to deliver oxygen to the tissues of the body not only depends on the flow rate in the capillary beds, cardiac output, and similar factors; it also depends on the oxygen-carrying capacity of the blood itself, which in turn depends on total blood volume, the number of red blood cells, their hemoglobin content, and the affinity of the hemoglobin molecule for oxygen.

OXYGEN CAPACITY

The blood volume of a bat is not especially large, ranging from 7 to 10 ml/100 g body weight. For an animal such as a bat that needs to minimize its body weight, it would not be practical to increase oxygen-carrying capacity simply by increasing blood volume. A different way to increase the oxygen-carrying capacity of the blood would be to increase the affinity of hemoglobin for oxygen. However, the oxygen affinity of bat hemoglobin is not significantly different from that of any ground-dwelling mammal. The measure of oxygen affinity, P_{50}, is the O_2 partial pressure at which 50% of the hemoglobin is saturated. In bats, P_{50} ranges from 29 to 50 mm Hg; in humans it is 30 mm Hg. The oxygen affinity of the blood in bat species that hibernate, like that of other hibernating animals, is less subject to temperature-dependent changes than that of homeothermic animals.

The blood of bats contains a higher number of red cells or erythrocytes than does human blood. The maximal value that has so far been measured in bats is 26 million erythrocytes/ml blood. In comparison, mice and insectivores have only 18 million red blood cells/ml. The resulting high hemoglobin concentration gives bats an oxygen capacity of 25–30%. Small ground-dwelling mammals of comparable body size have an oxygen capacity of only about 18%. During periods of inactivity, bats can store up to half of their red blood cells in the spleen. Thus, in inactive animals, the spleen is twice as large as in flying ones.

Erythrocytes in bats are smaller than those of other mammals. The consequent reduction in volume of individual red blood cells improves oxygen binding and thus enhances their ability to transport oxygen. Due to the small size and large number of red blood cells, the total surface area for oxygenation is larger. A measure of this is the hematocrit, or amount of blood cells as a percentage of total blood volume. As shown in table 2.2, the hematocrit, which is determined primarily by the volume of red blood cells, is clearly higher in bats than in other animals. Because hemoglobin is so densely packed into each red blood cell of bats, the total hemoglobin content of their blood is 18–24 g/100 ml blood—higher even than the hemoglobin levels of hummingbirds (18 g/100 ml), which are noted for their extraordinarily high oxygen consumption. The highest hemoglobin level ever measured in any mammal was in a species of pipistrelle bat (24 g/100 ml). This level exceeds that found in Wedell seals, diving animals that remain under water for long periods of time and thus require high oxygen-carrying capacity of the blood.

The oxygen-carrying capacity of the blood is thus able to meet the high demands of flight by means of an above-average hemoglobin content and recruitment of red blood cells from the spleen. The heart is powerful enough to ensure that the circulation remains highly efficient even at heart rates up to 1000/min. The blood circulation in the wing membranes functions not only as a transport system, but also as a reserve area where blood can be stored. In addition, the vasculature of the wing membrane plays an important role in thermoregulation. The arteriovenous shunts allow the circulation in the flight membrane to adapt to the instantaneous needs of the tissue. Thus, in parallel to the ability to fly, bats have developed a circulatory system that is highly specialized to meet the needs of flight. To date, however, the physiology of this system remains largely uninvestigated.

Table 2.2 Blood profile of different bat species compared with small mammals, birds, and humans

Species	Hematocrit (%)	RBC (10^6/µl)	Hb conc. (g/100 ml)	O_2 capacity (vol %)
Pteropus poliocephalus			18	
Rousettus aegyptiacus	57	15.4	17	
Cynopterus brachyotis	63			
Epomoophorus wahlbergi	63			
Phyllostomus hastatus	60			27.5
Antrozous pallidus	65	12.2	18	
Cheiromeles torquatus	64			
Tadarida mops	50			
Miniopterus schreibersi	57	12.8	14	
Miniopterus minor	75	12.8	14	
Pipistrellus pipistrellus	65	14.5	20	
Plecotus austriacus	48	12.1	15	
Myotis myotis	43	9.3	16	
Myotis natteri	60	12.6	20	
Myotis daubentoni (hibernating)	49	11.3	16	
Average value for bats	59	12.5	17	
Small mammals	45	7–8		18
Humans	42	4.5–6.0	16	20
Pigeon	51		17	20
Seagull	43		14	17
Domestic chicken	28		9	12

Data from Maina et al. (1989); Thomas and Suthers (1972); Arevalo et al. (1987); Wrightman et al. (1987); Wolk and Bogdanowicz (1987); Maina and King (1984); Bassett and Wiederhielm (1984); van der Westhuyzen (1988); Biscor et al. (1985).

RBC = red blood cell count; Hematocrit = Volume % of red blood cells per unit blood volume.

2.5 RESPIRATION AND GAS EXCHANGE

Oxygen and carbon dioxide are passively exchanged between the air and the body according to the physical laws of gas diffusion. The driving force is not the concentration gradient, but rather the difference in partial pressure between the air and the organism. In addition, the volume of gas exchanged depends on the surface area across which diffusion can occur, the permeability of the surface to the gas, and the thickness of the surface (see box 2.1). Gas exchange occurs in the alveoli of the lungs, which are highly vascularized by a dense capillary network. The oxygen that is taken in, as well as the carbon dioxide that is produced when oxygen is consumed, is transported from the blood to the surface of the alveoli where gas exchange takes place.

Box 2.1 The Law of Diffusion

$$dM / dT = \alpha d (F \Delta P / \Delta x)$$

d = Diffusion constant for a given membrane
F = Surface area of the membrane
Δx = Thickness of the membrane
ΔP = Difference in partial pressure across the membrane
α = Solubility of a gas in a medium (blood)

GAS DIFFUSION CAPACITY

The alveoli of the mammalian lung provide a huge surface area for gaseous exchange within a small area. For example, an Egyptian tomb bat (*Rousettus aegyptiacus*), weighing 160 g, has a lung surface area of 1 m^2. The membrane through which gas exchange occurs consists of two layers, the lung epithelium and the capillary wall. This two-layered membrane is only 1–2 μm thick.

The lungs of bats have undergone several adaptations that allow them to meet the high oxygen demand during flight:

- The lung volume is greatly increased relative to body weight,
- The alveoli are small, leading to a surface area that is nearly twice as large relative to body weight as that of the shrew, and six times as large as that of the domestic chicken,
- The alveoli are highly vascularized, with a high capillary volume relative to alveolar surface area.

These lung specializations in bats result in a gas diffusion capacity that is higher, relative to body weight, than that of any other mammal or any bird.

OXYGEN EXTRACTION

When mammals breathe, the direction of air flow in the lungs changes as the animal inhales and exhales. In contrast, the lungs of birds are constructed with a system of air sacs and intercommunicating parabronchi so that air always flows from the rear to the front regardless of whether the bird is inhaling or exhaling. This unidirectional respiratory flow has been interpreted as an adaptation for high oxygen requirements during flight. As would be expected, oxygen extraction in the lungs, measured as the fraction of total O_2 that passes into the blood, is greater in birds than in mammals.

Among those animals with a conventional lung, bats are the only ones that achieve oxygen extraction values comparable to those of birds. During flight, bats can transport up to 200 ml O_2/min-kg bw. In comparison, a vigorously exercising man can transport only 60 ml O_2/min-kg bw.

Oxygen extraction during flight. Based on the values for oxygen extraction measured at rest, it would be predicted that during flight the animal would extract 300–400 ml O_2/min-kg bw.

RATE OF RESPIRATION AND MINUTE VOLUME

As shown in fig. 2.7, in a resting bat, respiratory rate and other measures of oxygen consumption are lowest when the ambient temperature is close to, but below, body temperature. The respiratory measure that most closely reflects oxygen consumption is the minute volume (fig. 2.7c), or the volume of air that is breathed in 1 min. The minute volume is the product of the respiratory rate and the inspired volume per breath. At rest, the minute volumes of bats are about the same as those of ground-dwelling animals of comparable size. However, within 3 s of the time the bat takes flight, the minute volume increases 10- to 17-fold. This enormous increase is due to the fact that respiratory rate increases 3- to 5-fold (fig. 2.8) and the inspired volume increases 2- to 4-fold. Depending on body weight and flight speed, respiratory rate during flight ranges from 600 to 100 breaths per minute (fig. 2.9). Thus, the respiratory rate of bats is without exception higher than that of birds. In resting *Plecotus auritus* (body weight 11 g), the respiratory rate increases with oxygen consumption from about 150 to 350 breaths per minute. When oxygen consumption sinks below 0.5 ml/min, bats may alternate between periods in which they breathe at the normal rate and periods without breathing (apnea). The periods of apnea allow the bat to lower the average respiratory rate while maintaining an economically optimal frequency during the periods when breathing does occur.

HYPERVENTILATION

If the oxygen extraction values measured at rest are extrapolated to flight, a flying bat should extract 300–400 ml O_2/min-kg bw—far more than it would need. Therefore, oxygen extraction actually decreases during flight, perhaps through the opening of anastomoses in some of the blood vessels that carry venous blood to the alveoli. This hyperventilation and oversupply of oxygen could be due to two different causes.

1. The respiratory rate in bats is limited by the wingbeat rate (fig. 2.9). Wingbeat and respiratory cycles are coupled in a 1:1 ratio. Because the contraction of the flight muscles contributes to respiration, this coupling is probably obligatory.
2. The lungs not only exchange oxygen and carbon dioxide, they also give off water vapor and heat. The large amount of work performed by the muscles during flight produces a great deal of heat, which must be dissipated. For example, the body temperature in *Phyllostomus hastatus* rises to 42°C during flight. Vigorous ventilation of the lung is one way to rid the body of excess heat. Thus, the respiratory rate during flight may be determined by factors other than oxygen demand.

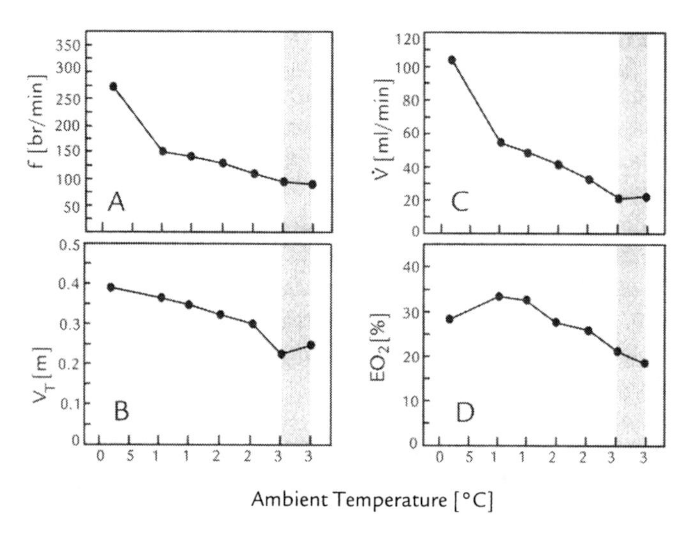

Figure 2.7 Ventilation in a Neotropical bat (*Noctilio albiventris*) as a function of ambient temperature. (A) Respiration frequency; (B) Tidal volume; (C) Minute volume; (D) Oxygen extraction. Shaded: thermoneutral zone (30–35°C). After Chappell and Roverud (1990).

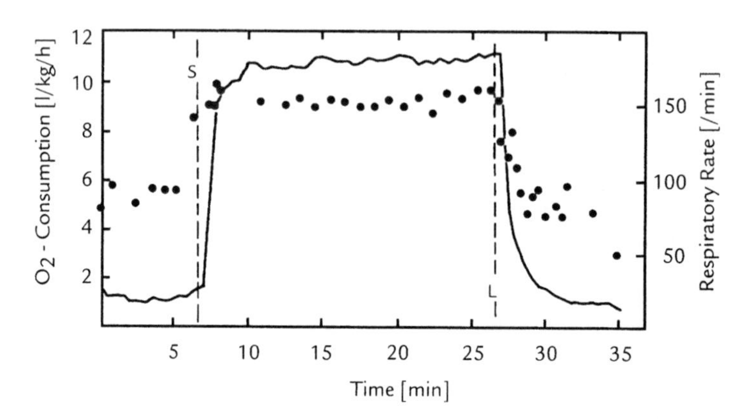

Figure 2.8 Respiration rate (filled circles) and oxygen utilization (solid line) of an 80-g flying fox (*Pteropus poliocephalus*) before, during, and after flight in a wind tunnel. Flight speed = 7 m/s. L, landing, S, take-off. From Carpenter (1985).

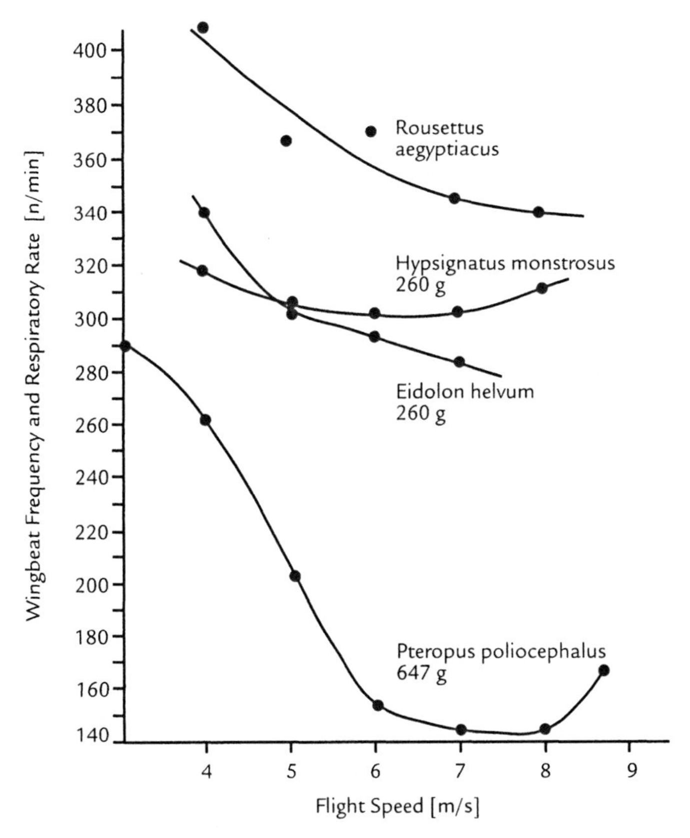

Figure 2.9 Respiration and wingbeat frequency as a function of flight speed in four species of flying foxes. The body weight in grams of each species is indicated next to the corresponding curve. Drawn from data in Carpenter (1985, 1986).

If the respiratory parameters of bats are compared with those of birds, it can be seen that bats are characterized by a high respiratory rate, whereas birds are characterized by a high respiratory volume. In bats, the resulting minute volume depends on the wingbeat frequency; in birds it depends on a large pulmonary dead space consisting of long bronchi and additional air sacs. The respiratory volume of bats is 1.75 times larger than that of ground-dwelling mammals. Based on the median relative lung volumes of 0.08 cm^3/g measured in the flying fox, *Pteropus gouldi*, it is possible to calculate a lung volume of 70 cm^3. With every breath, the flying fox takes in 58% of its total lung capacity. This is high when one considers that humans seldom take in more than 40% of their total lung capacity and that breathing more deeply than this quickly exhausts the respiratory muscles. It is likely that in the case of the flying fox, the flight musculature contributes some of the energy necessary for breathing and thus allows a larger volume per breath.

DO BATS GO INTO OXYGEN DEBT?

During vigorous exercise, mammals require more energy than can be produced by oxidative metabolism, even when the lungs are functioning at maximal capacity. Under these circumstances, there is an alternative energy source available through anaerobic conversion of glucose and glycogen to lactic acid. However, the energy produced through anaerobic glycolysis is only about 6% of that produced through oxidative metabolism. The body develops a short-term "oxygen debt," which can be accurately gauged by the concentration of lactic acid in the blood. When lactic acid concentrations were measured in flying foxes after short flights of less than 2 min, an oxygen debt of 6–7 ml O_2 was measured immediately after flight, but was fully compensated for by 24–28 s after landing.

This result seems to be contradicted by the finding that bats hyperventilate during flight and even decrease the amount of oxygen extraction from alveolar air. However, oxygen debt has only been measured after short flights and not after long ones, so it is possible that this deficit occurs at the start of flight but is compensated for during sustained flight. In any case, bats probably do not utilize anaerobic metabolism any more than do other mammals.

GAS EXCHANGE THROUGH THE SKIN

The naked flight membranes of bats are four to eight times as large as the parts of the body surface that are covered with fur. The skin is highly vascularized, so that heat and water vapor, as well as gases, can diffuse across the skin. There has to date been only one experiment on this topic, showing that *Eptesicus fuscus* and the flying fox *Pteropus poliocephalus* gave off an average of 2–4% of dissipated CO_2 via the skin, but did not take up any oxygen. For comparison, in humans, the fraction of CO_2 dissipated through the skin is 1.4%. The fraction of CO_2 diffusion across the skin is temperature dependent. In *Eptesicus fuscus*, it increases from 0.4% at 18°C to 11.5% at 37.5°C. This temperature dependence is related to the skin circulation, which is maximal at high temperatures in order to dissipate heat from the body.

References

Heart, Circulation, Blood

Agar NS, Godwin IR (1992). Erythrocyte metabolism in two species of bats: common bent-wing bat and red fruit bat. Comp Biochem Physiol 101B:9–12.

Arevalo F, Perez-Suarez G, Lopez-Luna P (1987). Hematological data and hemoglobin components in bats (Vespertilionidae). Comp Biochem Physiol 88A:447–450.

Bassett JE, Wiederhielm CA (1984). Postnatal changes in hematology of the bat *Antrozous pallidus*. Comp Biochem Physiol 78A:737–742.

Bhatnagar KP, Spoonamore BA (1979). Ultrastructure of the atrioventricular node of the big brown bat, *Eptesicus fuscus*. Acta Anat 105:157–180.

Carpenter RE (1985). Flight physiology of flying foxes, *Pteropus poliocephalus*. J Exp Biol 114:619–647.

Condo SG, EL Sherbini S, Shehata YM, Corda M, et al. (1989). Hemoglobins from bats (*Myotis myotis* and *Rousettus aegyptiacus*): A possible example of molecular adaptation to different physiological requirements. Biol Chem Hopp 370:861–867.

Cook RA, Halpern D, Tast J, Dolensek EP (1987). Electrocardiography of the shorttailed, leafnosed bat *Carollia perspicillata*. Zoobiology 6:261–263.

Davis MJ (1988). Control of bat wing capillary pressure and blood flow during reduced perfusion pressure. Am J Physiol 255:H1114.

Davis MJ, Shi X, Sikes PJ (1992). Modulation of bat wing venule contraction by transmural pressure changes. Am J Physiol 262:H625–H634.

Filho IPT (1990). Venular vasomotion in the bat wing. Microvasc Res 39:246–249.

Hafferl A (1933). Das Arteriensystem. In L. Bolk, E. Göppert, E. Kallius, W. Lubosch, eds., Handbuch Vergleichende Anatomie der Wirbeltiere, Vol. 6, 563–684. De Gruyter, Berlin.

*Kallen FC (1977). The cardiovascular systems of bats: structure and function. In W.A. Wimsatt, ed., Biology of Bats, Vol. 3. Academic Press, New York.

Maina JN, King AS (1984). Correlations between structure and function in the design of the bat lung: a morphometric study. J Exp Biol 111:43–61.

Maina JN, King AS, Settle G (1989). An allometric study of pulmonary morphometric parameters in birds, with mammalian comparisons. Trans R Soc Lond B 326:1–57.

Mathieu-Costello O, Agey PJ, Szewczak JM (1994). Capillary-fiber geometry on pectoralis muscles of one of the smallest bats. Resp Physiol 95:155–169.

O'Shea JE (1970). Temperature sensitivity of cardiac muscarinic receptors in bat atria ventricle. Comp Biochem Physiol 86C:365–370.

O'Shea JE (1993). Adrenergic innervation of the heart of the bat, *Miniopterus schreibersi*. J Morphol 217:301–312.

Slaaf DW, Reneman RS, Wiederhielm CA (1987). Pressure regulation in muscle of unanesthetized bats. Microvasc Res 33:315–326.

Speakman JR (1988). Position of the pinnae and thermoregulatory status in brown longeared bats. J Therm Biol 13:25–30.

Starck D (1982). Vergleichende Anatomie der Wirbeltiere, Vol. 3. Springer Verlag, Heidelberg.

Studier EH, O'Farrell MJ (1976). Biology of *Myotis thysanodes* and *M. Lucifugus* (Chiroptera: Vespertilionidae)—III. Metabolism, heart rate, breathing rate, evaporative water loss and general energetics. Comp Biochem Physiol 54:423–432.

Thomas SP, Suthers RA (1972). The physiology and energetics of bat flight. J Exp Biol 57:317–335.

Van der Westhulzen J (1988). Haematology and iron status of the Egyptian fruit bat *Rousettus aegyptiacus*. Comp Biochem Physiol 90A:117–120.

Viscor G, Marques MS, Palomeque J (1985). Cardiovascular and organ weight adaptations as related to flight activity in birds. Comp Biochem Physiol 82A:597–599.

Wahlström A (1978). Ionic content and ionic fluxes in the smooth muscle of the metacarpal vein of the flying fox, *Pteropus giganteus*. Comp Biochem Physiol 59A:21–26.

Wightman J, Roberts J, Caffey G, Agar NS (1987). Erythrocyte biochemistry of the greyheaded fruit bat, *Pteropus poliocephalus*. Comp Biochem Physiol B88:305–308.

Wolk E, Bogdanowicz W (1987). Hematology of the hibernating bat, *Myotis daubentoni*. Comp Biochem Physiol 88A:637–640.

Respiration

Carpenter RE (1985). Flight physiology of flying foxes, *Pteropus poliocephalus*. J Exp Biol 114:619–647.

Carpenter RE (1986). Flight physiology of intermediatesized fruit bats (Pteropodidae). J Exp Biol 120:79–103.

Chappell MA, Roverud RC (1990). Temperature effects on metabolism, ventilation and oxygen extraction in a neotropical bat. Resp Physiol 81:401–412.

Condo SG, El-Sherbini S, Shehata YM, Corda M et al. (1989). Hemoglobins from bats (*Myotis myotis* and *Rousettus aegyptiacus*): A possible example of molecular adaptation to different physiological requirements. Biol Chem Hoppe Seyler 370:861–867.

Herreid CF, Bretz WL, Schmidt-Nielsen K (1968). Cutaneous gas exchange in bats. Am J Physiol 215:506–508.

Jürgens KD, Bartels H, Bartels R (1981). Blood oxygen transport and organ weights of small bats and small nonflying mammals. Resp Physiol 45:243–260.

Maina JN, King AS, King DZ (1982). A morphometric analysis of the lung of a species of bats. Resp Physiol 50:1–11.

*Maina JN, King AS (1984). Correlations between structure and function in the design of the bat lung: a morphometric study. J Exp Biol 111:43–61.

Maina JN, King AS, Settle G (1989). An allometric study of pulmonary morphometric parameters in birds, with mammalian comparisons. Phil Trans R Soc Lond B 326:1–57.

Maina JN, Nicholson T (1981). The morphometric pulmonary diffusing capacity of a bat, *Epomophorus wahlbergi*. J Physiol 325:36.

*Speakman JR, Racey PA (1991). No cost of echolocation for bats in flight. Nature 350:421–423.

Thomas SP (1982). Ventilation and oxygen extraction in the bat *Pteropus gouldii* during rest and steady flight. J Exp Biol 94:231–250.

*Thomas SP, Lust MR, Riper van HJ (1984). Ventilation and oxygen extraction in the bat *Phyllostomus hastatus* during rest and steady flight. Physiol Zool 57:237–250.

Thomas SP, Suthers RA (1972). The physiology and energetics of bat flight. J Exp Biol 57:317–335.

3

HEAT AND WATER BALANCE

THE FOLLOWING TYPES of heat exchange occur between an animal and its environment.

Heat conduction. Conduction occurs when there is a temperature difference between the body temperature (T_b) and the ambient temperature (T_a). The amount of heat transferred depends on the size of the surface over which heat is exchanged and the specific heat conduction properties of the media involved. Heat conduction capacity for a living organism can be expressed in terms of O_2 utilization, and is measured as ml O_2/h/°C/g needed to maintain a given temperature difference between the body and the environment.

Convection. If there is movement of the media through which heat transfer occurs, equilibrium will never be reached, and more energy will be transferred than if the media were static. Because air and water are usually in motion, animals in nature lose the largest amount of heat through convection.

Radiation. During the day many animals gain heat from sunshine, but on clear nights when there are no clouds to reflect heat back to the earth, animals experience a net loss of heat.

Cooling through evaporation. The transformation of water from the liquid to the gaseous state requires energy. To transform 1 g of water from the liquid to the gaseous phase, 2.45 kJ of energy are needed. Evaporation through breathing and through the skin is unavoidable, so terrestrial animals continually lose large amounts of heat.

The poor insulating properties of water prevent vertebrates with gills from maintaining a high body temperature. Therefore, aquatic animals cannot maintain their body temperature above that of the environment for extended periods of time. Animals with gills are "cold-blooded" animals, or poikilotherms. Their body temperature, and consequently their metabolic rate, is completely dependent on the temperature of their surroundings. Because air has relatively good insulating properties, land-dwelling animals are able to maintain a high body temperature. The high body temperature of birds and mammals is achieved through an increase in the mitochondrial surface area within the cells. As a consequence, they are the "high performance models" of the animal kingdom. Within certain limitations, mammals and birds can continuously maintain an optimal body temperature of 35–39°C, thus they are "warm-blooded" animals, or homeotherms.

Homeothermy consumes a significant proportion of an animal's energy. The lower the body weight, the higher the ratio is of body surface area to metabolically active tissue. Small animals must therefore sacrifice a large proportion of their energy intake to compensate for heat loss (fig. 3.1). In temperate zones, during the cold seasons of the year, animals can suffer a negative energy balance. One solution to this problem is heterothermy. Heterothermic animals can facultatively, and in a controlled way, reduce their body temperature to save energy. They can also facultatively return to the normal homeothermic state in which their body temperature is 35–39°C without requiring any external energy input.

The large lungs and naked flight membranes of bats result in heat loss from a surface area 6 times larger, and a thermal conductance 1.5–4 times greater, than that of wingless mammals. Thus, bats that live in temperate zones and vespertilionids and molossids that live in the tropics have in varying degrees the ability to lower body temperature and go into a state of torpor (diurnal lethargy) or hibernation. For instance, a torpid long-eared bat (*Plecotus auritus*) at an ambient temperature of 5°C expends only 0.7% of the energy it requires when in a homeothermic state at the same ambient temperature.

3.1 HOMEOTHERMY

In their active state, all bats maintain their body temperature between 35° and 39°C and are homeotherms like other mammals. The fixed point for body temperature is controlled by neurons in the preoptic nuclei of the diencephalon, groups of nerve cells that lie just above the point where the optic nerve enters the brain. As long as the core temperature does not deviate greatly from the set point, the diencephalon induces only minor regulatory measures. However, if there is a threat of overheating, there is an increase in blood flow to exposed surfaces such as the flight membrane, ears, and scrotum, thereby increasing heat loss. If the body becomes too cool, the peripheral circulation shuts down and the hairs that cover the body stand on end, resulting in a thicker layer of stationary air surrounding the body. Because air is not a good conductor of heat, it provides good insulation.

In nursery colonies, pregnant and nursing females often hang together in large clumps. This type of social interaction is also an effective way of saving energy. A female *Myotis lucifugus* hanging alone under the roof of a barn would have a body temperature of 32.2°C and would use 32.5 ml O_2/h/animal. In contrast, an animal hanging in a group in the same building under the same conditions would have a body temperature of 36.9°C and use only 19.9 ml O_2/h/animal.

THE THERMONEUTRAL ZONE

The ambient temperature range within which a resting animal that is not digesting food consumes the least amount of oxygen is called the thermoneutral zone (TNZ) (figs. 3.2 and 3.3). Within this range, an animal can maintain a constant body temperature without additional energy cost. The range of the thermoneutral zone is bounded by the upper and lower critical temperatures and narrows with de-

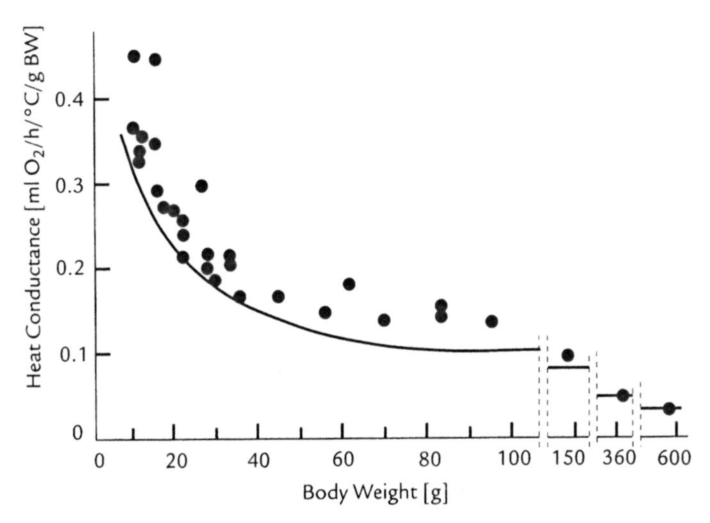

Figure 3.1 Heat conductance as a function of body weight in bats (filled circles) and other mammals (solid curve). From Hill and Smith (1984).

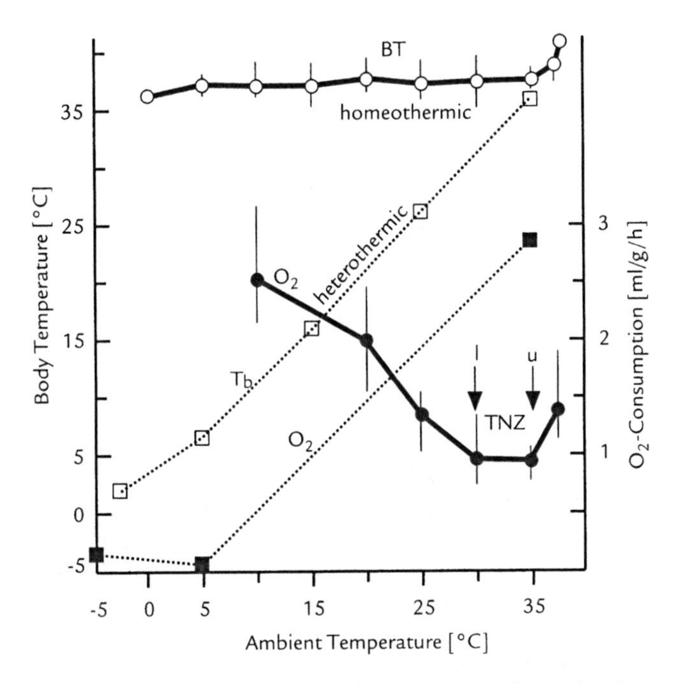

Figure 3.2 Homeothermy in the tropical species *Macroderma gigas* (open circles, solid line) and heterothermy in the hibernating species *Myotis sodalis* (open squares, dotted line). Homeothermy: the tropical bat maintains its body temperature (T_b) constant at 35°C throughout the range of ambient temperature. The range of minimal oxygen consumption (O_2) lies in the thermoneutral zone (TNZ), bounded by the upper (u) and lower (l) critical temperatures. Outside of the thermoneutral zone, O_2 consumption and energy expenditure increase steeply (filled symbols). In the heterothermic bat, *Myotis*, body temperature and oxygen consumption decrease when the ambient temperature decreases. From Yalden and Morris (1975).

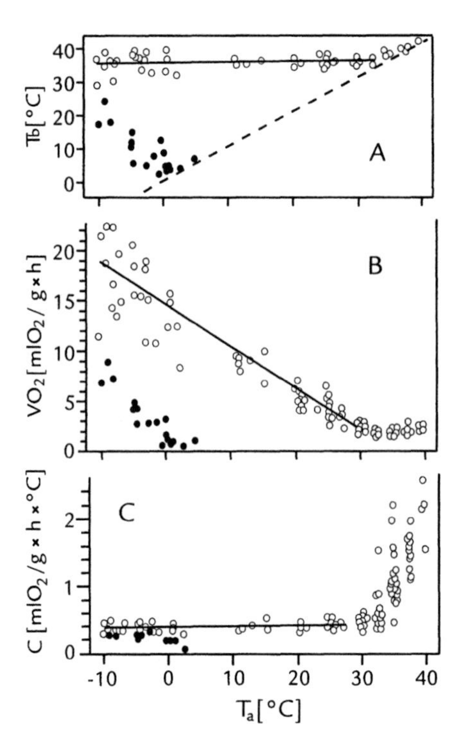

Figure 3.3 Thermoregulation in a subtropical tree bat, *Lasiurus seminolus*, as a function of ambient temperature, T_a. (Open circles) Bats that maintained normothermia; (closed circles) bats that went into torpor. (A) Body temperature, T_b; horizontal line = mean; (B) oxygen consumption with regression line; (C) thermal conductance; horizontal line = mean. After Genoud (1993).

creasing body size. For large flying foxes, the thermoneutral zone lies between 24° and 35°C, whereas for a lightweight bat such as *Saccopteryx bilineata*, weighing only 7 g, it is restricted to the range 30–35°C. Outside this zone, increased energy expenditure is needed to maintain a constant body temperature different from the ambient temperature (figs. 3.2 and 3.3).

BEHAVIORAL STRATEGIES FOR TEMPERATURE CONTROL

Animals are constantly generating heat. Because the normal body temperature is close to the lethal value of 44–45°C, they are less able to protect themselves from overheating than from overcooling. If the ambient temperature is above body temperature, bats quickly die. When body temperature was measured in a *Saccopteryx bilineata* placed in a water-saturated atmosphere at an air temperature of 37.2°C, its body temperature was found to rise from its normal value of 35.5°C to 43.1°C within 2 h. The only means of lowering the body temperature are evaporative cooling and the creation of convection currents by movement of the air. Flying

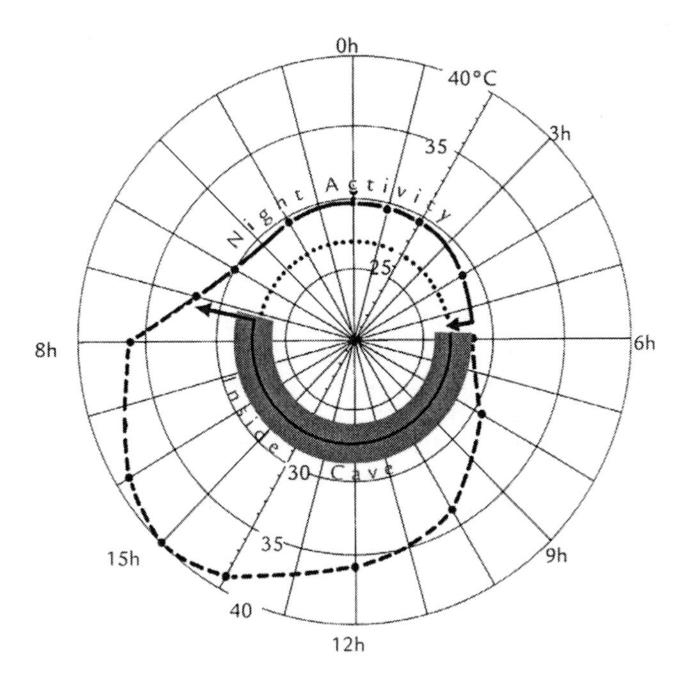

Figure 3.4 Temperature stabilization by protective day roosts. Temperature to which *Hipposideros speoris*, a tropical bat that normally spends its day roosting in a cave in Madurai, south India, is exposed during the course of a day. Outer circle = time of day; solid line = ambient temperature experienced by the bat over a 24 h period. Arrows mark the times when the bats fly out of the cave and when they return. The cave temperature remains constant at 27°C (solid and dotted half circles), whereas the outside temperature (dashed and solid line with points) varies between 40°C at 3:00 PM and 28°C at 4:00 AM.

foxes of the genus *Pteropus* spend their days hanging in large trees, often in full sun. No flying fox ever moves from its own place within the hierarchical and highly territorial colony. When the temperature begins to rise in the middle of the day, the animals start fanning the air with bent wings. If the ambient temperature approaches body temperature, the animals wet their fur with saliva and pant to further increase the evaporative effect of the fanning. Like most mammals, bats cannot sweat.

Microchiroptera and the smaller flying foxes avoid intense heat through their choice of day roosts. Even in the tropics, caves and cracks in rocks afford an ideal, thermally stable environment with temperatures in or near the thermoneutral zone (fig. 3.4). Because the bats only fly out at night, they never run the risk of overheating. Microchiroptera that roost in trees during the day have clever ways of avoiding the midday heat. The African bat *Lavia frons* (Megadermatidae) spends the day in pairs in the crowns of acacia trees. During the cool morning hours, the bats sunbathe with their bellies facing the sun to store as much warmth as they can before retreating into the shade of the foliage. On some days the temperature rises above 37°C. Then the pairs of bats leave their 6-m high roosts in the trees and crawl into low bushes where the temperature does not exceed 32°C. Thus, behavioral reactions are the simplest and most effective defense against overheating and restricting the temperature range experienced by bats in the tropics.

SHIVERING

Whenever the temperature sinks below the critical point at the lower limit of the thermoneutral zone, animals can generate heat by increasing their metabolism and thus maintain their body temperature. A significant amount of heat comes from muscle contraction. When the ambient temperature is cool, muscle tone increases up to the point at which shivering begins. In shivering, the skeletal muscle fibers contract rhythmically but asynchronously so that they generate heat without causing body movement. The generation of heat consumes a large amount of energy; small bats have only limited energy resources. Therefore, their response to cool temperatures, like that of many other mammals, is to enter a state of torpor.

3.2 TORPOR

When the ambient temperature sinks below the thermoneutral zone, bats can decide whether they will use their energy reserves to maintain their normal body temperature or whether they will allow their body temperature to drop close to or equal to the ambient temperature and thereby enter a state of diurnal lethargy, or torpor, allowing them to minimize their metabolic needs. This decision depends not only on the ambient temperature, but also on the availability of food and on reproductive condition. For this reason, it is possible to find both active and lethargic animals in the same colony on the same day (fig. 3.5). For example, in a colony of *Eptesicus fuscus* at an ambient temperature of 11°C, all of the males were torpid, while the nursing mothers had gathered together into a cluster and remained active. When the temperature sank below 9°C, these females also became torpid.

Bats, like all small mammals, have a high mass-specific basal metabolic rate. Homeothermic bats need a continuous daily food intake because they do not have significant fat stores. In insectivorous bats, temporary food scarcities may induce torpor in resting bats even at high ambient temperatures. For instance, fasting causes the vespertilionid bat *Nyctophilus gouldi*, an insectivorous bat native to Australia, to enter torpor even at an ambient temperature of 30°C. In the laboratory, decreases in ambient temperature cause these bats to respond by decreasing their body temperature, which has been observed to drop as low as 10°C. However, at ambient temperatures below 20°C, these bats increase their metabolic rate considerably to maintain body temperature. The result is that, at ambient temperatures below 15°C, energy consumption in torpid bats begins to increase. Under natural conditions, bats avoid such unfavorable conditions by selecting protective roost sites.

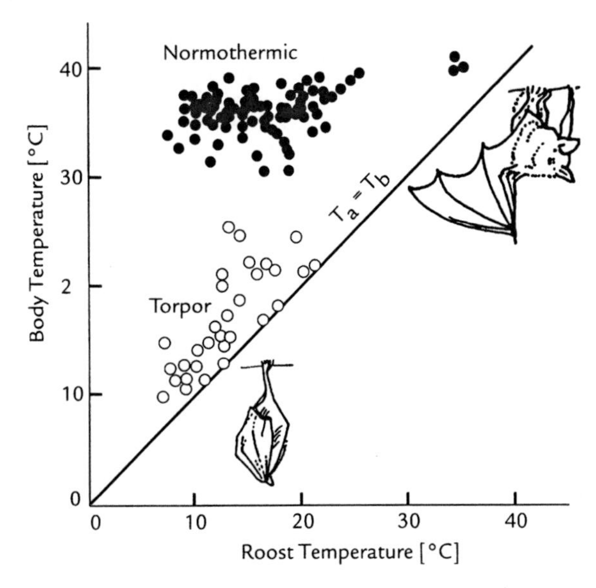

Figure 3.5 Facultative normothermy and heterothermy in *Plecotus auritus*. One group of animals maintains its body temperature within the normal range (solid circles), while another group goes into a state of torpor when the outside temperature sinks below 25°C (open circles). T_a = ambient temperature, T_b = body temperature. From Speakman (1988).

During torpor, bats enter into an arrythmic pattern of respiration in which periods of apnea alternate with brief bouts of ventilation. At lower ambient temperatures, the duration of the periods of apnea increases. In torpid *Pipistrellus pipistrellus*, for instance, the duration of apnea was observed to increase from 1.5 min at an ambient temperature of 12°C to 19.5 min at an ambient temperature of 0°C. The result was a drop in the mean ventilation frequency from 6.8 breaths/min at 12°C to 0.81 breaths/min at 0°C.

During apnea in torpid bats, CO_2 is stored in the blood. The excess CO_2 is released at the beginning of each ventilatory bout, as indicated by high and fluctuating respiratory exchange rates (CO_2 out/O_2 in). Initiation of a ventilatory bout may be triggered by a critical level of CO_2 in the blood. Apnea not only reduces the overall gaseous exchange rate, but it also effectively minimizes evaporative water loss during torpor. However, the mechanisms that control the onset and arousal from torpor have yet to be discovered.

In northern temperate zones, there are many extended periods of cold weather even in summer. These cold snaps threaten the development of young bats because more energy is needed to maintain normal body temperature, but at the same time the insect supply decreases and the mothers' milk supply is no longer sufficient. During critical periods of low temperatures, the mothers do not warm their young, but instead fly out of the nursery colony in the evening to hunt and do not return. The abandoned pups, and presumably the mothers in their alternative roosts, decrease their body temperature and enter a state of torpor. In this way, both the mothers and the pups are able to minimize energy consumption. Whenever the cold period ends, sometimes after many days, the mothers return to the nursery colony. This behavioral pattern effectively minimizes the death rate of pups under unfavorable temperature conditions.

RELATIONSHIP BETWEEN TORPOR, BASAL METABOLIC RATE, AND TERRITORY

The ability to enter a state of torpor varies according to climatic zone and feeding habits and is also influenced by the basal metabolic rate (see box 3.1). When body weight is taken into account, the basal metabolic rate of frugivorous bats is about the same as that of other mammals. Large species of *Pteropus* have no need to become torpid and are unable to do so. Small species of flying foxes, however, are able to reduce their body temperature to 25°C, even though they normally maintain their body temperature at 5°C above the ambient temperature. Nectar-feeding bats have a higher metabolic rate than other mammals of comparable size; they too are unable to become torpid.

Insectivorous bats have a basal metabolic rate that is about 35% lower than that of other mammals of similar size, and correlated with this low metabolic rate is the ability to enter a state of torpor. When *Eptesicus fuscus* returns to its day roost in the morning, its body temperature can drop by 16°C within 30 min. At temperatures below 15°C, all tropical bats, even those that are torpid during the day, raise their body temperature up to the normal value, suggesting that such low temperatures are life threatening.

Experiments have shown that tropical and subtropical bats cannot maintain their body temperature below 17°C for more than 1–2 h. At lower temperatures, they are no longer able to generate sufficient energy to return their body temperature to its normal value and consequently die. In contrast, insectivorous bats of the temperate zone can continue to reduce their body temperature until it reaches 11°C, at which point they enter a state of hibernation.

The limiting body temperature for torpor lies somewhere between 17°C and 11°C. During torpor, the body temperature is maintained approximately 5°C above the ambient temperature. During hibernation, however, the body temperature can drop to just above freezing. Thus, torpor and hibernation are different physiological processes.

3.3 HIBERNATION

Depending on climatic conditions, hibernation can last for days, weeks, or months, up to a maximum of 7 months. Much time goes into preparation for hibernation. Animals store fat reserves and may migrate for long distances to find appropriate winter quarters. In addition, their reproductive cycles are determined by the hiber-

Box 3.1 Energy Balance and Basal Metabolic Rate

Every animal must ensure that its energy balance—the difference between energy intake and energy output—remains neutral or positive. To calculate energy turnover, it is necessary to know what type of foodstuff is being metabolized. An approximate measurement of energy turnover is the respiratory quotient (RQ = CO_2 output/O_2 intake). It is known that 36.9 ml CO_2 is approximately equal to an energy value of 1 kJ, and 1 l O_2 releases 20.1 kJ at an RQ of 0.79 (Kurta, 1986). The different classes of foodstuffs have the following caloric equivalents:

1 g Carbohydrate (RQ = 1) provides 17.7 kJ
1 g Protein (RQ = 0.8) provides 23.6 kJ
1 g Fat (RQ = 0.7) provides 39.5 kJ
1 g Chitin (RQ = ?) provides 21.2 kJ

There are three fundamental measures that can be used to describe energy consumption over a period of hours or days, for a whole animal or as a function of body weight (e.g., kJ/day-animal or kJ/h-kg bw):

1. Basal metabolic rate (BMR) is the energy utilization in a resting animal at a temperature in the thermoneutral zone, when no food is being digested. In frugivorous bats, the basal metabolic rate is about the same as that of other mammals of comparable size. In insectivorous bats, the BMR is on average only about 65% of what would be expected based on body weight, probably due to these animals' ability to go into a state of torpor. The BMR in nectar-feeding bats is higher than would be predicted from their body weight, due to the large amounts of energy used in hovering flight.
2. Resting metabolic rate (RMR) is the energy utilization in a resting animal at a temperature in the thermoneutral zone. RMR is approximately equal to 1.25 BMR.
3. Field metabolic rate (FMR) is the metabolic rate of a freely behaving animal engaging in normal species-characteristic activities for 24 h in the field. FMR is about equal to two to three times the BMR.

In all animals, metabolic rate is a power function of body weight. This means that on a double-log plot of this relationship, the slope of the line is equivalent to the exponent in the equation for calculating FMR. For bats, the following average value has been found:

$$FMR = 184.5 \cdot bw^{0.767} \text{ kcal/day or } 771.9 \cdot bw^{0.767} \text{ kJ/day}$$

The bat exponent (0.767) is considerably higher than that for insectivores and rodents, presumably because of the high energy utilization during flight. However, it is also higher than the exponent calculated for birds. If energy utilization is calculated for a month or a year instead of for a day, the exponent for bats in temperate regions would probably not be higher than that for birds because bats conserve energy during periods of torpor and hibernation.

nation period. Therefore, it is not just the cold that causes bats to hibernate; instead, it is a yearly cyclic event initiated by the animal's own body processes. Virtually nothing is known about the mechanisms that regulate hibernation, although the substances vasoactive intestinal polypeptide and hibernation-inducing trigger seem to be somehow related to hibernation.

PHYSIOLOGICAL STATUS

During hibernation, bats hang stiff and immobile from the ceiling or in crevices in caves, or in hollow trees. As long as the cave temperature remains above freezing, the bats maintain a body temperature of about 1°C above the ambient temperature, but usually not lower than 6°C. Oxygen consumption falls from a rate of 3 ml/h/g during the active state to 0.02 ml/h/g. Heart rate drops from about 400/min to 11–25/min, and the peripheral circulation practically shuts down. The respiratory quotient (see box 3.1, p. 71) is between 0.6 and 0.7, indicating that nearly all the energy used is derived from the metabolism of fat. The blood sugar level drops from 155 mg/dl during the active state to 28 mg/dl during hibernation. Cells of the intestinal epithelium contain crystalline inclusions of unknown origin and function. The weight of the pancreas decreases by about 30%. During hibernation, bone tissue is broken down, and bone marrow, the source of new blood cells, is partially replaced with fat. When the bat awakes from hibernation, osteoblasts, the cells responsible for producing new bone tissue, reappear and new hemopoeitic cells squeeze the fat from the bone marrow. During hibernation, water and ionic balance remain intact. It appears that the small amount of water produced through metabolism is sufficient to replace the water lost through evaporation. Urine production decreases to 1% of the volume during the active state.

It is not clear which physiological changes are directly responsible for triggering hibernation. It is possible that hibernation is brought about by an autonomically controlled decrease in heart rate because, unlike those of other mammals, the cardiac ventricles of bats are densely innervated by cholinergic parasympathetic fibers. This innervation is capable of reducing the heart rate by up to 50%. It is interesting to note that in bats, the vagus, a large parasympathetic nerve, remains active even at low body temperatures. In other mammals, if the heart is cooled, activity in the vagus nerve is no longer effective because the affinity of muscarinic receptors for acetylcholine decreases at low temperatures. In contrast, it was found that in the vespertilionid bat *Miniopterus schreibersi*, the affinity of muscarinic cholinergic receptors remained unchanged over a temperature range of 37–12°C.

A hibernating bat may consume 0.02 ml/g/h of O_2. A small bat weighing 7 g would have an O_2 uptake of 3.36 ml per day. Assuming that fat is oxidized during hibernation, and that under conditions of lipid oxidation, 1 ml of O_2 provides energy equivalent to 19.8 J, the bat would consume energy equivalent to 66.5 J/day. Because 1 mg of fat provides energy equivalent to 39.41 J, the bat would deplete its fat stores by only 1.7 mg per day and would lose only 0.024% of its total body weight. Thus, in a bat weighing 13 g, just 100 mg of stored fat is sufficient to fuel a full month of continuous hibernation.

Table 3.1 Dependence of metabolic rate on body temperature

Body temperature (°C)	Q_{10}
30–20	4.11
20–10	3.71
10–0.1	2.85

Is metabolism actively suppressed? During torpor, the Q_{10} value (i.e., the change in metabolic rate for each 10° change in temperature) lies between 2.2 and 2.4 so that if the temperature changes by 10°C, metabolic rate would change by a factor of 2.2–2.4. During hibernation, metabolic rate is even more dependent on temperature, especially in the range around normal body temperature (table 3.1). From the values in table 3.1, one must conclude that metabolism is not simply slowed down at low temperatures, it is also actively suppressed. For example, a torpid bat at 20°C uses twice as much energy as a hibernating bat at the same temperature. Normally, even during periods of reduced metabolic activity, waste products are generated, and these must be excreted from time to time. In hibernation, this does not happen. It seems probable that the ability to hibernate for long periods of time is made possible by active suppression of metabolic activity. At present, it is only possible to speculate about the mechanisms responsible for this suppression. The following mechanisms have been proposed:

- *Acidosis.* When hibernation begins, CO_2 is stored, causing pH values to drop. The resulting acidosis suppresses glycolysis and enzyme activity. Upon awakening, CO_2 is depleted through increased respiration, and acidosis is reversed.
- *Thyroid hormone.* During hibernation, the concentration of thyroid hormone decreases.
- *Fatty Acids.* In other hibernating mammals it has been shown that prior intake of polyunsaturated fatty acids causes metabolic rate to decrease during hibernation.

The dilemma. The fact that energy requirements are minimal at low temperatures coupled with the need to find winter quarters that have a stable temperature and are protected from outside influences raises a conflict for the bat. Winter quarters deep in the interior of a cave have a stable temperature and are likely to be undisturbed, but the temperature is between 9° and 12°C. Only bats with good fat reserves can afford to winter at such warm temperatures. For animals with smaller fat reserves, especially juveniles born the previous summer, lower temperatures are more favorable. Although areas near cave entrances are likely to have the appropriate temperature range, they are also subject to larger temperature fluctuations when the temperature outside changes. If the temperature at the roost drops below freezing, the bats wake up and fly deeper into the cave or look for new quarters. However, waking uses a great deal of energy and results in severe depletion of the animals' reserves. If the daytime temperature rises above 10°C, the bats may become aroused and fly out to hunt for insects.

The solutions. In England, native horseshoe bats have found a solution to this dilemma. In the spring, summer, and autumn months, the males and juveniles hang in the cave entrances. They are seldom inactive for more than 6 days at a time, even into December. Because of the extra food they obtain, their weight decreases very slowly during the early winter period. In January and February, however, their weight decreases at a faster rate than that of females with their larger fat reserves, who winter deep inside the cave under constant temperature conditions. However, the males and juveniles make up for this weight loss when the weather begins to warm up again in late winter and they can forage occasionally. By the end of the hibernation period some animals have actually gained weight.

Many other species of bats have adopted a similar strategy. They select their winter quarters according to the size of their fat reserves and according to weather conditions. Thus, most bats change their quarters several times during the winter (fig. 3.6). Bats do not necessarily sleep continuously through the entire winter under natural conditions. At any time they may alternate between long and short periods of sleep. The lower the body weight, the more common it is to find active individuals in the winter quarters (fig. 3.7). Under laboratory conditions, bats can sleep continuously for up to 200 days. In nature, it is uncommon for a bat to sleep more than 80 days at one time. The length of the sleep period does not depend on body temperature. Bats have been known to sleep for 15 days at a body temperature of 12°C, but to awake after only a few days sleep at a body temperature of 2°C. On average, *Myotis lucifugus* sleep for 8–9 days at a time, *Pipistrellus flavus* 14–17 days, and *Eptesicus fuscus* 5–8 days. There is a relationship between the longer sleep periods and the ambient temperature. In the mild winters of Scotland, the long-eared bat, *Plecotus auritus*, may fly almost daily. Only during prolonged periods with night temperatures below 4°C might the bats remain in hibernation.

ONSET OF HIBERNATION

Little is known about the factors that bring about the onset of hibernation, especially in bats. It could be that hibernation depends on internal factors such as fat reserves, hormonal balance, and neuromodulators, as well as external factors such as changes in the quantity and quality of food available. It is certain that a prolonged period of cold weather can bring about hibernation. Under experimental conditions, bats kept for several days at low ambient temperatures go into hibernation, regardless of the time of year.

The first measurable sign that hibernation is about to begin is a cyclic change in heart rate that occurs approximately every 20 min. The heart rate changes from 600/min to 300/min several times, and within about 2–3 h has decreased to 10–80/min (fig. 3.8). This alteration in heart rate is followed by a change in respiratory rate, a decrease in body temperature, and an overall reduction in the metabolic rate. The peripheral circulation is reduced by a narrowing of the blood ves-

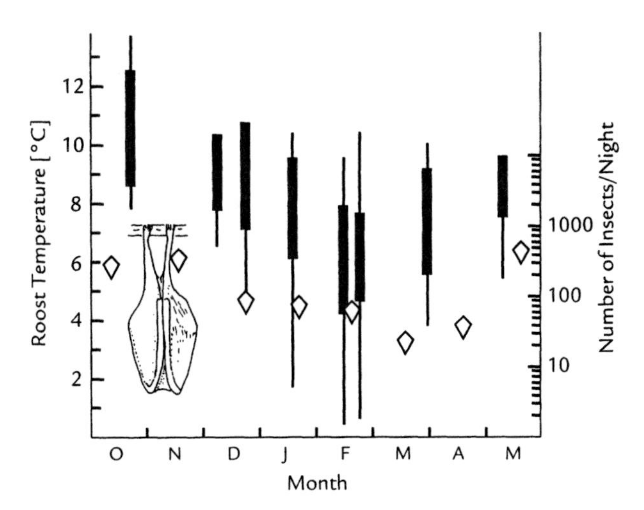

Figure 3.6 Choice of roosting place during the hibernation period depends on the outside temperature and availability of insects. This figure illustrates temperatures in the winter quarters of the horseshoe bat, *Rhinolophus ferrumequinum*, in England. Thin vertical lines = temperature range within the winter quarters; thick bars = temperature range preferred by the bats. In autumn and spring, the bats move to warmer roosting places. Open diamonds = number of insects caught in light traps per night. From Yalden and Morris (1975).

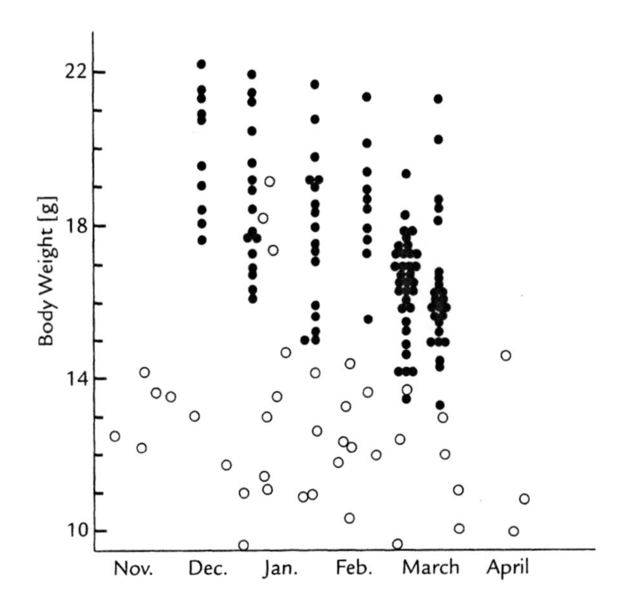

Figure 3.7 Difference between the body weight of hibernating *Eptesicus fuscus* (filled circles) and that of animals found actively moving about in the winter quarters (open circles). The clear boundary between active and hibernating bats at 14 g suggests that exhaustion of fat reserves acts as a stimulus for arousal. From Brigham (1987).

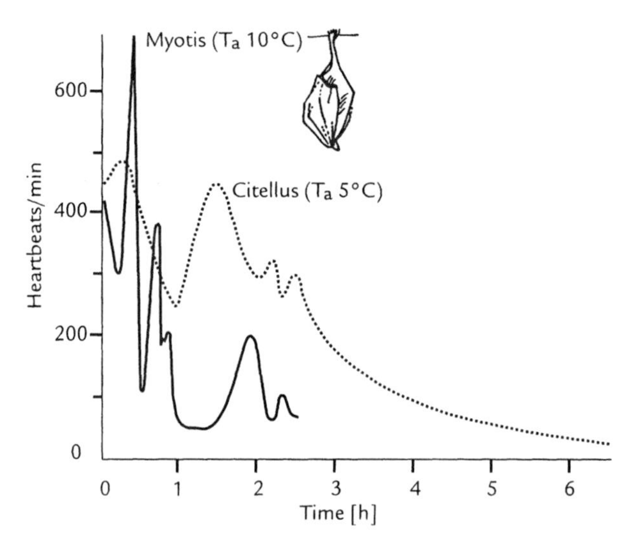

Figure 3.8 Fluctuations in heart rate upon entering a state of hibernation: *Myotis myotis* (ambient temperature of 10°C) compared to the rodent, *Citellus tridecemlineatus* (ambient temperature of 5°C). From Raths and Kulzer (1976).

sels, and some of the red blood cells are stored in the spleen. At body temperatures below 12°C, it is no longer possible to record neural activity in response to sound at the inferior colliculus. Nevertheless, stimuli such as touch, light, and dryness can cause the bat to awaken at any time.

These observations suggest that it is the reduction in heart rate that brings about the onset of hibernation. However, it is also quite possible that the change in heart rate is the result of some previous event that has not yet been measured, such as decreased metabolic activity in the brain.

AROUSAL FROM HIBERNATION

The process by which animals awaken from hibernation has been more thoroughly studied than the events leading up to hibernation. One of the most important factors in the arousal process is the ambient temperature. The lower critical temperature is the freezing point. At 0°C, a hibernating bat can maintain its body temperature at 2°C for a short time by shivering, but after some time, the bat awakens so that it can move to warmer quarters. The upper critical temperature for arousal varies from species to species, but is on average about 10°C. An extreme case is seen in *Lasiurus borealis*, a vespertilionid bat that spends the winter outdoors in the branches of fir trees until the temperature reaches 13°C. This bat is well-protected from the cold by its long, thick fur and does not awaken from hibernation until the ambient temperature has reached at least 16°C.

Apart from temperature, stimuli that have been suggested as possible triggers for arousal from hibernation are energy depletion and evaporative water loss. Measurements of oxygen consumption in torpid bats make it seem unlikely that

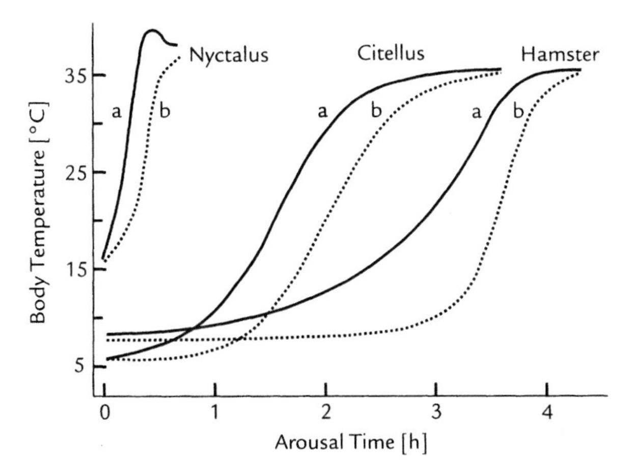

Figure 3.9 Rise in body temperature during arousal in three species of hibernating animals: bat (*Nyctalus noctula*), rodent (*Citellus undulatus*), and field hamster (*Cricetus cricetus*). a=Temperature of the front part of the body; b=rectal temperature. From Raths and Kulzer (1976).

energy supply is a major limiting factor controlling the duration of continuous hibernation periods. It seems that water loss by evaporation is a more critical factor. Water loss is directly correlated with the difference in water vapor pressure between the body surface and the ambient air. At air temperatures of 2–4°C and relative humidities between 90% and 98%, a bat may daily lose about 0.1–0.4% of its body weight through evaporation. Because respiration is greatly reduced during hibernation, water is almost exclusively lost through the body surface. Metabolic production of water through lipid oxidation is insufficient to compensate for the unavoidable water losses that occur under the air conditions that prevail in most hibernation roosts. Therefore, the need to drink water may be the main reason most bats interrupt longer hibernation periods at irregular intervals. The durations of hibernation periods that have been observed in the field closely match those calculated assuming that water loss is the limiting factor.

During hibernation, the sympathetic nervous system becomes spontaneously active at irregular intervals, quickly causing heart rate to increase. Arousal begins with a volley of discharges in sympathetic nerve fibers and a consequent increase in heart rate. This is followed by an increase in the rate of respiration, and a dilatation of selected blood vessels, which restores circulation to the front part of the body. It is only after these events have taken place that body temperature begins to rise (fig. 3.9).

The process of arousal consists of a slow first phase and a rapid second phase. In bats, the entire process of arousal can be completed within half an hour, much faster than in other hibernating animals. The time course of heat generation during the arousal process is shown in fig. 3.9. A horseshoe bat requires 2 kJ to raise its body temperature from 6°C to 37°C (fig. 3.10). How is it that bats can generate so much heat in so short a time?

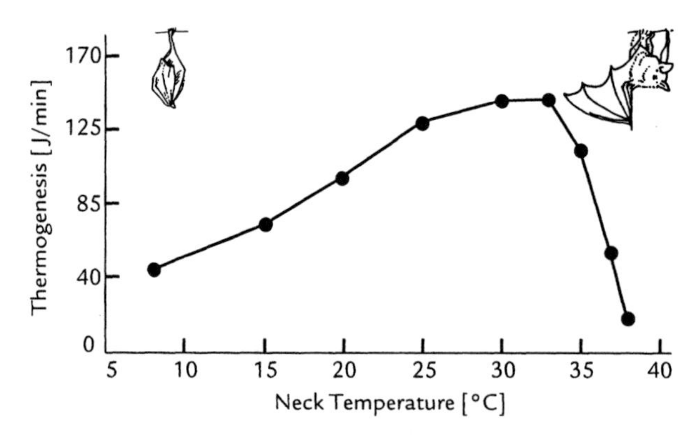

Figure 3.10 Thermogenesis during arousal in *Myotis myotis*. From Heldmaier (1969).

Phase I. The first phase of warming takes place slowly, and lasts until the body temperature reaches about 15°C. Most of the energy used during this phase comes from brown adipose tissue (see box 3.2 and fig. 3.11). The brown fat acts as a "furnace" which, like the heart, is switched on by the release of norepinepherine by the sympathetic nervous system. The heat produced is immediately carried via the blood circulation to the heart, head, and liver; subsequently, it is distributed throughout the rest of the body.

It has been shown that in dwarf hamsters and other small mammals, a low molecular weight "uncoupling protein" is necessary for the oxidation of brown fat. This protein allows the oxidation process to proceed without ATP production. This protein is produced in large amounts and stored in the mitochondria under conditions of cold ambient temperatures, short day lengths in autumn and winter, or through the influence of melatonin, a hormone that controls circadian rhythms

BOX 3.2 Brown Adipose Tissue

The most important source of warmth in hibernating animals is the brown adipose tissue (BAT), a type of fat specialized for heat production. BAT is mainly found between the shoulder blades (fig. 3.11) of hibernating animals, and in infants and juveniles of other mammals, which also have problems maintaining a body temperature of 37°C. The cells of BAT contain many large mitochondria and small fat droplets. The mitochondria burn fat and thereby produce only heat. This form of heat production through fat metabolism is called "nonshivering thermogenesis" because it takes place in the absence of muscular contraction. Because the fat content of BAT is low, it is thought that fatty acids are transported from the white adipose tissue and glucose is transported from other sources so that it can be metabolized by the BAT to produce heat. During arousal, the BAT is the warmest part of the bat.

BOX 3.3 The "Exact Diurnal Clock" of Bats in Contrast to the Circadian Clock of Other Mammals

Bats kept in constant and complete darkness and shielded from all outside stimuli awake spontaneously between 2:00 PM and 8:50 PM every day. The persistence of this circadian rhythm, which is maintained in the absence of any recognizable *Zeitgeber* (resetting stimulus), and at brain temperatures as low as 2°C, clearly differentiates it from other circadian phenomena such as cyclic changes in body temperature, which in darkness become free-running with a period of 26.5 h. This highly precise (not "circadian") internal clock in bats is unique within the animal kingdom.

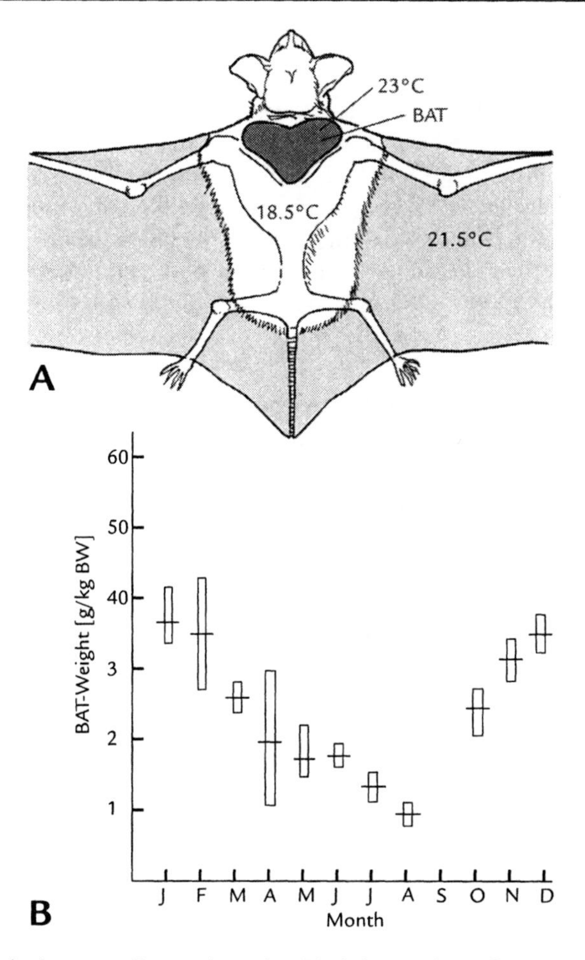

Figure 3.11 The brown adipose tissue (BAT). (A) Location of BAT on the neck and comparison of the temperature of the BAT to that of other parts of the body during arousal. From Lyman (1982). (B) BAT volume throughout the year in *Myotis californicus*. The bars show the range and the horizontal lines the median weight of the BAT. From O'Farrell and Schreiweis (1978).

(box 3.3). Because of the action of the uncoupling protein no chemical energey is produced, and the only metabolic product is heat. However, the uncoupled oxidation of brown fat delivers only 30–70% of the heat produced during nonshivering thermogenesis. It is thought that the remaining heat comes from the muscular activity of the heart. The cardiac muscle is the first organ that increases its activity during arousal. Like all muscular contractions, those of the heart muscle produce heat, which contributes to the warming of the whole body.

Phase II. At the beginning of the rapid second phase of arousal, the animal begins to shiver, and this provides an additional source of heat. This is especially true during the beginning of phase II, when the body temperature is raised from 15°C to 20°C.

There is some evidence that still other sources of energy may be mobilized during arousal. Glucose oxidation is much greater during arousal than during hibernation. However, because glycogen reserves are limited, it may be that some of the energy needs are also supplied through the breakdown of proteins. The metabolism of protein does, in fact, increase during arousal, and the weight of the pectoralis muscle decreases slightly. Bats, unlike other hibernating animals, store insulin in the pancreas. During arousal, the blood insulin concentration increases by one-third. Presumably the insulin is quickly removed upon the animal's awakening to prevent excessively high blood sugar levels and hyperosmolarity.

The duration of the arousal process. The time for arousal to take place depends on the temperature differential that must be overcome to restore the body temperature to normal. The arousal process is considered to be complete when the bat is able to fly. To raise the body temperature to a normal level starting from 3.5°C, *Myotis myotis* requires 49 min and *Plecotus auritus* requires 39 min. Starting at 9°C, only about 30 min is required. The rapidity with which this process takes place is due mainly to the bat's "central heating system," the brown adipose tissue, and the distribution of the heat produced via the blood vessels. Appropriately, the brown fat is highly vascularized by branches of the dorsal aorta.

REGULATION OF HIBERNATION

How is it that a bat can awaken from its state of diurnal lethargy every evening, but sleep for days or weeks during hibernation? As mentioned before, the reduction in body temperature during hibernation is due to a drop in the set point of the "thermostat" in the brain. Thus, hibernation is under neural control as well as indirect hormonal control. Most of what is known about the mechanisms that regulate hibernation comes from experiments on species other than bats. According to these experiments, two neuromodulatory peptides, vasoactive intestinal peptide (VIP) and hibernation-inducing trigger (HIT) play a role in the regulation of hibernation.

VIP is found not only in the digestive tract, but also in the brain of mammals. It is known to play a role in the regulation of processes that influence metabolism, such as body temperature, heart rate, and brain circulation. In bats, the highest concentrations of VIP are found in the anterior hypothalamus. As the body temperature drops, the concentration of VIP drops along with it. This relationship may be due to the shutting down of the brain circulation and brain metabolism as well as temperature reduction.

HIT was first found in certain species of fish that survive during dry periods in the summer by entering a state of estivation. It has been reported that when ground squirrels are injected with HIT, they go into hibernation. However, this finding is still controversial. The blood plasma and urine of bats contain HIT.

Endorphins. Recent experiments have shown that endogenous opiates (endorphins and enkephalins), which also function in pain regulation, play an important role in the reduction of body temperature. If hibernating bats are injected with the opiate antagonist naloxone (50 mg/kg body weight), they awaken immediately. However, the awakening is incomplete. Naloxone increases heart rate and respiration in hibernating mammals, but not in animals that are active. Administration of opiates causes heart rate and respiration to decrease and lowers body temperature. Thus it is possible that an increased brain concentration of opiates or an increased affinity of the opiate receptors could serve to maintain a state of hibernation. When the body temperature of a bat is lowered from 30°C to 4°C, the affinity of opiate receptors for naloxone increases by 80% in the hypothalamus but remains unchanged in cortex.

These findings suggest that the endorphin-containing autonomic pathways of the central nervous system induce and maintain hibernation, while another pathway that uses a neurotransmitter similar to naloxone is responsible for arousal. It is also possible that these neuromodulatory systems act on the autonomic nervous system because norepinepherine release by the sympathetic nervous system cannot be detected until arousal is in progress. The question of what stimuli or physiological events activate or suppress the neuromodulatory system remains unanswered.

The ability of bats to go into torpor or hibernate has been just as important a factor in their evolution as has flight and echolocation. This is convincing evidence that adaptation and taking over an ecological niche are multifactorial processes and can only be understood when all the different body systems of an animal and their interdependencies are taken into account.

3.4 WATER TURNOVER

For land animals, water is the critical limiting factor for life, just as oxygen is for aquatic animals. Water is necessary for life because all the functions of the cell must take place in an aqueous milieu. Water metabolism plays a role in three vital physiological functions:

- *Homeostasis*, or maintenance of a constant, optimally formulated ionic balance. A global measure of homeostasis is the osmotic pressure of the blood.
- *Thermoregulation.* Evaporation of water from the skin surface and lungs is the only means the body has for cooling.
- *Excretion.* Waste products and ions are water soluble. It is vital that toxic nitrogen-containing wastes such as urea or uric acid be removed from the blood and excreted through the urine.

WATER BALANCE

Animals obtain water through their food and by drinking. In addition, water becomes available as a by-product of metabolism. The body loses water by excretion in urine and feces, by evaporation from the lungs and skin surface, and through specialized secretions such as milk. These forms of water loss are unavoidable but can be minimized through appropriate behavior and adaptations of the body to the specific environment.

Evaporation plays an important role in the water balance of bats. Because bats have unusually large lungs and 80% of their body surface consists of the naked flight membrane, they can lose large amounts of water through evaporation. Evaporative water loss (EWL) is especially great during flight because the wings are constantly surrounded by convection currents of air. Just as in the case of energy balance, water turnover is related to body weight (fig. 3.12). As would be expected, mass-specific water loss in bats is higher than in other mammals and approaches the high values found in songbirds.

It is possible to measure energy and water turnover in free-living animals using the "double-labeled water method" (radioactive tracers). Data are available on four different species of bats. Table 3.2 shows water turnover rates for two insectivorous vespertilionid bats. As expected, there is no net gain or loss of water. The daily water turnover of lactating *Plecotus auritus* (body mass 8 g) in the wild is 5.36 g, or 67% of body mass. This daily water expenditure could be fully covered by water obtained from food if the bat ate 3.75 g wet mass every day. However, to satisfy energy demands, the bat would need an intake of only 1.44 g wet mass. The additional water requirement could be met either by drinking water or by intake of excess food. Actual demands on drinking may vary among individuals because water turnover also depends on evaporative water loss by flight and general level of activity. In the physiological range, ambient temperature plays a negligible role (fig. 3.13). The amount of time spent in flight determines how much water is lost by evaporation and how much through the urine.

WATER INTAKE

Do bats need to drink water? A critical factor in determining the ecological adaptability of a species is whether it needs to drink water. As shown in table 3.3, not all species of bats drink water. This is especially true for species that feed on nectar or fruit. The nectar-feeding species *Anoura caudifer* (Phyllostomidae), which weighs only 11.5 g, was observed to ingest nectar with a water content of 13.4 ml

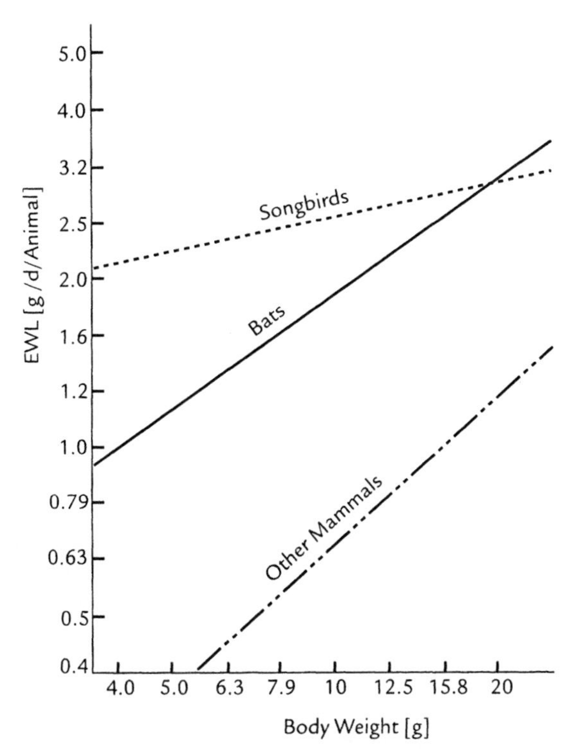

Figure 3.12 Relationship between evaporative water loss (EWL) and body weight. Bats: log EWL = log 0.389 + 0.672 log bw; songbirds: log EWL = log 1.563 + 0.217 log bw; other mammals: log EWL = log 0.087 + 0.883 log bw. The functions plotted are intended only to demonstrate their relationship with one another and are not necessarily accurate in absolute terms. From Studier (1970).

Table 3.2 Water balance in two insectivorous bats

Animal	Body weight (g)	Water balance (ml/day) gain	loss	Water balance (ml/day-kg bw) gain	loss
Myotis lucifugus					
Pregnant	9.02	6.16	6.27	691	702
Lactating	7.88	6.91	7.07	878	899
Eptesicus fuscus					
Pregnant	20.84	8.47	8.47	406	407
Nursing	17.37	17.13	17.00	997	989

The differences between gain and loss are insignificant, but those between pregnant and lactating females are highly significant. From Kurta et al. (1989, 1990).

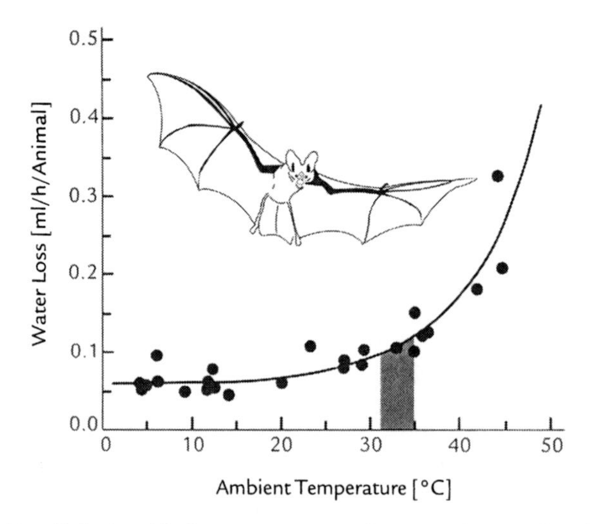

Figure 3.13 Relationship between evaporative water loss and ambient temperature in *Macrotus californicus*. Shaded area = thermoneutral zone. This graph clearly shows that at temperatures below the thermoneutral zone, there is little change in evaporative water loss. From Bell et al. (1986).

per day. This means that every night the bat takes in 116% of its own body weight in water. This estimate seems reasonable because a similarly high water intake of 120% body weight has been measured in nectar-feeding hummingbirds.

Under certain conditions, insectivorous species can also go without drinking water. *Macrotus californicus* spends the winter in geothermally warmed caves and only spends about 2 h each evening hunting. Individuals of this species have a water turnover of only 2.66 ml/day (or 200 ml/day-kg bw) and can survive without ever drinking water. In geothermally warmed caves, the temperature is about 29°C and evaporative water loss amounts to about 18% of body weight. Tropical populations of the same species do drink water, and their kidneys produce a less concentrated urine. Insectivorous bats in the deserts of Namibia have been observed to survive for 72 days without water at an ambient temperature of 40°C (11% humidity), simply by eating their normal diet of insects. During this stressful period, some animals actually put on weight, and one was able successfully to nurse a pup.

Species such as *Myotis lucifugus* that are not adapted for life in the desert die quickly when water is not available. Death generally occurs within 12 h of water deprivation, but the bat can survive for up to 24 h if it is also food deprived. This suggests that death does not occur due to dehydration, but rather due to uric acid poisoning. This could also explain why in summer, desert bats live off of their fat reserves, thus killing two birds with one stone: They save water by minimizing urine excretion and by avoiding the evaporative water loss that would necessarily occur if they went out to hunt the few available insects; in addition, they obtain water as a by-product of fat metabolism.

Table 3.3 Water balance in water-deprived bats

Species	Body weight (g)	Diet	Water intake from food (ml/day)	Water loss				Total loss	Water balance
				EWL in flight	EWL at rest	Urine	Feces		
Eptesicus	16.5	Insects	5.67 (34)[a]	2.08 (13)	1.98 (12)	5.06 (31)	0.59 (4)	9.71 (60)	-4.04 (-26)
Tadarida	11.2	Insects	3.84 (34)	1.32 (12)	1.35 (12)	2.53 (23)	0.41 (4)	5.61 (51)	-1.77 (-17)
Pizonyx	25	Marine crustaceans	12.0 (48)	2.8 (11)	1.2–3.7 (5–15)	6.3 (25)	1.3 (5)	11.6–14.1 (46–56)	+0.4–-2.1
Leptonycteris	22	Nectar	16.78 (76)	3.44 (16)	1.67 (8)	9.67 (43)	2.0 (9)	16.78 (76)	+2–-8 (0)

The balance is calculated as the water intake minus the total water loss. EWL, evaporative water loss. The wide range of EWL in *Pizonyx* is due to variability in the quality of its daytime roosts. From Carpenter (1968a,b).

[a]Numbers in parentheses indicate percentage of body weight.

Because of the high evaporative water loss through the lungs and the flight membranes, the nightly hunt for insects accounts for the greatest water loss (fig. 3.14). Soon after feeding, much water is lost through urine excretion, but water loss then decreases greatly during the day and remains low until the time when the bats' nightly hunting activities resume. Interestingly, the amount of water available does not seem to influence evaporative loss or urine volume. In species that drink during the day, the urine is less concentrated.

Despite all the possible flaws in the measurements of water balance that have been made in the laboratory and in the field, it is still clear that the only means by which bats can control their water balance is through behavioral adaptations and through the function of the kidneys.

EXCRETION

The mammalian kidney. More than almost any other organ, the structure and function of the kidney reflect the ecological conditions under which an animal lives. Comparative studies of the kidney have shown that, in addition to genetically determined differences, kidney function is astonishingly adaptable, resulting in a high degree of plasticity in the regulation of water balance.

The mammalian kidney is highly efficient at filtering, reabsorption, and concentration of solutions. It filters the entire volume of extracellular fluid 10–12 times each day. The functional unit of the kidney is the nephron. It begins with the glomerulus, which extracts a filtrate from the blood that passes through. The composition of the glomerular filtrate is about the same as that of blood plasma minus the proteins, since molecules larger than 4–5 nm cannot pass through the glomerular filter. Subsequently, the filtrate passes into the proximal and distal convoluted tubules of the nephron, where all of the metabolically useful molecules in the filtrate, such as glucose, amino acids, bicarbonate, and other ions, along with water, are selectively reabsorbed. The reabsorption of ions and water, which is under hormonal control, regulates the internal milieu and maintains homeostasis. What remains in the collecting tubules of the nephron after this thorough and precisely controlled reabsorption process is the urine, which contains only substances that are harmful or useless to the organism.

The long nephron tubules of birds and mammals contain a special "concentrating machine," the loop of Henle. This structure has a hairpin shape and runs through the medulla, or inner layer, of the kidney parallel to the collecting tubules. The glomeruli and reabsorbing convoluted tubules are located in the cortex, or outer layer, of the kidney. The loop of Henle employs active ion transport and selective water permeability to establish a countercurrent mechanism that exerts an osmotic pressure gradient throughout the entire medulla. This gradient arises at the outer border of the medulla and increases progressively up to the innermost part of the medulla. The longer the loop of Henle, the greater the ionic concentration in the medulla, and the greater the maximal urine concentration that can be achieved.

The length of the loop of Henle can be determined from the thickness of the renal medulla in histological sections. Commonly, only the thickness of the inner

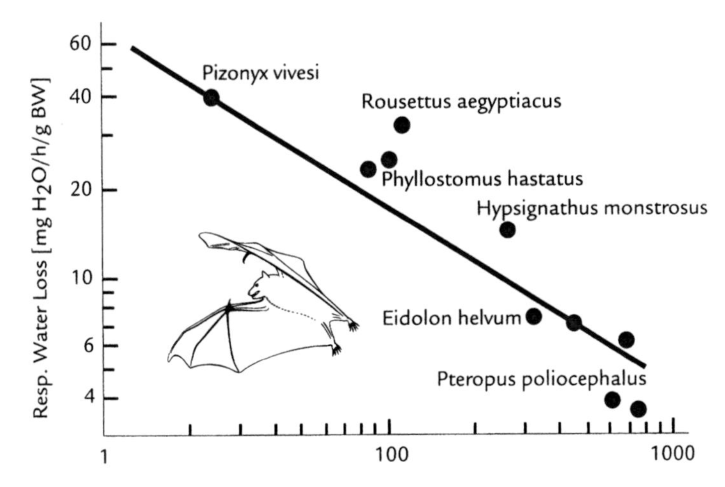

Figure 3.14 Water loss through respiration in flying bats as a function of body weight. From Carpenter (1986a).

medullary zone is measured. Thus, the ratio of the thickness of renal medulla to renal cortex provides a measure of the maximal urine-concentrating power of the kidney. Without the loop of Henle, the kidney could not produce urine with a concentration higher than that of blood plasma. A physiological measure of the concentrating power of the kidney is the ratio of the osmotic pressure in urine to that in blood plasma (U/P). Due to the evolutionary adaptation of the loop of Henle, this ratio in birds and mammals is greater than 1.0. This ability to form concentrated urine enables body water to be conserved and represents an important step in the establishment of life on land because it reduces the dependence of animals on external water supply.

The length of the loop of Henle is not the only factor that determines the urine-concentrating ability of the kidney. The faster the filtrate passes through the loop, the lower the concentration of the urine. Because the flow rate depends on the renal blood pressure, the blood pressure indirectly influences urine concentration. In small species of bats, the nephron tubules have very narrow lumens, which limits the flow rate. Another important factor controlling urine concentration is the hypophyseal hormone antidiuretic hormone, or vasopressin. This peptide causes increased water reabsorption from the collecting tubules as they exit the kidney in the renal medulla. Urea also plays a role in regulating the concentration of the urine in the renal medulla. Thus, the urine concentration also depends on the concentration of urea in the blood and therefore indirectly depends on the protein content of the diet.

Kidney function in insectivorous Chiroptera. The indirect relationship between dietary protein content and urine concentration by the kidney can clearly be seen in insectivorous chiropterans. After a meal of insects, both the amount of urine and the urine concentration increase abruptly. The osmotic pressure doubles

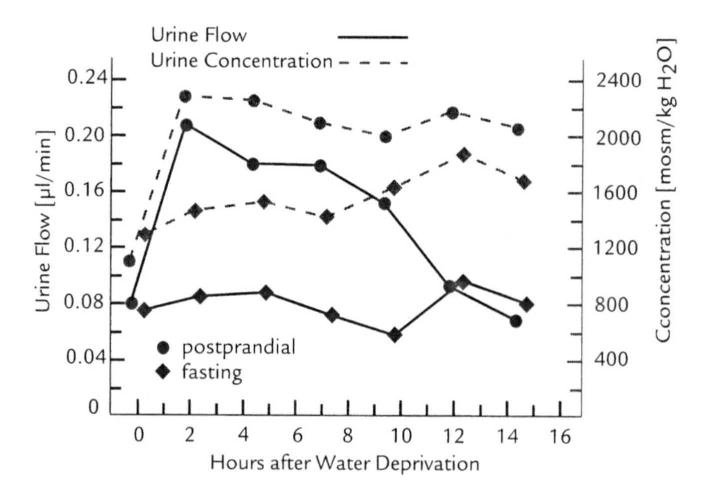

Figure 3.15 Daily pattern of urine production and urine concentration in the insectivorous bat *Myotis lucifugus* (Vespertilionidae). Comparison with a fasting bat shows that the high output of concentrated urine after a meal of insects is related to protein metabolism. From data in Bassett and Wiebers (1979).

to a value of 2400 mosmol/l (fig. 3.15). In fasting and water-deprived animals, urine flow is low. Although urine concentration increases slowly throughout the day, it never reaches the high values seen after an insect meal. The large output of highly concentrated urine is due to protein metabolism after feeding on insects, which leads in turn to a high blood urea concentration, four to five times as high as in omnivorous mammals. Nevertheless, the urine-concentrating ability of the kidney in insectivorous species is about the same as in other mammals, with a U/P relationship of 3.5–6.5. For comparison, that of humans is 4.0.

The vespertilionid bat *Pyzonix vivesi* (body weight ca. 25 g) lives on an island in the Gulf of California where there is neither fresh water nor insects. The extraordinary urine-concentrating ability of this bat's kidneys allows it to obtain all of its water requirements from the water contained in the shrimp and fish on which it preys, and from sea water. It is, in fact, possible to maintain these bats in the laboratory on a diet of frozen shrimp, without drinking water. *P. vivesi* turns over about 12 g of salt water obtained from prey every day (table 3.4). This water contains 540 mEq/l of Cl^-. *Pizonyx* can excrete 615 mEq/l of Cl^-, resulting in a net water gain of 0.11 ml of fresh water/ml salt water consumed. The more the animal drinks, the greater the amount of urine that must be excreted (fig. 3.16). To replace the maximal daily water loss of 2.1 ml, as shown in table 3.3, *Pizonyx* would have to drink 19 ml of sea water, amounting to three-fourths of its body weight, or else eat an additional 7 g of shrimp. The desalination of sea water results in a high level of chloride excretion by the kidney. The urine of *Pizonyx* contains 615 mEq/l Cl^-; in comparison, the Cl^- concentration in other bats is only 15–22 mEq/l. Thus, the ionic regulation in the kidneys is highly adaptable.

Table 3.4 Water balance in insectivorous bats

	Myotis lucifugus				*Eptesicus fuscus*			
	Pregnant		Lactating		Pregnant		Lactating	
	ml/day	% bw	ml/day	% bw	ml/day	% bw	ml/day	% bw
Water gained from								
Food	3.85	63	4.69	68	5.60	66	12.04	71
Metabolism	0.71	11	0.60	9	1.02	12	1.61	9
Drinking	1.60	26	1.62	23	1.85	22	3.42	20
Total	6.16	100	6.91	100	8.47	100	17.07	100
Loss from (calculated fractions)								
Urine	2.91	46	2.47	35	6.07	72	9.63	56
Evaporation	2.78	45	2.01	28	1.52	18	1.80	10
Feces	0.58	9	0.74	11	0.84	10	1.81	11
Milk	–	–	1.85	26	–	–	3.83	23
Total	6.27	100	7.07	100	8.43	100	17.07	100

From Kurta et al. (1994).

Within a single species, kidney function depends on the environment in which the population lives. In three populations of North American bats living in the dry climate of New Mexico, the loop of Henle is longer, and the maximal urine concentration higher than in populations of the same species in the wetter climate of Indiana. Thus, the length of the loop of Henle and consequently the urine concentration can be modified according to environmental conditions. It would be interesting to identify the key stimulus that triggers these modifications and to know how the environmental information is translated into physiological changes.

The dependence of kidney structure on environment means that urine concentration plays an important role in the regulation of water balance. Because of the need to maintain water balance, it could be argued that the unavoidable evaporative water loss through the lungs and the wings forced flying mammals to restrict their period of activity to the night, when it is cool. This argument may sound rather far-fetched, but it points out how many interrelated causes there may be for a specific behavior.

The widely held view that the kidneys cease to function during hibernation is contradicted by a series of measurements in *Myotis velifer*. Using glass capillaries, small amounts of urine were removed from the hibernating bat's ureter. At an ambient temperature of 39.4°C, the urine output in *Myotis* is 0.206 μl/min-g bw, but at 8°C it drops to 0.001 μl/min-g bw. This minimal amount of urine is very dilute (fig. 3.17). During hibernation mainly fat and protein are metabolized, so that the result may even be a positive water balance, resulting in urine production. Therefore, under normal conditions, hibernating bats do not suffer from dehydration. Furthermore, they use the frequent arousal phases during the winter to urinate and to drink.

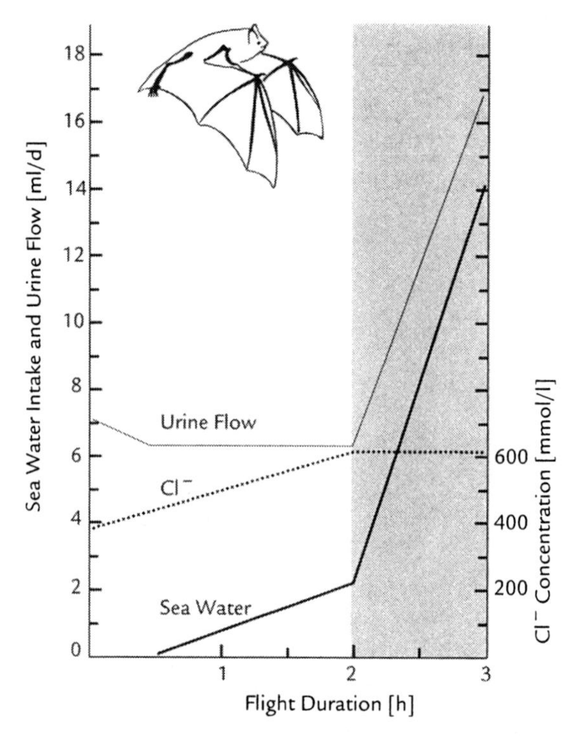

Figure 3.16 The effect of evaporative water loss during flight on sea water consumption and chloride concentration in the urine of *Pizonyx vivesi*, a vespertilionid bat that feeds on shrimp and drinks sea water. It can be seen that as water loss increases, water intake increases. Because of the desalination process, urine production also increases once the maximal rate of Cl^- excretion of 615 mEq/l is reached (dotted line). From Carpenter (1968b).

Kidney function in the vampire. When vampires ingest blood, they take in an unusual combination of high water content and high protein concentration. The water needs to be quickly excreted in order to minimize the animal's weight during flight. As shown in figure 3.18, in the first hour after feeding, vampires excrete a large volume of dilute urine (diuresis). The dense capillary bed that lies beneath the stomach epithelium takes up water from the ingested plasma and transports it rapidly to the kidney. Within the next 6 h, urine production falls from the highest relative volume measured in any mammal (0.24 g/h-g bw) to a minimum of 0.01 g/h-g bw, or 4% of the maximal rate. At the same time, the osmotic pressure of the urine increases 10-fold to about 3416 mosmol/l due to the breakdown of protein and production of urea. Thus, diuresis and urine concentration alternate due to urea production. The high urine concentration in the vampire is normally due to protein metabolism, while the high osmotic pressure of the urine in *Pizonyx vivesi* is due to its ionic content. During diuresis, the high osmotic pressure of the urine in the vampire bat is also due to a high NaCl concentration.

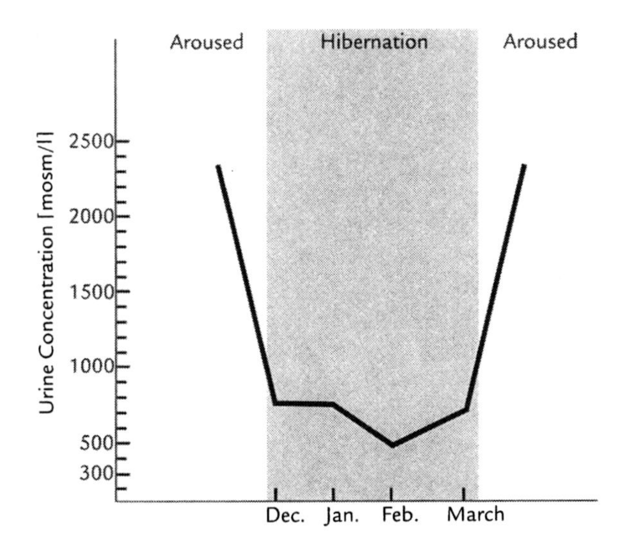

Figure 3.17 Average urine concentration in a population of 52 hibernating *Myotis velifer* (Vespertilionidae). From Caire et al. (1982).

Kidney function in bats that feed on fruit juice. The lowest demand for urine concentration by the kidney is found in frugivorous bats because their diet is high in water and low in protein. The Egyptian fruit bat, *Rousettus aegyptiacus* (bw ca. 150 g), daily excretes a volume of about 19 ml of a dilute urine (113 mosmol/l). Under desert conditions, however, this fruit bat reduces its urine volume to 1/20 of the normal value and increases its concentration fivefold. In frugivorous Phyllostomid bats, the urine-concentrating ability of the kidney, measured as the maximal osmotic pressure attainable, is only 1181 mosmol/l. This value is only reached under conditions of stress, when animals have been deprived of food and water for 21–29 h. The normal value is around 700 mosmol/l. As shown in fig. 3.19, urine concentration decreases rapidly and markedly when the bat is feeding on fruit. This rapid diuresis is related to the glucose content of fruit juice. Glucose is quickly absorbed from the gut, so that what remains is a hypotonic solution with an osmolarity of only about 120 mosmol. This pressure gradient between the intestinal lumen and the blood then leads to transport of a large volume of water. The excess water is eliminated in the urine. Due to the nature of the diet, the urine concentration of frugivorous bats decreases during feeding, while that of insectivorous bats increases. Because fruits contain large quantities of K^+, much K^+ is eliminated over a period of many hours after feeding. At the same time, Na^+ excretion drops. This relationship suggests that ions are transported by means of a Na^+/K^+ pump in the nephron.

The relationship between diet and the urine-concentrating ability of the kidney is clearly illustrated by comparing the maximal urine concentration in 33 different species of bats found in one region of Panama (fig. 3.20). The urine of several

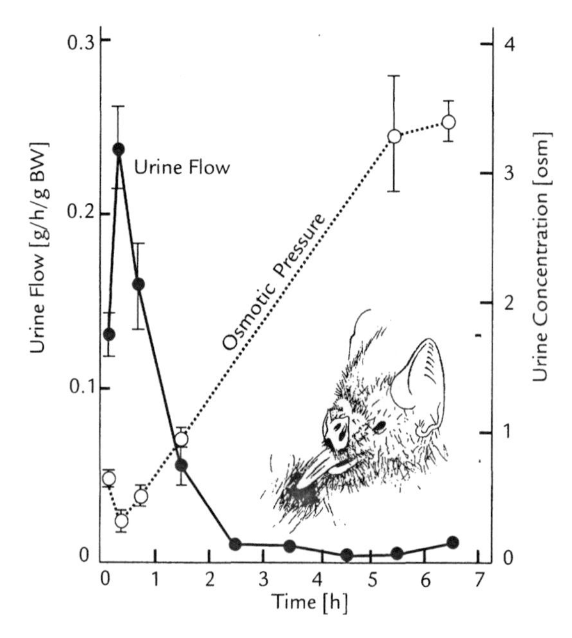

Figure 3.18 Urine excretion in vampire bats during and after feeding (0 = time when the bat begins to lick blood). Solid line, urine output; dotted line, osmotic pressure of the urine; error bars, standard deviations. The low osmotic pressure at the beginning of urine flow is due to NaCl concentration and the high osmotic pressure at the end of the curve is due to urea concentration. From Bush (1988).

insectivorous species that sometimes spend time in desert regions reaches an osmotic pressure as high as 3500 mosmol/l. Such a high concentration is otherwise found only in animals that are strictly desert dwellers. The correlation between the maximal urine concentration and the length of the loop of Henle, measured as the ratio of the thickness of the inner medullary layer to that of the renal cortex, indicates that the active processes that take place in the loop of Henle are a decisive factor in determining urine concentration.

Ammonia tolerance. Ammonia is extremely toxic to cells. Fish can excrete ammonia because it is diluted by the surrounding water as soon as it is produced. Terrestrial animals, unlike fish, do not excrete ammonia because there is no efficient way to prevent it from adhering to their body. Humans can tolerate only trace amounts of ammonia in the air because in high concentrations it destroys the alveolar cells of the lungs. Bats, on the other hand, can live quite well in air that contains high concentrations of ammonia, as can be seen from the many colonies that live in caves under such conditions. For example, in caves occupied by huge colonies of *Tadarida brasiliensis*, their feces form a thick carpet on the cave floor. As the feces decompose, ammonia is produced, and the ammonia concentration in

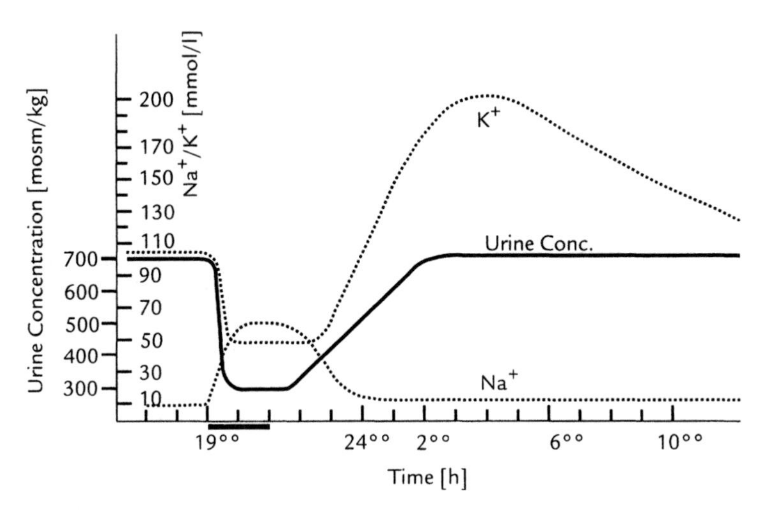

Figure 3.19 Urine excretion in the fruit-eating Phyllostomid bat *Artibeus jamaicensis*. Solid line, urine concentration. The bar beneath the horizontal axis indicates the feeding period. From data in Studier et al. (1983).

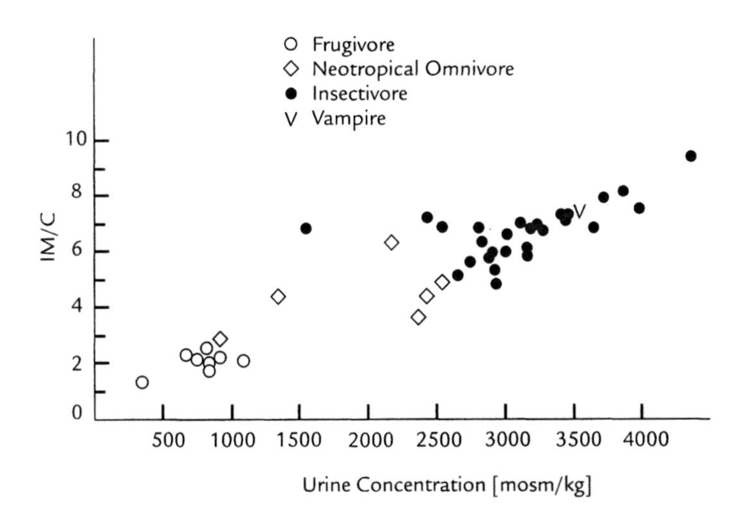

Figure 3.20 Relation between the maximal urine-concentrating ability of the kidney and diet. The urine-concentrating ability of the kidney is proportional to the length of the loop of Henle, and therefore proportional to the thickness of the inner zone of the renal medulla (IM), where the loop of Henle is located. The ratio IM/C (thickness of the inner zone of the renal medulla/thickness of the renal cortex, C) is a measure of the concentrating ability of the kidney. The data are from a population of 33 bats found in a single geographical region in Panama. From data in Studier and Wilson (1983), Studier et al. (1983), and Geluso (1978).

air of the poorly ventilated caves reaches concentrations of 100–1800 ppm. For humans, it is lethal to breathe an ammonia concentration of 500 ppm for an hour. In contrast, under laboratory conditions, *Tadarida brasiliensis* can tolerate an ammonia concentration of 5000 ppm for many days. However, when the ammonia concentration reaches 7000 ppm, even *Tadarida* die after 2–3 h. In bats, the ammonia content of the blood is 14 μmol/ml compared to 0.03–0.08 μmol/ml in dogs and in humans.

There have been no studies to investigate the protective mechanisms that enable bats to tolerate highly toxic concentrations of ammonia. It is thought that ammonia in the blood may be buffered by an unusually high CO_2 content. Then when the bats fly out in the evening, the excess ammonia is breathed out along with the CO_2. It is also thought that the lungs of cave-dwelling bats are protected from the toxic effects of ammonia by a neutralizing layer of mucus.

Urea and life span. All bats, regardless of whether they are homeothermic or hibernate in winter, have an unusually long life span given their body size, reaching an age of 20 years or more. There have been many speculations regarding the reasons for their longevity, but virtually no scientific investigations. In humans and in elephants, two long-lived large mammals, it has been suggested that longevity is related to high blood concentrations of urea. One experiment has shown that the blood urea concentration of insectivorous bats is four to five times as great as that of omnivorous mammals.

References

Temperature Regulation

Audet D, Fenton MB (1988). Heterothermy and the use of torpor by the bat *Eptesicus fuscus* (Chiroptera: Vespertilionidae): A field study. Physiol Zool 61:197–204.

Bauman WA (1990). Seasonal changes in pancreatic insulin and glucagon in the little brown bat (*Myotis lucifugus*). Pancreas 5:342–346.

Bonaccorso FJ, Arends A, Genoud M, Cantoni D, Morton T (1992). Thermal ecology of moustached and ghost-faced bats (Mormoopidae) in Venezuela. J Mammal 73: 365–378.

Brack V, Twente JW (1985). The duration of the period of hibernation of three species of vespertilionid bats. I: Field studies. Can J Zool 63:2952–2954.

Brigham RM (1987). The significance of winter activity by the big brown bat (*Eptesicus fuscus*): The influence of energy reserves. Can J Zool 65:1240–1242.

Doty STB, Nunez EA (1985). Activation of osteoclasts and the repopulation of bone surfaces following hibernation in the bat, *Myotis lucifugus*. Anat Rec 213:481–495.

Erkert HG, Rothmund E (1981). Differences in temperature sensitivity of the circadian system of homoiothermic and heterothermic neotropical bats. Comp Biochem Physiol 68:383–391.

Genoud M (1993). Temperature regulation in subtropical tree bats. Comp Biochem Physiol 104A:321–331.

Gustafson AW (1979). Male reproductive patterns in hibernating bats. J Reprod Fertil 56:317–331.

Hays GC, Speakman JR, Webb PI (1992). Why do brown long-eared bats (*Plecotus auritus*) fly in winter? Physiol Zool 65:554–567.

Hays GC, Webb PI, Speakman JR (1991). Arrhythmic breathing in torpid pipistrelle bats, *Pipistrellus pipistrellus*. Resp Physiol 85:185–192.

Heldmaier G (1969). Die Thermogenese der Mausohrfledermaus (*Myotis myotis*) beim Erwachen aus dem Winterschlaf. Z Vergl Physiol 63:59–84.

Hill JE, Smith JD (1984). Bats, a Natural History. British Museum, London.

Kurta A (1986). Factors affecting the resting and postflight body temperature of little brown bats, *Myotis lucifugus*. Physiol Zool 59:429–438.

Kurta A, Johnson KA, Kunz TH (1987). Oxygen consumption and body temperature of female little brown bats under simulated roost conditions. Physiol Zool 60:386–397.

Kurta A, Kunz TH (1988). Roosting metabolic rate and body temperature of male little brown bats (*Myotis lucifugus*) in summer. J Mammal 69:645–651.

Laburn HP, Mitchell D (1975). Evaporative cooling as a thermoregulatory mechanism in the fruit bat. *Rousettus aegyptiacus*. Physiol Zool 48:195–202.

Lyman CH (1982). Hibernation and Torpor in Mammals and Birds. Academic Press, New York.

McNab BK (1989). Temperature regulation and rate of metabolism in three Bornean bats. J Mammal 70:153–161.

Morris S, Curtin AL, Thompson MB (1994). Heterothermy, torpor, respiratory gas exchange, water balance and the effect of feeding in Gould's long-eared bat, *Nyctophilus gouldi*. J Exp Biol 197:309–335.

Noll UG (1979). Body temperature, oxygen consumption, noradrenaline response and cardiovascular adaptations in the flying fox, *Rousettus aegyptiacus*. Comp Biochem Physiol 63A:79–88.

O'Farrell MJ, Schreiweis DO (1978). Annual brown fat dynamics in *Pipistrellus hesperus* and *Myotis californicus* with special reference to winter flight activity. Comp Biochem Physiol 61:423–426.

O'Shea JE (1970). Temperature sensitivity of cardiac muscarinic receptors in bat atria ventricle. Comp Biochem Physiol 86C:–365–370.

*Raths P, Kulzer E (1976). Physiology of Hibernation and Related Lethargic States in Mammals and Birds. Bonner Zoological Monograph No. 9, Museum Alexander König, Bonn.

Shump KA, Shump AU (1980). Comparative insulation in vespertilionid bats. Comp Biochem Physiol 66A:351–354.

Speakman JR (1988). Position of the pinnae and thermoregulatory status in brown longeared bats. J Therm Biol 13:25–30.

Szewczak JM, Jackson DC (1992). Ventilatory response to hypoxia and hypercapnia in the torpid bat, *Eptesicus fuscus*. Resp Physiol 88:217–232.

Thomas DW, Cloutier D, Gagné D (1990). Arrhythmic breathing, apnea and non-steady-state oxygen uptake in hibernating little brown bats, *Myotis lucifugus*. J Exp Biol 149: 395–406.

Thomas DW, Cloutier D (1992). Evaporative water loss by hibernating little brown bats, *Myotis lucifugus*. Physiol Zool 65:443–456.

Twente JW, Twente J, Brack V (1985). The duration of the period of hibernation of three species of vespertilionid bats. II: Laboratory studies. Can J Zool 63:2955–2961.

Twente JW, Twente J (1987). Biological alarm clock arouses hibernating big brown bats, *Eptesicus fuscus*. Can J Zool 65:1668–1674.

Vaughan TA (1987). Behavioral thermoregulation in the African yellowwinged bat. J Mammal 68:376–378.

Wilkinson M, Buchanan GD, Jacobson W, Younglai EV (1986). Brain opioid receptors in the hibernating bat, *Myotis lucifugus*: Modification by low temperature and comparison with rat, mouse and hamster. Pharmacol Biochem Behav 25:527–532.

Wolk E, Bogdanowicz W (1987). Hematology of the hibernating bat, *Myotis daubentoni*. Comp Biochem Physiol 88A:-637–640.

*Yalden DW, Morris PA (1975). The Lives of Bats. David and Charles Newton Abbot, London.

Water Balance and Excretion

Bassett JE (1980). Control of postprandial water loss in *Myotis lucifugus lucifugus*. Comp Biochem Physiol 65A:497–500.

Bassett JE (1982). Habitat aridity and intraspecific differences in the urine concentrating ability of insectivorous bats. Comp Biochem Physiol 72A:703–708.

Bassett JE, Wiebers JE (1979). Urine concentration dynamics in the postprandial and the fasting *Myotis lucifugus*. Comp Biochem Physiol 61A:373–379.

Bell GP, Bartholomew GA, Nagy KA (1986). The roles of energetics, water economy, foraging behvior, and geothermal refugia in the distribution of the bat, *Macrotus californicus*. J Comp Physiol B 156:441–450.

Bush C (1988). Consumption of blood, renal function and utilization of free water by the vampire bat, *Desmodus rotundus*. Comp Biochem Physiol 90A:141–146.

Caire W, Haines H, McKenna TM (1982). Osmolality and concentration in urine of hibernating *Myotis velifer*. J Mammal 63:688–690.

Carpenter RE (1968a). Structure and function of the kidney and the water balance of desert bats. Physiol Zool 42:288–302.

Carpenter RE (1968b). Salt and water metabolism in marine fisheating bat. Comp Biochem Physiol 24:951–964.

Carpenter RE (1986). Flight physiology of intermediate-sized fruit bats (Pteropodidae). J Exp Biol 120:79–103.

*Geluso KN (1978). Urine concentrating ability and renal structure of insectivorous bats. J Mammal 59:312–323.

Kurta A, Bell GP, Nagy KA, Kunz TH (1989). Water balance of freeranging little brown bats (*Myotis lucifugus*) during pregnancy and lactation. Can J Zool 67:2468–2472.

Kurta A, Kunz TH, Nagy KA (1990). Energetics and water flux of freeranging big brown bats (*Eptesicus fuscus*) during pregnancy and lactation. J Mammal 71:59–65.

Studier EH (1970). Evaporative water loss in bats. Comp Biochem Physiol 35:935–943.

Studier EH, Boyd BC, Feldman AT, Dawson W (1983). Renal function in the neotropical bat, *Artibeus jamaicensis*. J Biochem Physiol 74:199–209.

Studier EH, O'Farrell MJ (1976). Biology of *Myotis thysanodes* and *M. Lucifugus* (Chiroptera: Vespertilionidae)-III. Metabolism, heart rate, breating rate, evaporative water loss and general energetics. Comp Biochem Physiol 54:423–432.

Studier EH, Wilson DE (1983). Natural urine concentrations and composition in the neotropical bat, *Artibeus jamaicensis*. Comp Biochem Physiol 75:509–517.

Studier EH, Wisniewski SJ, Feldman AT, Dawson RW (1983). Kidney structure in neotropical bat. J Mammal 64:445–452.

Webb PI, Speakman JR, Racey PA (1993). Defecation, apparent absorption efficiency, and the importance of water obtained in the food for water balance in captive brown long-eared and Daubenton's bats. J Zool 230:619–628.

4

DIET, DIGESTION, AND ENERGY BALANCE

TEMPERATURE REGULATION AND water balance are closely related to energy balance. Environmental conditions that affect one of these processes also affect the others.

Among all the different mammalian orders, the one with the largest variety of diets is the bats. Although most are insectivores (see table 4.1), a significant number of bats feed exclusively on fruit, nectar, and pollen, a few feed on small vertebrates, and three species of neotropical vampires feed exclusively on the blood of mammals and birds. Thus, following a general description, it will be useful to consider separately feeding, digestion, and energy balance in insectivores, frugivores, nectar and pollen-feeding species, and vampires.

4.1 THE TEETH

Mammals have an adaptation that gives them a decided advantage over other vertebrates: Their teeth are not only used to seize their food so that they can swallow it; they also serve to break it up into small pieces before it is swallowed. The process of chewing makes food more digestible and increases the availability of nutrients. This mammalian adaptation is due to two innovations.

First, the development of a secondary joint in the jaw has caused the jaw muscles to be displaced outward. Due to this adaptation, the lower jaw not only can move up and down, but it can also move from side to side so that it provides a crushing and cutting motion. In parallel with this skeletal adaptation, an appropriate system of muscles for chewing also developed (fig. 4.1d). Second, the teeth of the upper and lower jaws are formed in such a way that they fit precisely against one another, or occlude. This makes the teeth well-suited for breaking apart and crushing food (fig. 4.2).

Bats have adapted this basic structural plan to fit their special requirements. Their shortened jaw (fig. 4.2) gives them a powerful bite. Because their short jawbone reduces the amount of space for teeth, bats tend to have a reduced number of teeth. Bats have, at most, 38 teeth (*Thyroptera*, *Myzopoda*, 4 genera of Natalids and 4 genera of Vespertilionids, including *Myotis*). The form of the jaw and teeth in each species of bat provide a perfect reflection of its diet.

Table 4.1 Dietary specializations in bats

Diet	Number of species	Percentage of all species
Insects	700 (approx.)	71
Fruit	230 (approx.)	23
Nectar/pollen	50 (approx.)	5.3
Meat	7	0.7
Blood	3	0.3

MILK TEETH

Like most mammals, newborn bats have a set of milk teeth that are the precursors of the incisors, canines, and premolars. The milk teeth are like fine hooks that point toward the back of the mouth and are used by the pup to grip the mother's nipple. The milk teeth perform a vital function for the pup because, if it were to lose its grip and fall, it would be unlikely to survive. The permanent teeth appear at about the same time that the pup begins to fly, at about 3–6 weeks after birth.

Horseshoe bats, hipposiderids, and megadermatids are not born with milk teeth; instead, their permanent canine teeth are already in place. Milk teeth are present in these species during development but are reabsorbed before birth. The young of these species use their canine teeth to grip not only the breast nipples, but also the pubic attachment nipples, located in the fold of the hip. Pubic nipples are found only in these three families of bats.

DIGESTIVE ORGANS

There have been no systematic physiological studies of digestion in bats, so knowledge about the function of the digestive system is incomplete. As far as is known, digestion in all Chiroptera follows the standard mammalian plan. The stomach serves as a storage receptacle for large amounts of food ingested over a short period of time, for the destruction of bacteria by the stomach acid, and for the initial breakdown of proteins by the gastric enzymes pepsin and cathepsin. True enzymatic digestion occurs in the small intestine through the action of enzymes secreted by the pancreas and intestinal epithelium, and fat is emulsified through the action of bile, making it available as a substrate for hydrophilic enzymes. The nutrients absorbed from the intestine enter the portal circulation and are transported to the liver, the most important metabolic organ. The large intestine and colon act mainly to reabsorb water and excrete the indigestible remains of the food.

The stomach of bats can be subdivided into the same compartments as in other mammals (fig. 4.3b). The cardiac region is a small area surrounding the entrance to the stomach, the fundus is a saclike region that extends above the cardiac region, the body is the section that lies between the entrance and exit of the stomach, and the pylorus is the region surrounding the junction with the duodenum. The stomach is highly distensible and capable of accommodating large quantities

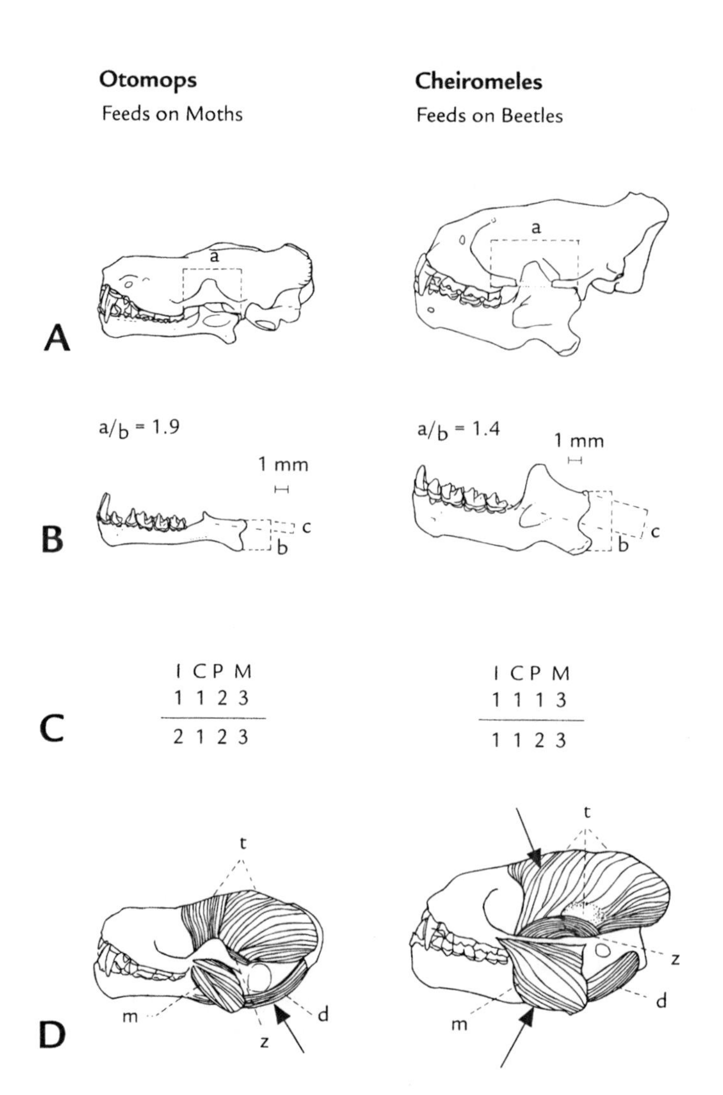

Figure 4.1 Teeth and muscles of mastication in two species of molossid bats. (A) Skull. a = distance from the joint to the insertion of the masseter muscle. (B) Lower left jaw. a/b = angle of jaw opening; c = distance from joint level to level of row of teeth. (C) Dental formulae. C, canines; I, incisors; M, molars; P, premolars. (D) Musculature. d, digastric muscle; m, masseter muscle; t, temporalis muscle; z, zygomaticomandibularis muscle. *Cheiromeles parvidens,* which feeds on insects with hard chitin exoskeletons, has a relatively short jaw with the mandibular joint located above the row of teeth. The mouth has a small angle of opening and powerful muscles for closing the jaw. This type of jaw is especially effective in producing pressure when the mouth is closed. *Otomops martienssi* feeds on insects with soft chitin exoskeletons. It is characterized by an elongated jaw, a mandibular joint that is nearly at the same level as the teeth, and a long digastric muscle. These features permit the mouth to open widely. This type of jaw exerts force through acceleration of the lower jaw as the mouth closes. From Freeman (1979).

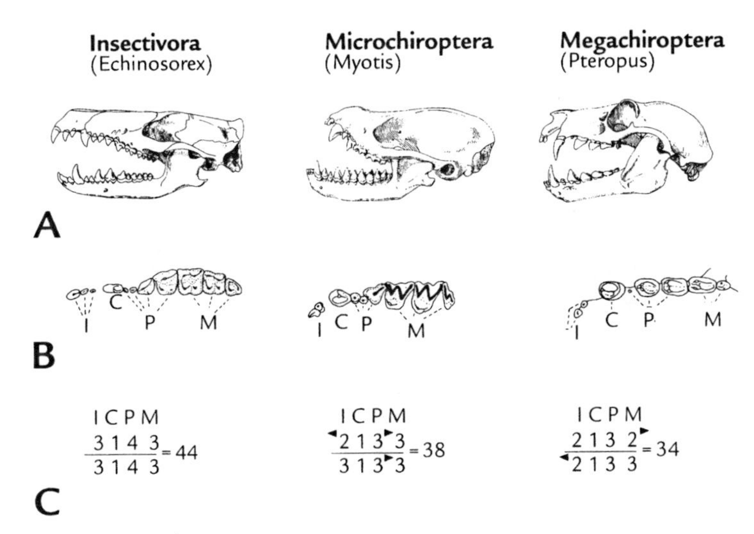

Figure 4.2 Comparison of the teeth in insectivores, Microchiroptera, and Megachiroptera. (A) Skull (not drawn to scale). (B) Top view of the teeth of the upper right jaw. (C) Dental formulas: number of teeth in the upper jaw is given above the line; number of teeth in lower jaw is given below the line. I, incisors; C, canines; P, premolars; M, molars. Triangles indicate reductions in the number of teeth. The anterior part of the skull of bats is considerably shortened compared to that of insectivores, and the occlusal surfaces of the teeth of insectivorous bats have sharp ridges arranged in a W-shaped formation. The occlusal surfaces of the teeth in frugivorous bats are flattened and well suited for crushing fruit. In insectivorous Microchiroptera, the jaw closes from back to front, while in frugivorous Megachiroptera, the premolars close first, followed by the molars. From Thenius (1989).

of crushed food due to its longitudinal folds or rugae. For example, during a half hour period, *Pipistrellus subflavus* can ingest food equivalent to one-quarter of its body weight and *Eptesicus nilssoni* can catch up to 20 insects in 1 min.

In Chiroptera, as in other mammals, the gut consists of the small intestine, which can be further subdivided into the duodenum, jejunum, and ileum, and the large intestine (fig. 4.3a). The large intestine is unusually short, consisting only of a descending segment. The ascending and transverse segments are absent, as is the caecum or appendix. In only one species, *Rhinopoma*, is there a short caecum at the junction between the large and small intestines. Phyllostomid bats have an ampulla filled with lymphoid tissue in the same location. The shortening of the intestine and the rapid passage of food through the digestive system of bats is likely related to the need to reduce weight during flight. To exploit the full nutritional content of the food during its rapid passage through the gut, the decreased time for digestion must be compensated for by increased enzymatic activity, but little is known in this regard.

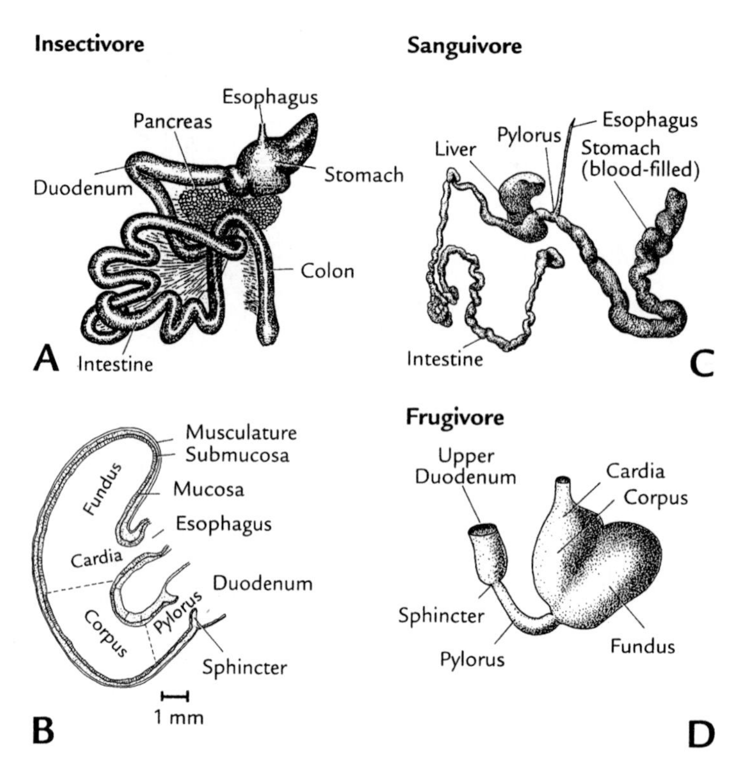

Figure 4.3 Stomach and intestine of bats. (A) The stomach and intestinal tract of *Myotis frater*, an insectivorous vespertilionid. From Ishikawa et al. (1985). (B) Schematic section through the stomach of *Myotis velifer*. From Rouk and Glass (1970). (C) The stomach and intestinal tract of the vampire, *Desmodus rotundus*. From Schmidt (1978). (D) The stomach of the frugivorous phyllostomid bat, *Ariteus flavescens*. From Mennone et al. (1986).

ENERGY BALANCE

Daily energy consumption depends on many different factors, including external conditions such as the time of year, availability of food, and pressure due to predators, as well as internal factors such as age, motivational state, and behavioral tendencies. Because the experiments that have been conducted to date on energy balance in bats have used very different methods and have been based on different assumptions, there are large variations in the results obtained even when studying a single species. For this reason it is difficult to compare the results from different experiments, and the data cited here can only serve as examples. On the basis of the available data, it is impossible to draw any general conclusions. Nevertheless, given these caveats, it is possible to make two general statements: Bats have a lower rate of energy turnover than birds of comparable size, and frugivorous and nectar-feeding bats tend to have a higher energy turnover than other bats and other small mammals of comparable size, up to about 1 kg.

4.2 INSECTIVOROUS AND CARNIVOROUS BATS

DENTITION IN INSECTIVOROUS BATS

As mentioned earlier, bats have a shortened jaw with a powerful bite. The shortening of the jaw restricts the amount of room for the teeth. The maximal number of teeth found in mammals is 44. In comparison, insectivorous bats lack the first upper incisor and the first upper and lower premolar, for a total of 38 teeth (figs. 4.1 and 4.2). In some species of bats—the molossids, for example—the necks of the canine teeth have taken on a cutting function. The premolars (P) and molars (M) of insectivorous bats are located far back in the jaw, just under the muscles of mastication, where the greatest force can be exerted. The upper cheek teeth are sharply serrated, with narrow triangular ridges and grooves into which the mirror-image structures on the surfaces of the lower teeth fit tightly. The teeth of bats function as a sort of mill with cutting and chopping surfaces that can break apart even the hardest chitin exoskeleton. As a result of this method of breaking up food, the masseter muscle, used for chewing, is smaller than the temporal muscle, which is used for biting (fig. 4.1). Forward and backward movements of the lower jaw aid in chopping up insect chitin. Special bony protuberances on the mandibular joint ensure that the lower jaw remains vertical during biting and prevent the jaw from twisting.

Many insectivorous bat species have a muscular, calluslike papilla on the gum at the level of the lower premolars. This papilla is thought to be used as a platform on which to break hard pieces of chitin and as an instrument for ejecting sharp chitin-covered parts of the prey that are too hard to eat.

The differentiation of the teeth and mandibular joint provide a guide to the type of diet of different species of bats as well as the general dietary preferences within a genus (fig. 4.1). Among the molossids, there are some species in which the jaw is short and the mandibular joint is located above the row of teeth. The teeth are strong and reduced in number. The temporal muscle is especially large and powerful, and it is located closer to the muzzle than usual (arrow in fig. 4.1d). These specializations are also found in cats. The result of this configuration is that when the mouth is closed, force is exerted across the entire length of the jaw. This type of jaw provides a static, high-pressure bite, and is especially well-suited for breaking apart hard-shelled prey. *Cheiromeles parvidens* (fig. 4.1) has this type of jaw and preys mainly on hard-shelled beetles.

Other species of molossid bats have long jaws with the mandibular joint located at the same level as the row of teeth so that the mouth can open wide (lever ratio a/b in fig. 4.1). The teeth are not as strong, and the second upper premolars are present. This type of jaw can close rapidly, and due to its mass and the long lever arm of the mandible, it produces a large amount of kinetic energy. This type of jaw provides a dynamic, acceleration-driven bite. It produces less biting power, but more movement-driven power, and because of the large size of the mouth, it is especially well suited for killing large, soft-bodied prey. *Otomops* has this type of jaw and preys mainly on moths. When a moth is caught, the large pointed canine teeth quickly penetrate its body and kill it.

THE TEETH OF CARNIVOROUS BATS

Carnivorous and omnivorous bats tend to have even fewer incisors. Megadermatid bats have no incisors at all, and lack a premaxillary bone as well. The scissorlike zigzag shape of the cheek teeth is flattened and laid out in a lengthwise orientation. This modification gives rise to oblong cutting edges of the same type as are found in cats, for breaking apart meat and bones. The first molar in *Megaderma* is similar in form to the fourth premolar of early carnivores.

DIGESTION

Chitin can be extremely hard. Even when it is crushed into small pieces it can still injure the mucosa. For this reason, the upper third of the esophagus of many insectivorous species is lined with keratinized epithelium. In *Eptesicus fuscus*, this part of the esophagus is lined with a thickened, calluslike layer. The esophagus empties into the single chamber of the stomach without an intervening sphincter. The gastric epithelium contains the ducts of many different glands of the type generally found in mammals. The cardiac region contains mainly mucus-producing glands. The fundus and body of the stomach contain not only mucus-producing glands but also parietal cells, which secrete hydrochloric acid, and chief cells, which secrete pepsinogen. The pyloric region contains mainly mucus-secreting cells. The gastric epithelium also contains various types of enteroendocrine cells. Based on their microscopic structure, these cells appear to secrete gastrin, which stimulates HCl secretion, or glucagon and other hormones that regulate digestion.

As a general rule, the intestinal tract of insectivorous bats is no longer than four times the length of the body. The minimal digestion time is 35 min in fully active animals, but can last as long as 170 min in animals that rest after feeding. As soon as 15 min after feeding, the partially digested food enters the duodenum. Several hours may elapse between the ingestion of a full insect meal and the elimination of the last remnants in the feces (fig. 4.4). Little is known about the digestion of insects. The amazingly short time for passage of food through the intestinal tract suggests a high rate of enzymatic activity. It has often been suggested that chitin is at least partly digestible. Although there is no conclusive evidence for or against this idea, the fact that well-preserved chitin fragments are found in the feces argues against it.

Although digestion ceases during hibernation, there are no obvious gross or microscopic changes in the gastric epithelium during this time. The only unusual observation is the finding of crystalline inclusions in epithelial cells of the duodenum, but nothing is known about their origin, composition, or function.

ENERGY BALANCE

The double-labeled water method was used to measure the field metabolic rate (FMR) in pregnant and lactating *Myotis* living freely in a nursery colony (table 4.2). When daily energy consumption was measured, it was found that 95% of the digestible nutrient energy was absorbed and assimilated, and that about 15% of the assimilated protein was lost in the form of urea. This energy loss has to

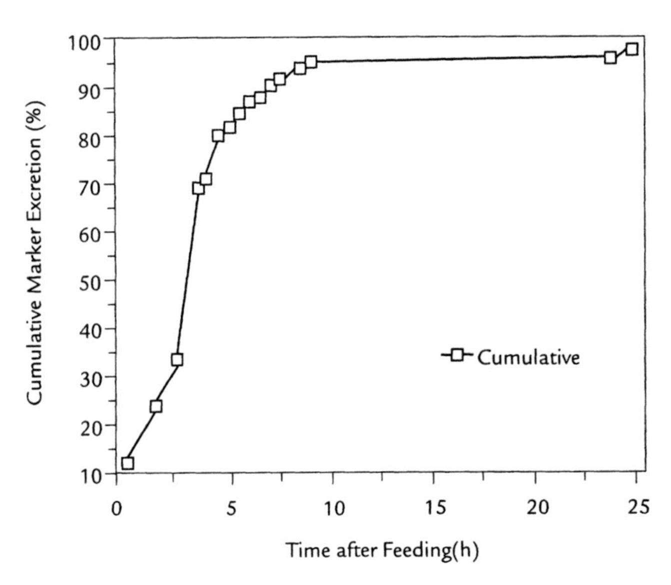

Figure 4.4 Food passage rate in mealworm-fed bats (*Nyctophilus gouldi*) as determined by a pollen marker with which the mealworms were dusted. After Morris et al. (1994).

be compensated by an equivalent energy intake (fig 4.5). An "average" insect is made up of 70% water, 17.8% protein, 4.5% fat, and 2.2% carbohydrate. The indigestible chitin exoskeleton makes up another 3.8%. Thus, 1 g of an insect diet provides 6 kJ of assimilable energy. A pregnant female *Myotis* thus requires 5.5 g of insects (61% of her body weight) every day, and a nursing female requires 6.7 g (85% of her body weight) to avoid using fat reserves to supply part of her daily energy requirements. Nursing females do appear to use fat reserves for energy because they lose weight during the suckling period. In *Eptesicus fuscus* there is an even greater difference in energy consumption between pregnant and nursing females. Pregnant females require 2274 kJ/day-kg bw, lactating females 4325 kJ/day-kg bw.

Table 4.2 Field metabolic rate in female *Myotis lucifugus*

	Average weight (g)	CO_2 Production (ml/day)	Assimilated energy	
			(kJ/day)	(kJ/day-kg bw)
Pregnant	9.02	1224 (47.3)	33.2 (1.3)	3681
Lactating	7.88	1036 (59.5)	28.1 (1.7)	3566

Standard deviations in parentheses. The bat's own daily energy consumption based on CO_2 measurement does not take into account energy for growth of the embryo (0.5 kJ/day) or milk production (13.2 kJ/day). From Kurta et al. (1989).

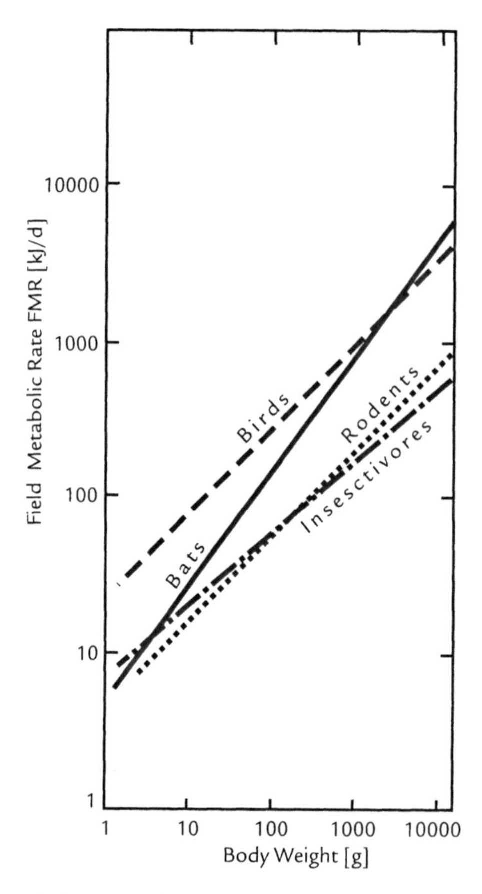

Figure 4.5 Metabolic rate of selected animals as a function of weight. Bats: FMR = $771.9 \cdot bw^{0.767}$; insectivores: FMR = $290.4 \cdot bw^{0.43}$; rodents: FMR = $371.5 \cdot bw^{0.54}$; birds: FMR = $851.9 \cdot bw^{0.605}$. From Helversen and Reyer (1984).

How is energy consumption related to daily activities? Measurements and estimation of time spent in various activities for a model animal (pregnant, 9 g body weight), indicate that the activity with the shortest duration, hunting in flight, accounts for nearly two-thirds of the daily energy consumption (table 4.3). Therefore, the costs of the daily "income" are the major factor that determines energy balance.

The insectivorous bat, *Macrotus californicus*, provides an impressive example of the sophisticated strategies that have been developed to maintain energy balance. Even though this species is a phyllostomid and incapable of hibernation, it spends the winter in the desert of Colorado, where nighttime temperatures drop to near freezing. Two special adaptations have made it possible for this species to inhabit the northernmost zone in which a phyllostomid is found. First, for their day roosts, they seek out open tunnels that are heated by geothermal energy to about

Table 4.3 Activity-specific energy expenditures in a pregnant *Myotis lucifugus* weighing 9 g

Activity	Duration (h)	FMR (kJ/h)	Fraction of total energy (%)
Daytime rest	15.6	0.52	27
Nighttime rest periods	4.3	0.82	11.7
Hunting insects	4.1	4.46	61

FMR, field metabolic rate. From Kurta et al. (1989).

29°C, close to the thermoneutral zone of 33–37°C. These animals avoid warmer caves, presumably because the higher temperature would lead to excessive water loss. Second, they seldom use echolocation when they hunt. Instead, they use vision and passive hearing to detect large prey such as locusts, *Nymphalidae* (brush-footed butterflies), sphynx moths, and cockroaches. This hunting strategy permits them to hunt during the first 2 h of the evening while the air is still relatively warm, between 15–20°C.

The FMR of these animals, with a body weight of about 13 g, is only 22.3 kJ/day, or 1715 kJ/day-kg bw. To maintain its FMR at this level, *Macrotus* must eat 3.7 g of insects every day (29% of its body weight). In February and March it is not possible to obtain this volume of insects. The deficit is compensated by mobilization of fat reserves which, during this period, decrease from 31% to 21% of the animal's body weight. Because *Macrotus* comes out of the cave for only about 2 h each day, 62% of the energy consumed is used during the time it is resting in the cave, and only 38% is used outside during hunting. Even during the winter *Macrotus* does not become torpid. The relation between the amount of energy used at rest and for hunting by *Macrotus* during the winter is the reverse of that seen in active *Myotis lucifugus*.

The model illustrated in figure 4.6 shows that the temperature of the day roost is a critical determinant of the energy balance typical of a region, and therefore also determines the chances of survival. Temperatures around 30°C minimize the energy-intensive hunting time necessary and therefore allow the animal to survive for an extra 2 weeks due to fat conservation. The availability of appropriate day roosts is a critical factor in determining whether an area will be inhabited by bats.

Optimization of energy balance is the credo of sociobiology, which has shown over and over again how animal populations develop evolutionarily stable strategies in which they optimize the exploitation of an energy source through specific patterns of behavior. An impressive model calculation shows how the collective behavior of a colony influences the energy gain of an individual under conditions of scarce resources. When a bat colony is occupying its winter quarters, individual bats wake up from time to time and fly out in search of food. In a colony of 100 bats, how frequently should an individual wake up if 100 insects are available each night, the energy cost for a night of foraging is 10 insects, and the rate of arousal

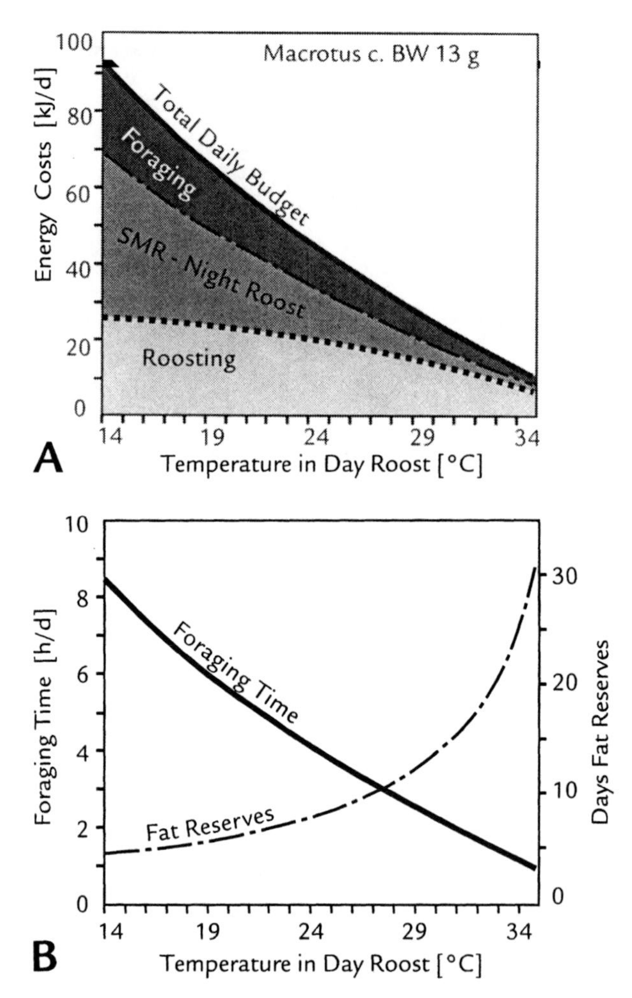

Figure 4.6 Dependence of daily energy expenditure on the temperature of the day roost in *Macrotus californicus* (Phyllostomidae). All values given are for an animal weighing 13 g. (A) The daily energy cost is the sum of energy consumption while resting in the day roost, the sustaining metabolic rate (SMR) outside the roost, and the energy used in hunting. (B) The energy required for hunting is inversely related to the temperature of the day roost. Survival increases as a function of the amount of fat reserve available for days when there is no food. From Bell et al. (1986).

in the colony is evenly distributed throughout the winter? As figure 4.7 shows, if all individuals awoke 4 times during the winter, each individual would have a net gain of 80 arbitrary units, whereas if the rate were higher, not all individuals in the colony would gain. The model shows that social behavior patterns and their distribution over time have decisive consequences for the development, selection, and optimization of strategies for obtaining energy.

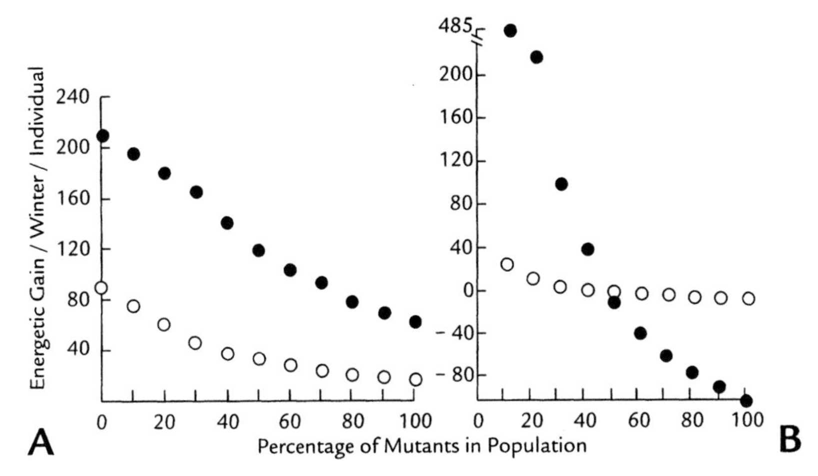

Figure 4.7 Influence of arousal rate in a hibernating colony on potential energy gain for individual bats (model calculations, ordinate in arbitrary units). (A) "Mutants" (filled circles) in the colony wake up four times during hibernation; "nonmutants" (open circles) wake up only once. (B) "Mutants" wake up 20 times; "nonmutants" wake up only once. It is clear from these calculations that with an arousal rate of four times per winter (filled circle at 100% in panel A) that every animal is assured of sufficient energy gain. From Avery (1985).

4.3 VAMPIRE BATS

Vampire bats live exclusively on blood. This highly specialized diet in three different vampire species has led to unique adaptations. For example, the cheek teeth are reduced in size and are nonfunctional. Vampires have only 20 teeth, the smallest number of any chiropteran species (fig. 4.8).

FEEDING

The most common vampire, *Desmodus rotundus*, lives on the blood of mammals, while the less common species *Diphylla* and *Diaemus* mainly feed on the blood of birds. The licking of blood by *Desmodus* is the most thoroughly studied feeding behavior.

Desmodus lands near its victim or directly on it. Vampires can run and hop rapidly supported on their wrists and legs, thereby avoiding defensive reflexes of their sleeping victims. The vampire locates a highly vascularized area, the neck and throat, for example, and wets a 10–15 mm area with its saliva. Then the vampire presses its protruding lower jaw and outwardly angled teeth against the skin so that they act as a chopping block. It then brings its upper teeth down in a rapid blow that cuts away the overhanging piece of skin (fig. 4.8a). The cut is razor-sharp because the incisors and canine teeth are broadened to take on a bladelike shape, and are equipped with slightly concave cutting surfaces (fig. 4.8c,d).

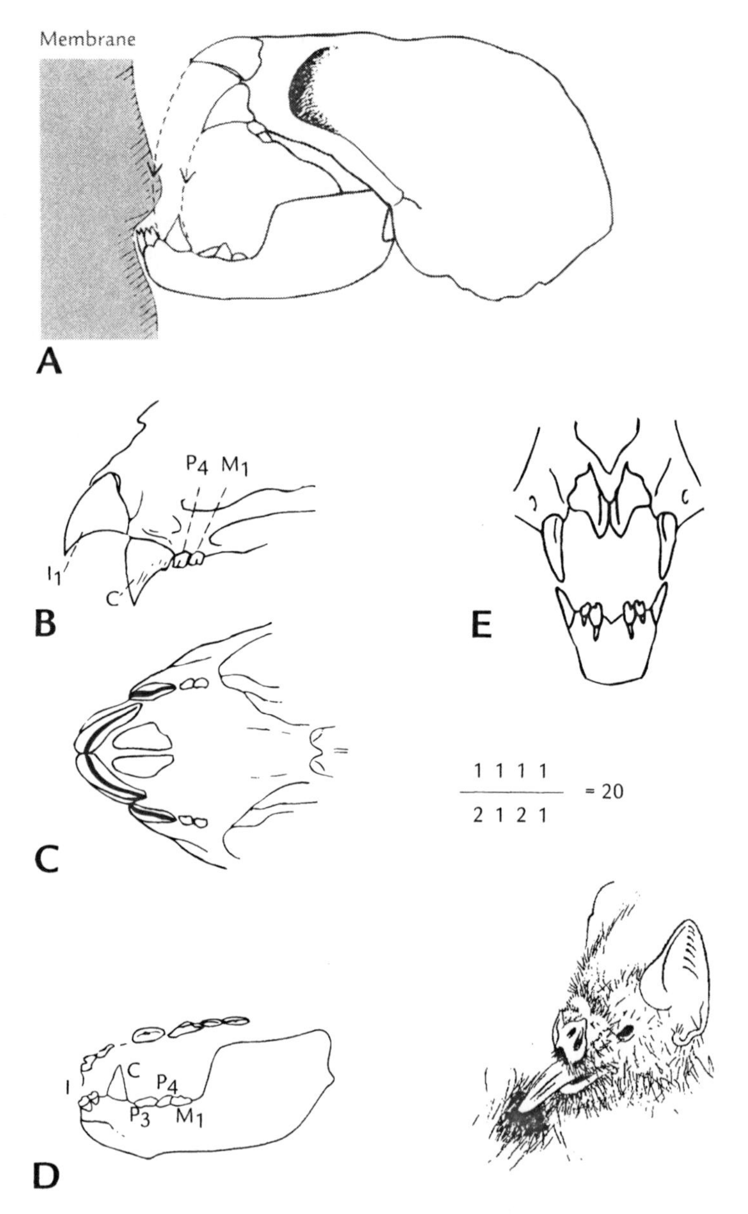

Figure 4.8 Teeth of the vampire bat, *Desmodus rotundus*, which feeds by licking blood. (A) Operation of the teeth in cutting the skin. (B) Outside view of the upper jaw. (C) Upper jaw as seen from below. Cutting edges are shown in black. (D) Inside view of the lower jaw, teeth of the lower jaw viewed from above. (E) View from straight ahead and dental formula. I, incisors; C, canines; P, premolars; M, molars. (A) From Storch (1968); (B)–(D) from Thenius (1989).

After the bite, the vampire jumps back and spits out the piece of skin. The wound, which reaches down into the dermis, continues to bleed for a considerable time because the vampire continually prods the wound with its rough keratinized tongue, opening new capillaries. The saliva of vampires contains substances that prevent clotting and make the wound continue to bleed. One of these substances is a plasminogen activator. Plasminogen prevents blood coagulation through fibrinolysis. The plasminogen activator in the saliva of vampires increases its activity 45,000-fold in the presence of fibrin I, and is therefore 220 times as selective as the plasminogen activator found in mammalian tissue. These properties make vampire plasminogen activator interesting as a potential treatment for thromboses in man. When the vampire licks, the divided lower lip rests against the edge of the wound. There is a groove on the left and right of the underside of the tongue, into which the blood flows through the split in the lower lip. It is thought that each time the tongue is extended the grooves are narrowed, pumping blood toward the rear of the mouth where it flows up over the side of the tongue into a channel in the center of the upper surface of the tongue that carries it to the pharynx. However, the course of the blood from the tip of the tongue to the back of the tongue is not fully understood.

A blood meal lasts about 25 min. However, the tongue actively licks up blood for a total of only about 12–13 min because short licking bouts of 1–22 s are interspersed with long pauses. During each licking bout, the tongue is rapidly dipped in the blood 2–4 times, with about 3 μl of blood ingested per lick. Vampires living in the wild ingest about 15 g of blood at each meal, equivalent to half their body weight. In extreme cases, the amount of blood consumed can amount to more than two-thirds of the bat's body weight.

DIGESTION

The numerous longitudinal folds in the fundus of the stomach make it highly distensible and well adapted for the rapid ingestion of large quantities of blood (fig. 4.3c). When empty, this portion of the stomach is only 2 mm wide, 62 mm long, and its volume is 0.8 ml. When full, it expands to a diameter of 8 mm, a length of 115 mm, and holds a volume of 23 ml. The blood meal is stored in the fundus and does not pass into the intestinal tract until a half hour after feeding. The wall of the stomach is thin, but supplied with a dense capillary bed through which blood plasma is quickly absorbed and water, which makes up 80% of the blood volume, is transported to the kidneys. It is thought that water absorption through the capillary network is regulated through distention of the fundus.

The thickened blood that remains consists mainly of cellular elements. These are digested enzymatically in the intestine. Because of the blood's hemoglobin content, a great deal of iron is present in the intestine. A vampire's daily meal of 15–18 g of bovine blood contains approximately 6.1 mg of iron—10 times the iron concentration in the vampire's own blood and 800 times the daily iron consumption in humans. Most of the iron is retained in the intestine through some as yet unknown mechanism, and only 4.2 μg is absorbed per day.

ALTRUISM

Vampires do not accumulate large fat reserves and can survive no more than 3 days without a blood meal. This negates some of the advantages of the easily digestible blood diet. Vampires reduce the danger of starvation through their impressive altruistic behavior. In response to a specific pattern of begging behavior engaged in by animals that have failed to obtain a meal, vampires that have fed will regurgitate a part of their blood meal not only for their young, but for other adult members of the colony.

ENERGY BALANCE

The nutritional value of mammalian blood does not vary. It consists of the following nutrients: 22% protein, 1% carbohydrate, 0.6% fat, and 75% water. Thus, blood provides about 5.6 kJ/g. At an ambient temperature of 22°C, a vampire weighing 42 g has a calculated FMR of 107.5 kJ/day, or 2560 kJ/day-kg bw. According to these figures, a vampire needs to take in 19g of blood every day, amounting to nearly half its body weight. This amount is in good agreement with the measured blood intake in vampires. At a temperature of 32°C, in the thermoneutral zone, the energy requirement decreases by about 25%.

4.4 FLOWER AND FRUIT FEEDERS

FEEDING FROM FLOWERS

Bats that are specialized for feeding from flowers, especially the New World *Glossophaginae* and the Old World *Macroglossinae*, obtain nectar by licking it up with their tongues. To better serve this purpose, the front part of the skull has been transformed into a tubelike structure that forms a channel to guide the elongated tongue. The only teeth that remain large and strong are the canines. The cheek teeth are small, rootless, and rounded and are no longer suited for chewing. The narrow jaws are greatly elongated, and the muscles of mastication are weak. The lower incisors are completely absent, and the upper incisors are pushed to the sides to form an opening through which the tongue can be extended.

The tongue of nectar-feeding bats is also specially adapted. In *Glossophaga soricina*, the tongue is at least three times as long as the muzzle. The longest tongue belongs to a bat that is highly specialized for feeding from flowers, *Choeronycteris harrisoni*. The tongue of this species measures 76 mm, nearly as long as the animal's entire head-to-tail length of 80 mm. The long tongue is well supplied with blood vessels. It is extended by the contraction of circular muscles; this muscular contraction appears to force blood toward the tip of the tongue, resulting in hydraulic pressure that further lengthens the tongue. The upper surface of the tongue is densely covered with large filiform papillae which form a spongelike layer into which nectar is sucked by capillary action. When the tongue is retracted, the nectar-filled "sponge" is squeezed out by pressing against the palate. The tongues of many nectar-feeders also have small channels that fill with nectar.

Glossophaga soricina hovers in front of a flower and dips its tongue in the nectar at a rate of 12 times a second. From one flower, the bat obtains about 0.5 ml nectar. A nectar-feeder weighing 20 g requires at least 4–5 g sugar every day.

FEEDING IN FRUGIVOROUS BATS

The Old World Megachiroptera and some Neotropical phyllostomids live on fruit. Thus, the teeth of flying foxes are adapted for fruit consumption (fig. 4.1). The strong, pointed canine teeth grip the fruit tightly and penetrate the skin. The wide, flattened cheek teeth act to crush the fruit, and the incisors have degenerated into small useless stumps.

With their wide, elastic cheeks, flying foxes can handle large pieces of fruit. The tongue is used to press the chewed fruit against the hard ridges of the palate, squeezing it to extract the juice. The fruit juice is swallowed and the remaining woody fibers are spit out. Thus, flying foxes do not actually eat fruit, they just drink the juice. Their esophagus is so narrow that it would be impossible for them to swallow bits of fruit. The pressed fibrous remains that are found lying around their roosting sites led to the curious suggestion that flying foxes did not have an anus.

DIGESTION

The intestinal tract of the vegetarian Megachiroptera and Microchiroptera is five to nine times as long as the body, and thus considerably longer than that of insectivorous bats. Nevertheless, passage of food through the digestive tract takes place even more quickly. The minimal passage times in the large flying fox *Pteropus* are only 12–34 min. Ten minutes after feeding, the ingested fruit juice enters the duodenum, and within another 11 min the duodenum has already emptied.

Frugivorous and nectarivorous species have a two-chambered stomach, consisting of an enlargement of the cardiac and fundic regions of the stomach pouch (fig. 4.3d). A weak sphincter at the entrance to the stomach has been described in several species of *Pteropus*. The presence of this sphincter could be related to the high internal pressure that results when the stomach is called upon to store large volumes of fruit juice taken in over a short period of time. The walls of the stomach are very elastic due to an extensive network of folds. In frugivorous phyllostomid bats, the duodenum widens just beyond its junction with the stomach to form an ampule or saclike structure (fig. 4.3d). The pouchlike structure of the stomach and the expansion of the duodenum suggest a high turnover of the stomach contents within a short time.

The rate at which glucose and fructose are absorbed in fruit-eating bats is many times higher than in mammals such as rats. The high rate of absorption is correlated with a large mucosal area, about three times as large as in the rat. Maltase and invertase, two enzymes responsible for the digestion of sugars, reach their highest levels of activity in the intestine. In contrast, amylase activity is highest in the esophagus, followed by the stomach, suggesting that this enzyme, which is responsible for digestion of polysaccharides, is secreted in the saliva of bats just as it is in other mammals. In fruit-eating bats, the glands of the gastric epithelium are flat and contain many cells that secrete HCl. As would be expected, many cells in the pyloric region secrete gastrin, which stimulates HCl production. Other cells in this region secrete somatostatin, which suppresses HCl secretion and glucagon production. The physiological function of the high HCl production in the stomach of frugivorous bats might be to sterilize fruit juice. Finally, cells have been found in the stomach of the fruit-eating bat *Artibeus* that secrete VIP. VIP acts to relax the smooth muscle in the stomach wall and make it more elastic.

What are the adaptations in protein metabolism that enable frugivorous bats to subsist on a diet that is poor in nitrogen (0.2–1.4% of the dry weight of fruit)? First, the maintenance nitrogen requirement (MNR) of flying foxes is unusually low, only about 457 mg/kg$^{0.75}$-day. This low nitrogen requirement can be explained by the low basal metabolic rate of bats and the consequent low rate of protein synthesis. Although fruits contain little protein, they do contain amino acids. The essential amino acids methionine and lysine are thought to be the limiting factors in this diet. Bats are capable of metabolizing 30–40% of the nitrogen that they take in. Theoretically, they are able to obtain all their nitrogen requirement from fruit juice or nectar through an appropriate choice of fruits or flowers, or through consumption of sufficiently large amounts. For example, *Artibeus* can consume approximately twice its own body weight in fruit every night.

Supplementing a fruit diet with animal tissue can also serve the same purpose. Thus, many phyllostomids eat insects as well as fruit, at least at certain times of year. Pollen provides yet another source of protein when it is consumed by nectar-feeding bats; the nitrogen content of pollen is about 1.5–7.1% of dry weight. Although a good source of protein, the digestion of pollen grains presents a problem because the outer surfaces of the pollen grains are coated with exines that protect them against mechanical and chemical influences. However, it is thought that the high sugar content of the stomach and the body heat of the bat cause the pollen grains to germinate and release their nutrients. The exines are excreted undigested by the bats.

ENERGY BALANCE IN NECTAR-FEEDING BATS

On average, the nectar of the flowers visited by bats consists of 82.6% water, 17% sugar, up to 4% protein, and 0% fat. One milliliter of nectar provides 3.26 kJ. Using the double-labeled water method, FMR was measured in a South American colony of *Anoura caudifer* (body weight 11–12 g) that spent the day under a bridge where the temperature was 25–27°C and flew out at night to a higher altitude region where the temperature was 15°C or less. The values found were as follows:

$$\text{FMR} = 46.44\text{–}56.9 \text{ kJ/day, or } 4184\text{–}4822 \text{ kJ/day-kg bw.}$$

Taking into account this result and the sugar concentration of nectar, it was calculated that to avoid any weight loss, each animal must have consumed about 15.5 g of nectar per night. This energy turnover is considerably higher than for insectivorous or vampire bats, and is probably due to the high energy costs of hovering in front of flowers. The authors estimated that every night an *Anoura* performs the equivalent of a 50-km flight at a speed of 4 m/s.

A study on *Leptonycteris sanborni*, a species that has an average body weight of 19 g and visits flowers in groups, calculated a much lower metabolic turnover based on the volume of nectar consumed and the estimated energy used in flight:

$$\text{FMR} = 40.4\text{kJ/day or } 2125\text{ kJ/day-kg bw.}$$

Within 20 min, each of these bats fills its stomachs with 4 g of nectar, and then spends another 20 min resting and digesting its meal. The animals thus take in 13 kJ of energy in 20 min. The flight energy required for feeding has been estimated at 2.8 kJ, resulting in a net energy gain of 10.2 kJ/20 min per animal. *Leptonycteris* flies a total of 3 h per night. If the above calculations are realistic, *Leptonycteris*, like the insectivorous *Myotis*, expends nearly two-thirds of its daily energy budget in 13% of the day.

ENERGY BALANCE IN FRUGIVOROUS BATS

The energy content of fruits varies considerably and depends on how much of the fruit is digestible. Flying foxes spit out the fibrous parts of the fruit they eat, thus reducing the volume actually consumed to about one-third of the fruit's original volume. Although a flying fox ingests only about one-third of the fruit's total energy content, this part represents four-fifths of the usable energy. In contrast, when a monkey eats the entire fruit, its stomach is filled with a pulp that is only 38% digestible. Flying foxes must spend large amounts of time feeding, but this is compensated by the rapid passage of food through their digestive tract, and by their rapid flight, which allows them to search large areas for fruit in a short time.

Using the double-labeled water method, the following FMR was measured in the South American fruit-eating bat *Carollia perspicillata*:

$$\text{FMR} = 58.2-94.3\text{ kJ/day or } 2900-4700\text{ kJ/day-kg bw.}$$

The individual variations are great, but the maximal values are in the same range of values measured by the double-labeled water method in the nectar-eating bat *Anoura*.

References

Avery MI (1985). Winter activity of pipistrelle bats. J Anim Ecol 54:721–738.

Beasley LJ, Leon M (1986). Metabolic strategies of pallid bats during reproduction. Physiol Behav 36:159–166.

Bell GP, Bartholomew GA, Nagy KA (1986). The roles of energetics, water economy, foraging behavior, and geothermal refugia in the distribution of the bat, *Macrotus californicus*. J Comp Physiol B 156:441–450.

Bush C (1988). Consumption of blood, renal function and utilization of free water by the vampire bat, *Desmodus rotundus*. Comp Biochem Physiol 90A:141–146.

Dorbat K, Peikert-Holle TH (1985). Blüten und Fledermäuse. Verlag Waldemar Kramer, Frankfurt.

Forman GL (1990). Comparative macro- and microanatomy of stomachs of Macroglossine bats (Megachiroptera: Pteropodidae). J Mammal 71:555–565.

*Freeman PW (1979). Specialized insectivory: beetle-eating and moth-eating molossid bats. J Mammal 60:467–479.

Freeman PW (1988). Frugivorous and animalivorous bats (Microchiroptera): dental and cranial adaptations. Biol J Linn Soc 33:249–272.

*Helversen von O, Reyer HU (1984). Nectar intake and energy expenditure in a flower visiting bat. Oecologia 63:178–184.

Howell DJ (1974). Bats and pollen: Physiological aspects of the syndrome of chiropterophily. Comp Biochem Physiol 46A:263–279.

Ishikawa K, Matoba M, Tanaka H, Ono K (1985). Anatomical study of the intestine of the insectfeeder bat, *Myotis frater kaguae*. J Anat 142:141–150.

Kurta A, Bell GP, Nagy KA, Kunz TH (1989). Energetics of pregnancy and lactation in freeranging little brown bats (*Myotis lucifugus*). Physiol Zool 62:804–818.

Kurta A, Junz TH (1988). Roosting metabolic rate and body temperature of male little brown bats (*Myotis lucifugus*) in summer. J Mammal 69:645–651.

Mennone AM, Phillips CJ, Pumo DE (1986). Evolutionary significance of interspecific differences in gastrinlike immunoreactivity in the pylorus of phyllostomid bats. J Mammal 67:373–384.

Morris S, Curtin AL, Thompson MB (1994). Heterothermy, torpor, respiratory gas exchange, water balance and the effect of feeding in Gould's long-eared bat, *Nyctophilus gouldi*. J Exp Biol 197:309–335.

Rouk CS, Glass BP (1970). Comparative gastric histology of five North and Central American bats. J Mammal 51:455–471.

Sazima I, Uieda W (1980). Feeding behavior of the whitewinged vampire bat, *Diaemus youngii,* on poultry. J Mammal 61:102–103.

Schmidt U (1978). Vampirrledermäuse. Neue Brehmbücherei Ziemsen Verlag, Wittenberg.

Storch G (1968). Funktionsmorphologische Untersuchungen an der Kaumuskulatur und an korrelierten Schädelstrukturen der Chiropteren. Abh Senckenberg Naturf Ges 517: 1–92.

Studier EH, O'Farrell MJ (1976). Biology of *Myotis thysanodes* and *M. Lucifugus* (Chiroptera: vespertilionidae)-III. Metabolism, heart rate, breathing rate, evaporative water loss and general energetics. Comp Biochem Physiol 54:423–432.

Thenius E (1989). Zähne und Gebiß der Säugetiere. Handbuch der Zoologie, vol. 8, Mammalia. Walter de Gruyter Verlag, Berlin.

Thomas ST (1975). Metabolism during flight in two species of bats, *Phyllostomus hastatus* and *Pteropus gouldii*. J Exp Biol 63:273–293.

5

CENTRAL NERVOUS SYSTEM

5.1 ENCEPHALIZATION

THE BRAIN, OR encephalon, is the body structure that determines the success of an animal species, and it is subject to the greatest evolutionary pressure. As a general rule, the relative mass of the brain is directly correlated with its performance. According to this assumption, the relative brain mass or "encephalization" of a phylogenetic group is a measure of its level of evolution. Thus, in evolutionary analyses, the relative brain weight or encephalization index (EI) is used. For each species, the log of the average brain weight is plotted as a function of the log of the average body weight. For each arbitrarily determined species grouping, a regression line is drawn through the values on the scatter plot to provide an estimate of the average ratio for that group. All species whose values lie on the regression line have an EI of 100%. Species whose values lie above or below the line have an EI above or below 100%. For example, a species with a value 0.301 log units above the line has an EI of 200% ($10^{0.301} = 2$); similarly, a species with a value 0.301 log units below the line has an EI of 50%. Among mammals, the "basal" insectivores, which include the hedgehog, shrew, and tenrec, are frequently chosen as a reference group because they have the smallest relative brain sizes.

The encephalization indices of bats lie far above those of the basal insectivores (fig. 5.1 and table 5.1), but below those of other mammals. In *Pteropus conspicillatus*, a flying fox with a body weight of 750 g, the brain weighs 8.3 g. In contrast, in a tenrec with a body weight of 830 g, the brain weighs only 2.6 g. The only part of the brain that is smaller in Chiroptera than in insectivores of comparable size is the olfactory bulb.

Compared to the basal insectivores, the mean EI values for the different families of bats range from 115% in the Natalidae to 247% in the Pteropodidae. As shown in table 5.1, Megachiroptera have a higher EI than Microchiroptera. The EI values of the different families of bats are correlated most closely with their diet, not with their position on the phylogenetic scale. The highest EI values are found in species that consume a diet rich in energy, but that is not available at predictable times and places. These species include fruit and nectar feeders, vampires, and fishing bats. Carnivorous bats have EI values in the middle range. The lowest EI values are found in insectivorous echolocating bats (fig. 5.1). Within this last group, the Thyropterids with their high EI are an exception. Some believe that this

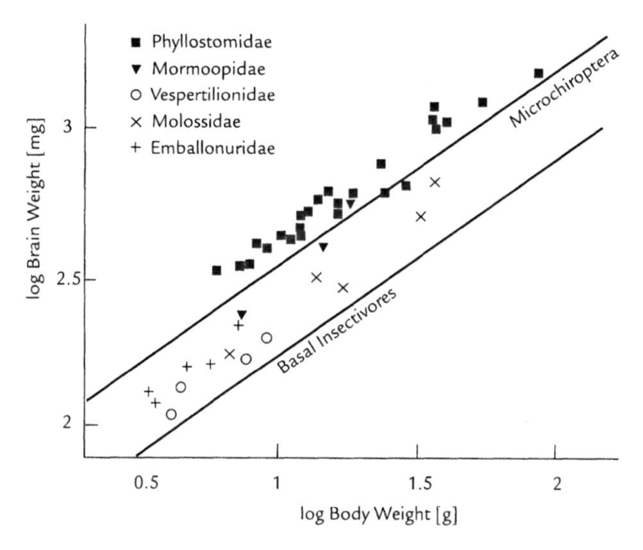

Figure 5.1 The degree of encephalization of bats relative to the "basal insectivores" (lower line) and expressed as the logarithmic ratio of brain weight to body weight (regression line). Insectivorous bats such as Mormoopidae, Vespertilionidae, and Emballonuridae all have a lower degree of encephalization than species that feed on nectar, fruit, or blood, represented by the Phyllostomidae. From Stephan (1977).

is due to the fact that every morning they must seek a new roost in a rolled-up banana leaf. Thus, the Vespertilionids, a family that is distributed worldwide, is highly successful, and that includes more species than any other, also occupies a position at the bottom of the EI scale. Because hunting for insects using echolocation is a highly specialized form of behavior, it can be concluded that the degree of encephalization is not determined by the degree of specialization, but rather by the degree of behavioral flexibility demanded of a species by its environment and daily activities.

The predictive value of the EI should not be overestimated because it is based strictly on the mass of the brain and not on the complexity of its circuitry or its degree of differentiation. Furthermore, the EI values are not completely independent of body weight; in many families, the largest species have the smallest EI values.

5.2 ANATOMY OF THE CENTRAL NERVOUS SYSTEM

Histological sectioning of a mammalian brain reveals a complex and tangled structure, made up of groups of neurons and fiber pathways, many of which have imaginative and highly descriptive names (box 5.1). There are many different strategies for bringing a logical order to this complex structure. The following description is based on a phylogenetic–functional organization similar to that developed by Starck in his *Comparative Anatomy of the Vertebrates*.

Table 5.1 Encephalization indices (EI) of the different families of bats, using the basal insectivores as a reference

	EI (%)		
Family/suborder	Minimum	Maximum	Average
Megachiroptera	177	340	247
Microchiroptera			
Phyllostomidae	148	284	235
Desmodontidae	216	243	232
Thyropteridae	230		
Megadermatidae	182	233	211
Nycteridae	178	216	196
Mormoopidae	171	203	180
Noctilionidae	128	220	174
Rhinolophidae	139	186	164
Hipposideridae	87	174	164
Furipteridae	—	—	152
Emballonuridae	132	169	151
Molossidae	107	166	137
Vespertilionidae	77	179	128
Microchiroptera that only hunt insects in flight			137
Microchiroptera that hunt in flight and glean from the substrate			156
Microchiroptera that glean only			186
Carnivorous microchiroptera			213

From Stephan (1977), Stephan and Nelson (1981), and Stephan et al. (1987).

OVERVIEW

The central nervous system (CNS) of vertebrates consists of the brain and spinal cord. It develops from the neural tube, which forms early in embryonic development over the notochord and encloses a fluid-filled central canal (fig. 5.2). At the rostral end, the canal widens to form one paired ventricle and two unpaired ventricles, the outer walls of which will become the brain. The long caudal portion of the tube develops into the segmentally organized spinal cord.

The phylogenetically oldest part of the brain can be divided according to its innervation patterns into the following parts:

- The spinal cord innervates the muscles and the skin of the trunk and extremities.
- The rhombencephalon, or hindbrain, innervates the head, the mouth, and the throat as well as the internal organs.
- The prosencephalon, or primitive forebrain, innervates the rostral part of the head, the eyes, and the mouth. Early in development the prosen-

Box 5.1 Definitions

Nerve (N)	A bundle of nerve fibers that connects the periphery with the central nervous system (CNS).
Afferent nerves	Nerve fibers that transmit information from the periphery to structures in the brain, such as sensory nerve fibers that transmit activity from the sense organs to the brain.
Efferent nerves	Nerve fibers that transmit patterns of excitation away from the neuron, such as motor fibers that transmit neural commands to the muscles.
Tract	Bundles of fibers that connect different CNS regions with one another.
Nucleus	A clearly defined group of nerve cells in the CNS.
Cortex	Layered arrangement of nerve cells in the CNS.
Ventricle	A hollow fluid-filled space in the brain.
Axon	Neuronal process that conducts electrical potentials away from the cell body.
Dendrite	Neuronal process that conducts action potentials toward the cell body.
Funiculus	Pathway connecting the spinal cord with the brain.
Fasciculus	Small bundle of nerve fibers.

cephalon differentiates into two parts, the diencephalon and the telencephalon.

Overlying this region of the brain are three integrative centers (fig. 5.2):

- The cerebellum, which lies above the rhombencephalon,
- The midbrain tectum (mesencephalon), located between the cerebellum and the diencephalon,
- The pallium, which overlies the telencephalon. The pallium and the basal telencephalon together make up the forebrain.

THE SPINAL CORD

The spinal cord (figs. 5.3 and 5.4) supplies the trunk and the four extremities with sensory and motor innervation. It is a center for neural control of bodily functions, and at the same time it is the gateway through which somatosensory information is transmitted to the telencephalon.

The neurons of the spinal cord are organized into a butterfly-shaped region of *gray matter*, which surrounds the fluid-filled central canal. The two ventral (or anterior) horns of the spinal cord contain motor neurons. The two dorsal (or posterior) horns contain sensory neurons, of which all bodies are located outside the spinal cord in the dorsal root ganglia on the spinal nerves that arise from each segment of the cord (fig. 5.4). The two nerve bundles that arise from the ventral and dorsal horns unite to form the *spinal nerves*. Each pair of spinal nerves innervates a topographically distinct myotome, or muscle segment, and a dermatome, or skin

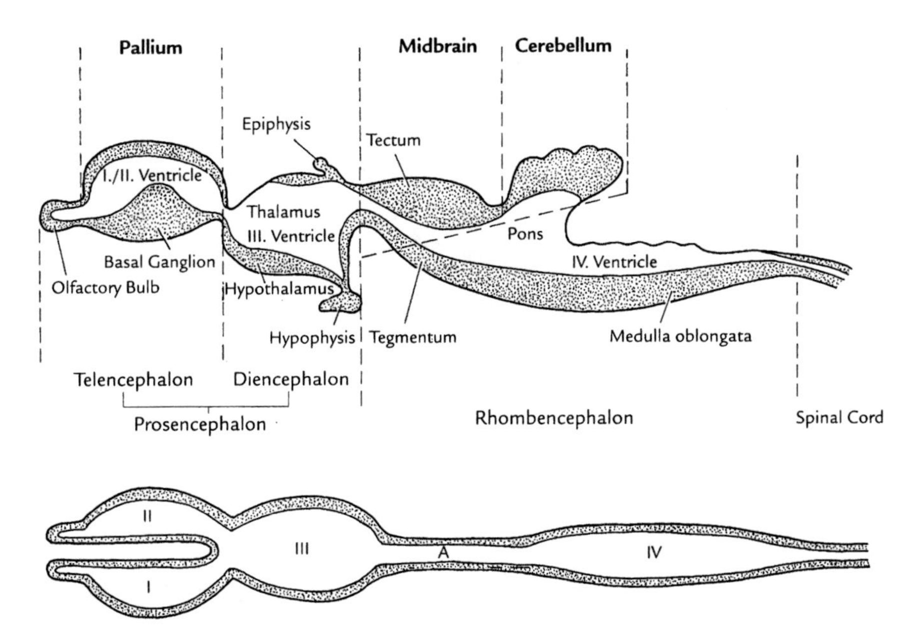

Figure 5.2 Schematic parasagittal section (above) and horizontal section (below) through the brain of a mammal. Bold print indicates major brain divisions. A, cerebral aqueduct; I–IV, ventricles.

region. The gray matter of the spinal cord is surrounded by afferent and efferent nerve fiber bundles called funiculli and fasciculi. Collectively, these fiber bundles are called the *white matter.*

The spinal cord of Microchiroptera is considered to be the most primitive in any mammal for several reasons. In Microchiroptera, the neurons in the gray matter have widespread dendritic branches that extend into the afferent and efferent fiber tracts of the white matter, forming a diffuse information processing network. Appropriately, the *fasciculi proprii* that connect the different segments of the spinal cord with one another are relatively well developed. In contrast, in Megachiroptera, the dendrites of neurons remain strictly confined to the gray matter as they do in other mammals. Thus, there is a clear separation between information processing and transmission of information.

Perhaps because of the diffuse internal connectivity between gray and white matter, the afferent and efferent tracts of the spinal cord are much less well developed than those of other mammals. The *posterior funiculus*, a sensory pathway that transmits information about touch and deep pressure to the brain, is so small that it is practically surrounded by the dorsal horns of the two sides (fig. 5.4a). The *funiculus gracilis*, a fiber tract of the dorsal white matter which transmits information from the lower half of the body to the brain, is nearly absent in many species of Microchiroptera. This is not surprising, considering the small size and relative unimportance of the legs in Microchiroptera. Flying foxes, on the other hand, must

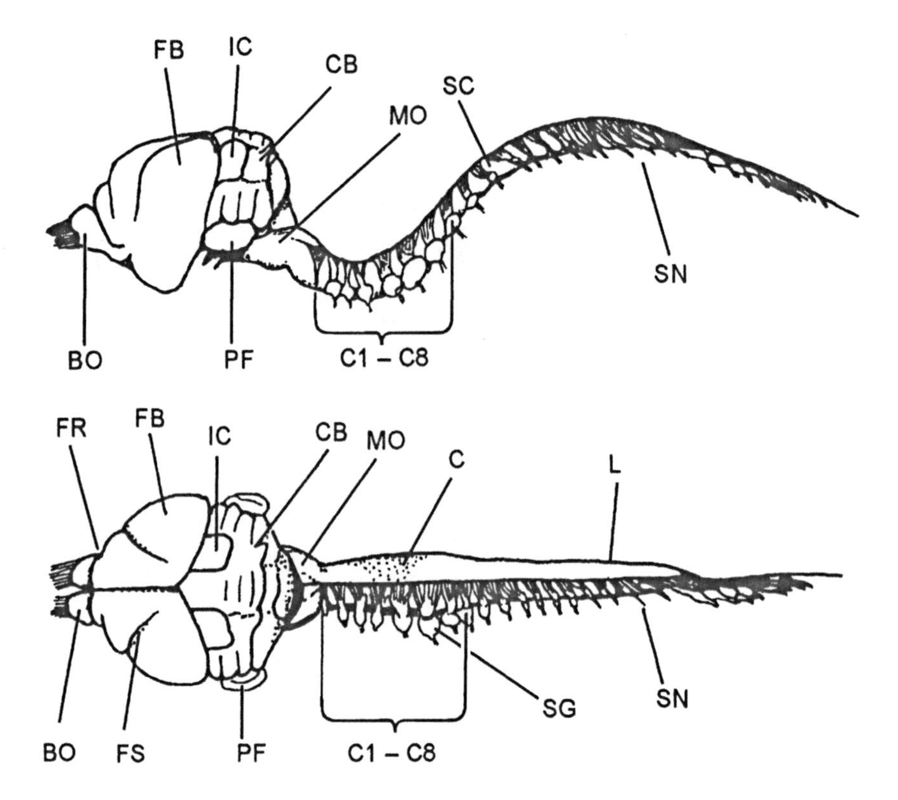

Figure 5.3 Central nervous system (CNS) of a bat. Side view (top) and dorsal view (bottom) of the CNS of *Pteronotus parnellii* (Mormoopidae). BO, olfactory bulb; C, cervical enlargement of the spinal cord; CB, cerebellum; IC, inferior colliculus (midbrain); C1–C8, spinal nerves of the first through eighth cervical segments; FB, forebrain; FS, Sylvian fissure; FR, rhinal fissure; L, lumbar enlargement of the spinal cord; MO, medulla oblongata (hindbrain); PF, paraflocculus of the cerebellum; SC, spinal cord; SG, spinal ganglion; SN, spinal nerve. From Henson (1970).

scramble along branches using their thumbs and their feet to reach fruit, so in these species both the wings and the legs have a relatively large representation in the brain. The cuneate and gracile tracts of the spinal cord are appropriately well developed (fig. 5.4b).

Due to the reduced number of fiber pathways in the spinal cord of Microchiroptera, the relative volume of the gray matter far exceeds that of the white matter. In Microchiroptera, the ratio of gray to white matter is twice as large as in the mouse, six times as large as in the cat, and nine times as large as in humans.

At the tip of the dorsal horns is a region filled with many small cells, the *substantia gelatinosa* (fig. 5.4a). In the cat, this region receives sensory information from the skin of the dorsal part of the body. The neurons in this region modulate the activity of the dorsal horn neurons. One function of the substantia gelatinosa

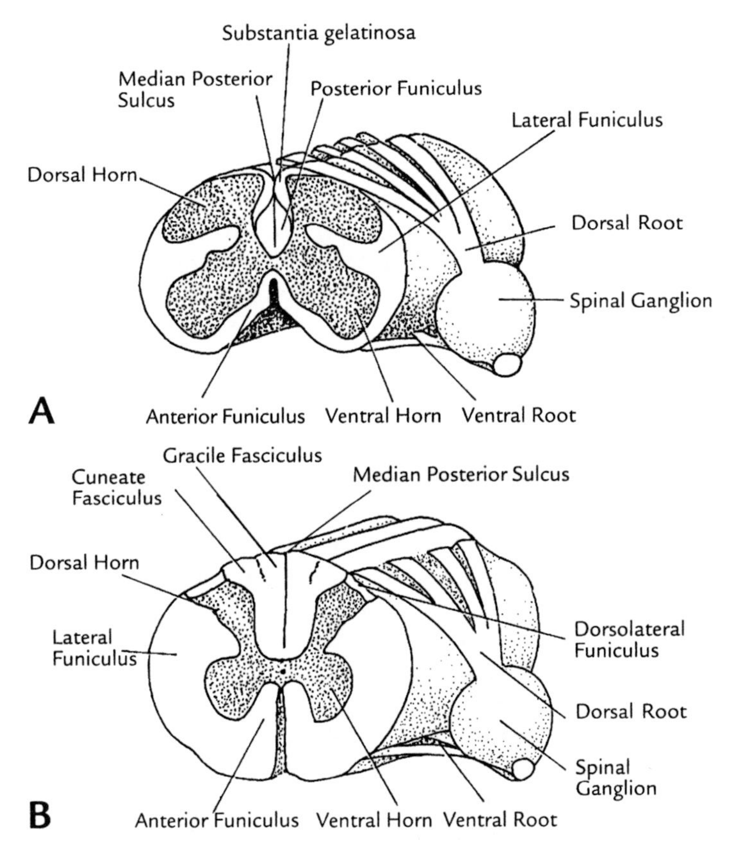

Figure 5.4 Cross sections through the spinal cord. (A) A microchiropteran bat, *Pteronotus parnellii*, and (B) a megachiropteran bat, *Eidolon helvum*. Shaded areas indicate gray matter; white areas indicate white matter. The anterior funiculus, cuneate fasciculus, and gracile fasciculus are sensory pathways from the lower body to the brain. The posterior funiculus transmits information about touch and deep pressure to the brain. From Henson (1970).

is to inhibit the transmission of sensory information from the skin to higher levels of the nervous system. In Microchiroptera the substantia gelatinosa is large and conspicuous; this is considered a primitive characteristic. The high degree of development of the substantia gelatinosa is thought to be related to the sensitivity of the flight membrane, although there is no experimental evidence to support this idea.

Finally, Microchiroptera possess a nucleus of the sympathetic nervous system, the *nucleus marginalis*, which is present in reptiles and birds, but absent from all other mammals.

Because the vertebral column of bats continues to grow postnatally but the spinal cord does not, the cord appears shortened. For example, the spinal nerves that innervate the trunk exit from the spinal canal at a considerable distance from the corresponding vertebrae. Thus, in *Pteronotus parnellii* (Mormoopidae), the nerves of the second lumbar vertebra through the third sacral vertebra exit the spine at the level of the twelfth thoracic vertebra; in *Artibeus jamaicensis* (Phyllostomidae), they exit at the level of the ninth thoracic vertebra.

The "wing brain." Unlike most other mammals, but similar to humans, the cervical segments of the spinal cord in bats are greatly hypertrophied (C1–C8 in fig. 5.3). This cervical enlargement is related to the innervation of the wings which, like the front limbs of all mammals, are innervated by the spinal nerves of the fourth cervical vertebra through the first thoracic vertebra. In bats, this cervical region represents nearly half of the entire length of the spinal cord. For this reason the enlarged cervical portion of the spinal cord that has developed in the bat is sometimes referred to as the "wing brain."

HINDBRAIN

The nomenclature used to describe the hindbrain is confusing. The rostral part that lies beneath the midbrain is called the *tegmentum*, but is also referred to by the descriptive anatomical term of mesencephalon (fig. 5.2). The caudal part is referred to as the *medulla oblongata* or myelencephalon (fig. 5.2). In mammals, in parallel with the development of the cortex and cerebellum, an additional connection region, the *pons*, has developed. Together, the tegmentum, pons, and medulla make up the hindbrain. To add to the confusion, the pons, and cerebellum together are referred to as the metencephalon.

Initially, the hindbrain is little more than a continuation of the spinal cord, and there is no clear boundary between the two regions. The hindbrain consists of columns of neurons surrounded by a loosely organized network of neurons and fibers called the *reticular formation*. The reticular formation extends through the midbrain up to the level of the diencephalon (fig. 5.5). During the course of evolution the dorsal sensory columns and the ventral motor columns have broken apart into local groups of neurons called *nuclei*. The hindbrain contains mainly the nuclei belonging to those cranial nerves (V, VII, X, and XI) that are embryologically derived from the branchial nerves, as well as the nuclei of the somatosensory and auditory nerves (see box 5.2 and fig. 5.5).

Sensory nuclei. The hindbrain contains the *cochlear nucleus*, the *nuclei of the trapezoid body*, the *superior olivary complex*, and the *nuclei of the lateral lemniscus*, which together make up a major portion of the auditory pathway. These nuclei and their functions are described in chapter 6, Echolocation. In echolocating bats, the ventral cochlear nucleus, the superior olivary complex, and the nuclei of the lateral lemniscus are hypertrophied. In *Pipistrellus*, the superior olivary complex is seven times larger than in the flying fox *Pteropus*. In echolocating Microchiroptera, the auditory nuclei together make up 14–23% of the hindbrain. In contrast, in nonecholocating Megachiropteran bats they make up only 5.3%. In *Rousettus*, a flying fox that echolocates, the auditory nuclei make up 8.5% of the hindbrain.

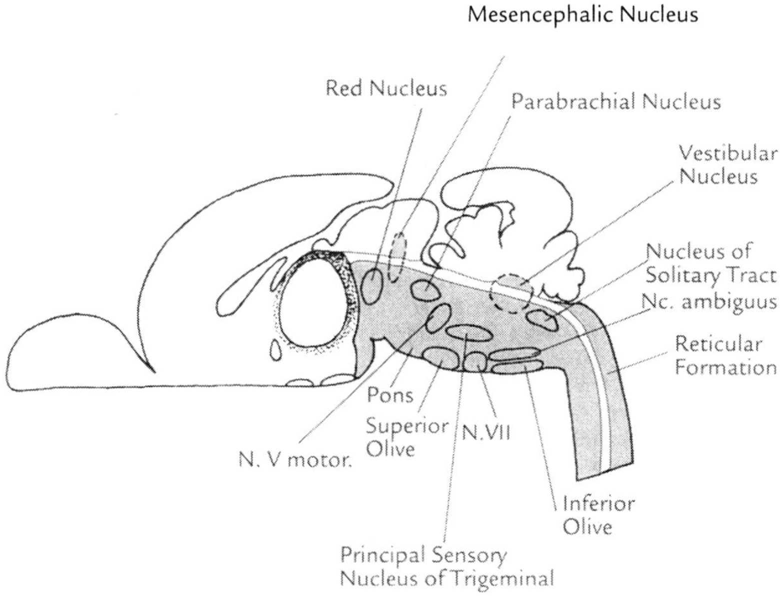

Figure 5.5 Schematic longitudinal section through the hindbrain showing selected nuclei. The shaded area indicates the reticular formation of the hindbrain. Vm, motor nucleus of the trigeminal nerve; VII, nucleus of the facial nerve.

The branches of cranial nerve VIII, the vestibulo-cochlear nerve, that originate in the organs of balance terminate in the *vestibular nuclei* (fig. 5.5). The vestibular nuclei project mainly to the cerebellum, to the spinal cord to regulate flight, and to the nuclei that govern eye movements. The vestibular nuclei of bats are larger than those of ground-dwelling mammals and are especially large in vampire bats which not only fly, but also run and jump with great agility.

In bats, the somatic motor tract that originates in the *spinal trigeminal nucleus* passes through the caudal part of the medulla oblongata. Because this tract develops from the substantia gelatinosa, it is hard to distinguish it from the rest of the spinal cord. The spinal trigeminal nucleus receives sensory information from the skin of the face related to touch, temperature, and pain. Sensory information related to pressure and touch in the snout region is transmitted via the ascending branch of the trigeminal nerve, which terminates in an extension of the spinal trigeminal nucleus, the *principal nucleus of the trigeminal* (fig. 5.5). This nucleus is especially large in species that have long whiskers. The *mesencephalic nucleus of the trigeminal* is actually a spinal ganglion that has migrated into the central nervous system. It receives sensory information via several different cranial nerves from the jaw muscles, teeth, and palate.

The first central station for gustatory information is the *nucleus of the solitary tract.* This nucleus is especially large in vampires and nectar-feeding bats.

Box 5.2 The cranial nerves of mammals

Nerves derived from the branchial arches (branchial nerves)

1. *Trigeminal* (cranial nerve V). Supplies sensory and motor innervation to the jaw musculature and face. Derived from the branches that supplied the most rostral (mandibular) branchial arches.
2. *Facial* (cranial nerve VII). Supplies sensory and motor innervation to the muscles of the face and ear, and the digastric muscle of the lower jaw. Transmits sensory information from the front part of the tongue and the throat. Derived from the nerve supplying the hyoid arch (blow hole). During mammalian development, the hyoid musculature enlarges to cover much of the head and becomes the facial muscles.
3. *Glossopharyngeal* (cranial nerve IX). Supplies both sensory and motor innervation to the pharynx, mouth lingual epithelium, and salivary glands. Derived from the nerve of the first branchial arch.
4. *Vagus* (cranial nerve X). Supplies sensory and motor innervation to the viscera, heart, lungs, larynx, etc., as well as parasympathetic innervation of the upper body. Derived from the nerves of the second through fifth branchial arches.
5. *Spinal accessory* (cranial nerve XI). Motor nerve that innervates the muscles of the neck and throat. Derived from the vagus nerve.

Somatic motor nerves

1. *Occulomotor* (cranial nerve III).
2. *Trochlear* (cranial nerve IV).
3. *Abducens* (cranial nerve VI).

These three nerves innervate the extrinsic muscles of the eye.

4. *Hypoglossal* (cranial nerve XII). Innervates the tongue muscles. Derived from a spinal nerve.

Sensory nerves

1. *Olfactory* (cranial nerve I). Projects from the olfactory epithelium to the olfactory bulb (paleocortex).
2. *Optic* (cranial nerve II). The correct term for this nerve is "optic tract" because it connects two parts of the brain, the eye, and the diencephalon. In development, the eye is derived from an outpouching of the central nervous system and is thus part of the brain.
3. *Vestibulo-cochlear* (cranial nerve VIII). Consists of two branches, one of which projects from the organs of balance to the vestibular nuclei of the brainstem; the other branch projects from the spiral ganglion of the inner ear to the cochlear nucleus, also located in the brainstem.

Motor nuclei. The muscle groups innervated by the cranial nerves are controlled by output from appropriately named nuclei (fig. 5.5). Of these motor nuclei, the ones that control respiration and movements of the larynx are especially important. Respiration is controlled by the *parabrachial nucleus* and poorly differentiated parts of the reticular formation. The *nucleus ambiguus* innervates the larynx and plays a key role in the production of sounds and echolocation calls. The *facial nucleus* (nucleus of cranial nerve VII) consists of four subdivisions. The lateral subdivision innervates the lips, and the medial subdivision has been shown to innervate the pinnae, at least in cats. The medial subdivision is especially large in bats. The nuclei that belong to the motor pathways of the pyramidal and extrapyramidal system and project to the spinal cord are thought to operate in conjunction with cortical systems to control movements.

Reticular formation. The nuclei and tracts of the brainstem lie within a diffuse network of cells and fibers (shaded area in fig. 5.5). This network extends throughout the entire brainstem from the spinal cord up to the level of the diencephalon. It is especially dense in the region of the mesencephalon. This network functions as a data bus that receives information from the entire body and distributes it to all parts of the brain. The reticular formation is especially important in controlling sleep and waking and in regulating attention. In addition, it is responsible for autonomic control. Over the course of mammalian evolution, there has been a tendency to form more and more nuclei in the reticular formation. Bats seem to occupy a position at the beginning of this progression because they have even fewer brainstem nuclei than does the hedgehog, an animal generally considered to be the lowest on the mammalian scale. However, there are two exceptions to this general rule: (1) The *lateral reticular nucleus* in bats is highly developed. In *Pipistrellus pipistrellus* its relative size is larger than in any other mammal. This nucleus receives sensory information via the spinal cord, from the principal nucleus of the tegmentum, the *red nucleus* (fig. 5.5), and from the cerebellum via the mossy fibers. Thus it is possible that the lateral reticular nucleus may be involved in the control of flight. (2) The hypertrophied *inferior olivary nucleus* (fig. 5.5) may also be involved in flight control. This nucleus is also connected with the cerebellum via the climbing fibers.

MIDBRAIN

From a phylogenetic and functional point of view, the midbrain tegmentum is a part of the brainstem. The higher-level midbrain center in lower vertebrates is the *tectum,* which corresponds to the collicular midbrain in mammals. The *inferior colliculi* are the central components of the mammalian auditory pathway. In echolocating bats they are hypertrophied. This can easily be seen by comparing the relative size of the inferior colliculi in Microchiroptera and Megachiroptera in figure 5.6. The function of the inferior colliculus is discussed in chapter 6. The two *superior colliculi* receive visual information via the optic nerve and from the visual cortex. The lower layers of the superior colliculi receive primarily somatosensory and auditory information. The superior colliculus plays an important role in controlling reflex orientation movements of the eyes and pinnae, whereby objects are brought into view on the fovea or sound sources are maintained in the most sensitive direction of hearing. The superior colliculus also provides a bidirectional pathway to and from the cortical visual fields in parallel with the primary visual pathway via the diencephalon.

The midbrain also contains the *substantia nigra,* an important nucleus of the extrapyramidal system. The cerebral aqueduct, a highly constricted portion of the

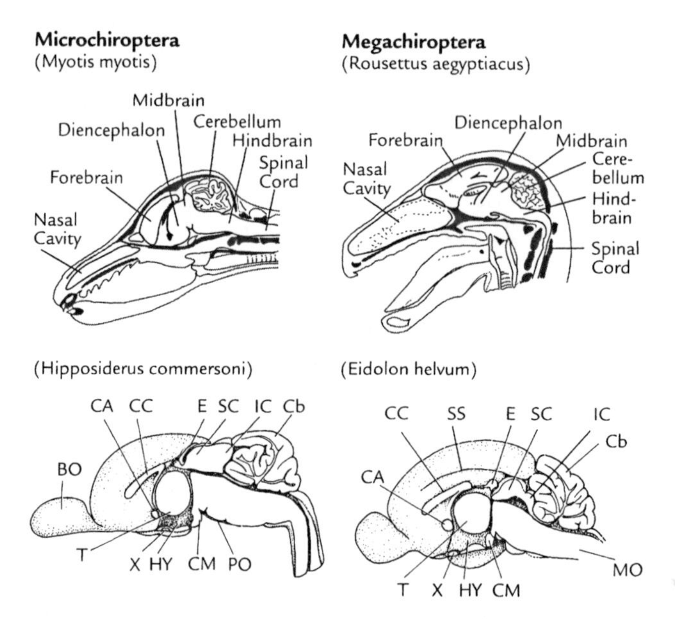

Figure 5.6 Brains of Microchiroptera (*Myotis myotis*, top left; *Hipposideros commersoni*, lower left) and Megachiroptera (*Rousettus aegypticus*, upper right; *Eidolon helvum*, lower right). Top: Location of the brain within the skull. Bottom: Sagittal section through the brain. BO, olfactory bulb; CA, anterior commissure; Cb, cerebellum; CC, corpus callosum; IC, inferior colliculus (midbrain); CM, mamillary bodies (diencephalon); HY, hypothalamus; MO, medulla oblongata; PO, pons; SS, splenial sulcus; T, thalamus; X, optic chiasm. From Henson (1970) and Schneider (1957).

ventricular system that lies within the midbrain, is surrounded by the *central gray*, a diffuse aggregation of neuron cell bodies and fibers. Electrical stimulation of the central gray can elicit vocalizations.

CEREBELLUM

The fact that bats cannot fly without a cerebellum points out the importance of this structure in vertebrates. The cerebellum is responsible for the coordination and regulation of complex movements in space. Although the cerebellum does not initiate movements, it functions as an "auxilliary computer" for the spatiotemporal coordination and fine control of muscle contractions.

The cerebellum is located just above the pons (fig. 5.6) and has a highly convoluted three-layered cortex as well as input and output nuclei. Cerebellar neurons have no direct access to motor neurons, but instead project out via the cerebellar nuclei. An important pathway from the cerebellum to the extrapyramidal system of the forebrain projects via the *dentate nucleus* of the cerebellum, the red nucleus in the brainstem, and the thalamus in the diencephalon. This pathway is thought to play an important role in sensory–motor coordination. The dentate nucleus is the most recently evolved of the cerebellar nuclei, and in bats it is relatively small.

The *interpositus nucleus* transmits information from the cerebellum to the red nucleus, which in turn projects to the thalamus and spinal cord. This pathway controls motor reflexes and muscle tone. The *fastigial nucleus* is intimately associated with the sense of balance. One function of this nucleus is to transmit information from the cerebellum to the vestibular nuclei in the brainstem, which in turn project to the spinal cord. The fastigial nucleus also integrates vestibular and auditory information as well as information about muscle tone relative to the earth's gravitational field. Clearly, these functions are of great importance for a flying mammal. Indeed, all of the cerebellar areas and nuclei associated with the sense of balance are conspicuously large in bats.

Information from the motor centers of the cortex reaches the cerebellum via the pontine nuclei and the mossy fibers from the pons. Two other important sources of information are the vestibular nuclei, as already mentioned, and the climbing fibers from the inferior olive.

The cortical areas represent the true switching centers of the cerebellum. The neurons of the cerebellar cortex are connected in an orderly way, heightening the impression that the cerebellar cortex functions as a computer. The part of the cerebellar cortex that is phylogenetically oldest, the archicerebellum, forms two lateral lobes, the *flocculus* and the *paraflocculus* (fig. 5.3). These lobes have connections with the vestibular nuclei. In adult bats the flocculus and paraflocculus are fused together and unusually large. On the basis of their size, their inputs, and their physiology, it seems likely that the flocculus and paraflocculus are important areas for converting input information from echoes into output to control flight.

The inputs and outputs described here provide an incomplete picture of the complexity of the connections between the cerebellum and other parts of the brain. The connections between the cerebellum, the diencephalon, and the telencephalon form nonhierarchical circuits. Such circuits allow cerebellar neurons, for example, to prepare for a movement that is subsequently executed via signals from the pyramidal neurons of the cortex.

The only physiological experiments on the cerebellum of bats have been concerned with processing of auditory information. The cerebellar cortex receives auditory input from the cortex and the inferior and superior colliculi by way of the pontine nuclei. The cerebellar cortex contains many neurons that do not differ appreciably in their responses to sound from those of the main auditory pathway. In *Eptesicus fuscus*, the majority of cerebellar neurons are most sensitive to sounds that originate from the direction in which the bat is flying. In horseshoe bats, the most sensitive cerebellar neurons are those that are tuned to the frequencies of the echolocation sounds.

THE DIENCEPHALON

The diencephalon is arranged in four divisions that are stacked one on top of the other. These divisions are the epithalamus, the thalamus, the subthalamus, and the hypothalamus (fig. 5.2). The diencephalon performs two important classes of functions. First, the thalamic nuclei provide a "gateway" to the cortex through which all consciously perceived sensory information must pass. The thalamic nuclei are under the control of the cortical areas responsible for the various sensory modalities. Second, nuclei in the diencephalon control vegetative functions such as body temperature, circulation, hunger, thirst, hormone concentrations, and sexual behavior and are also involved in controlling emotional states such as aggression and fear.

Epithalamus. The *epiphysis*, or pineal gland, develops from the roof of the diencephalon. The pineal, through the secretion of melatonin, plays a role in regulating body functions that are subject to light-triggered circadian rhythms. Below the roof of the diencephalon lies the *habenula*. The habenula gives rise to a fiber bundle that projects to the raphe nuclei of the midbrain, thereby influencing the serotonergic system. The habenula is often described as part of an olfactory–somatic reflex pathway and is thought to be especially well developed in bats that have sensitive olfaction.

Thalamus. The nuclei of the thalamus are conspicuous and highly variable among species. The mammalian thalamus is usually divided into anterior, medial, ventral, lateral, posterior, and intralaminar cell groups, as well as the lateral and medial geniculate. The anterior and medial cell groups contain regions that modulate autonomic functions and emotional states.

The large nuclei of the *ventrolateral* thalamus represent a relay station through which sensory information is transmitted to the cortex. The *ventral nucleus* is the cortical "gateway" for sensory information from the skin, including pressure, temperature, and pain, for proprioceptive input from the muscles, and for sensory input from the internal organs of the body. The visual sensory pathway projects to the cortex via the *lateral geniculate*, and the auditory sensory pathway via the *medial geniculate*. Some thalamic nuclei such as the pulvinar are thought of as sensory integration centers that are also involved in motor pathways that connect the cortex and cerebellum. However, the pulvinar is thought to be absent in horseshoe bats and perhaps in other Microchiroptera.

Subthalamus. The subthalamus is a continuation of the roof of the midbrain. It contains nuclei belonging to the extrapyramidal motor system and can therefore be considered the motor area of the diencephalon.

Hypothalamus. The hypothalamus contains a large number of different nuclei that participate in the control of vegetative body functions. It is difficult to assign specific functions to individual nuclei because in most cases a number of different nuclei are involved in an interactive network. The hypothalamic nuclei can be divided into groups based on their location. The *preoptic nuclei* lie in front of the optic chiasm and the *supraoptic nuclei* lie above it. The preoptic nuclei are involved in temperature regulation. The paraventricular and supraoptic nuclei produce neurohormones that regulate water balance, such as antidiuretic hormones. These neurohormones are transported to the neurohypophysis and stored there. Other nuclei produce hormone-releasing or -inhibiting factors that control the anterior lobe of the hypophysis.

The *mamillary complex* forms a bulge on the caudal floor of the hypothalamus (fig. 5.6). The nuclei of the mamillary complex have strong connections with the anterior thalamic nuclei, the hippocampus, and the fornix of the forebrain. For this reason they are considered part of the limbic system. The mamillary nuclei of bats are well developed. One consequence of lesions of the mamillary complex in humans is disorientation, so it is possible that the mamillary complex–hippocampal network could play a role in spatial memory. For this reason it would seem especially fruitful to explore these brain structures in echolocating bats.

TELENCEPHALON

Neopallium. The prosencephalon, olfactory bulb, and pallium together make up the forebrain. Because neurons of the pallium are arranged in layers, it is also called cortex. The basal ganglia and septum are the phylogenetically oldest regions of the forebrain. These regions, together with the overlying cortex and paired ventricles, make up the cerebral hemispheres (fig. 5.7). In amphibians, the original structure of the cerebral hemispheres can be clearly seen. In the ventral prosencephalon, the medial quadrant contains the septum, and the lateral quadrant contains the basal ganglia (fig. 5.7). In the dorsal part, the medial quadrant contains the archicortex and the lateral quadrant the paleocortex. In mammals, a new cortical area has evolved, located between the archicortex and paleocortex. This area is called the neocortex (fig. 5.7). During evolution, the neocortex of higher mammals has grown larger than any other part of the brain, and covers most of the lower parts. In humans, the neocortex is highly convoluted. These new expanded areas provide ample room for additional input and output fibers, greatly increasing the brain's capacity for analysis and integration. Generally, the neocortex consists of six layers, in contrast to phylogenetically older cortical areas, but the connectional basis underlying the novel and highly specialized properties of neocortical processing is not clearly understood.

The development of the neocortex was and is the most outstanding event in mammalian evolution. In echolocating bats, the development of neocortex is still in a relatively primitive stage. The neocortex represents less than 50% of the total cortical area in many species of bats. The surface of the cortex is smooth, or lissencephalic, with few convolutions. Even in flying foxes there are only three visible sulci: the lateral fissure in the dorsal third of the cortex; the Sylvian fissure (fig. 5.3), which divides rostral from temporal cortex; and the rhinal fissure (fig. 5.3), which divides the olfactory cortex from the occipital region. All bats have a hippocampal fissure on the medial surface of the hemisphere. Flying foxes also

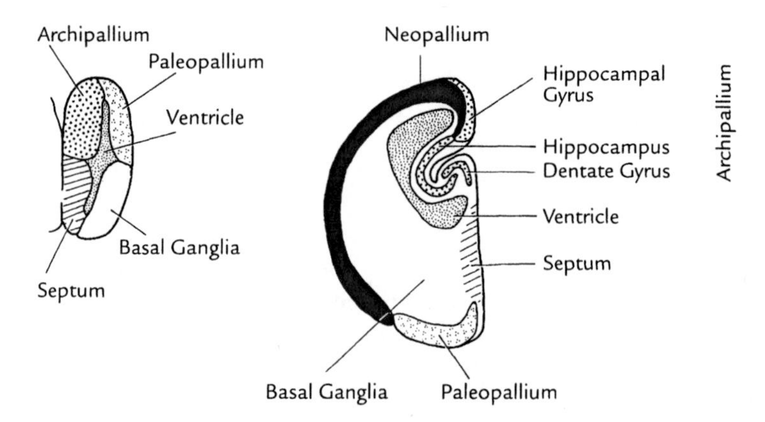

Figure 5.7 The mammalian cortex. *Left*: Original organization of the vertebrate telencephalon, as illustrated by the frog brain (schematic right hemisphere). *Right*: Organization of the mammalian telencephalon (schematic left hemisphere). From Starck (1982).

have a splenial fissure that divides the archicortex from the neocortex. The sulci are all shallow in bats, the Sylvian fissure is absent in many species, and the rhinal fissure is often the only one that is clearly recognizable.

As in the case of total brain weight, the proportion of the brain occupied by neocortex increases starting with the insectivorous echolocating bats and progressing through the flower-feeding and frugivorous Microchiroptera to the Megachiroptera. In parallel to the neocortex, there is a progressive increase in the size of the diencephalon and striatum, the highest subcortical station in the extrapyramidal motor system. Among the Microchiroptera, vampires appear to be the most "intelligent" based on the observation that next to the Pteropids and frugivorous phyllostomid *Artibeus*, they have the largest neocortex. In all bats the neocortex is the most progressive area of the brain in that it is two to six times larger than in basal insectivores (table 5.2).

On the basis of cytoarchitecture, the mammalian cortex is divided into as many as 52 different areas. The majority of these areas contain topographic sensory or motor representations of specific body areas. For other cortical areas, especially those in frontal cortex, there is no obvious single function that can be assigned to them; for this reason they are termed "association" areas. In Megachiroptera, only 26 of the 52 areas known from the human cortex appear to be present; in Microchiroptera, the number of cortical areas is even smaller. Because it is difficult to establish homology on the basis of cytoarchitecture alone, these numbers must be regarded with a certain amount of caution. Functional analysis is necessary to characterize the differentiation of the cortex, as exemplified by the studies of auditory cortex in *Pteronotus*.

In mammals, the frontal cortex (frontal lobe) is the most progressive region. It is largest in humans and is the basis for personality, judgment, and planning for the future. In insectivorous bats the frontal cortex appears to be small or absent, and in Megachiroptera only area 12 appears to be present. As would be expected

Table 5.2 Median progression indices (encephalization index values for selected brain structures) for bats compared to basal insectivores (100%).

Olfactory bulb	70–80%
Paleocortex	90%
Hippocampus, septum, hindbrain	140%
Midbrain, diencephalon, striatum	220%
Cerebellum	290%
Neocortex	380%
Vampire	620%
Fisherman bat, *Noctilio leporinus*	530%
Megachiroptera	500%
Frugivorous phyllostomids	400%
Nectar-feeding phyllostomids	360%
Myotis spp., rhinolophids, molossids, hipposiderids	180–240%

From Stephan (1970).

for animals with large eyes and good nocturnal vision, the visual (occipital) cortex of the Megachiroptera is large. In echolocating bats, the visual cortex is small. Based on cytoarchitecture, the nonprimary visual cortical areas are absent. In contrast, the auditory cortex (temporal lobe) occupies nearly half of the total cortical area (see fig. 6.22).

In flying foxes, the somatosensory field is differentiated into five distinct areas, each one with a complete somatotopic representation. Neurons in the largest area, the primary somatosensory cortex, have been shown to respond to stimulation of peripheral body parts, with very narrow receptive fields. Neurons in a neighboring area quickly habituate to repetitive stimulation. Neurons in a small ventral area were found to have large receptive fields and to respond to both somatosensory and auditory stimuli.

The representation of the body surface in the somatosensory cortical areas of bats is especially interesting because the topography of these areas provides insight into how an animal perceives its own body. In other words, the topographic organization of somatosensory projection fields is the basis for the subjective perception of where the various parts of the body are located in space. In ground-dwelling animals, the representation of the snout and the limbs face in the same direction so that parts of the body that lie below the head point toward the rostral part of the cortex (fig. 5.8, rat). In contrast, in bats the representation of the arm and wing are rotated 180° so that they point caudally and in the same direction as the back. This organization suggests that the wings are perceived as being above the body whether the bat is flying or at rest (fig. 5.8, *Macroderma* and *Pteropus*).

The size of the cortical area devoted to the representation of a given part of the body provides information about the density of sensory innervation and importance of that part of the body for the animal. When the somatosensory cortical representation of a nonecholocating flying fox is compared with that in an echolocating bat (*Pteropus* and *Macroderma* in fig. 5.8), the following differences are conspicuous:

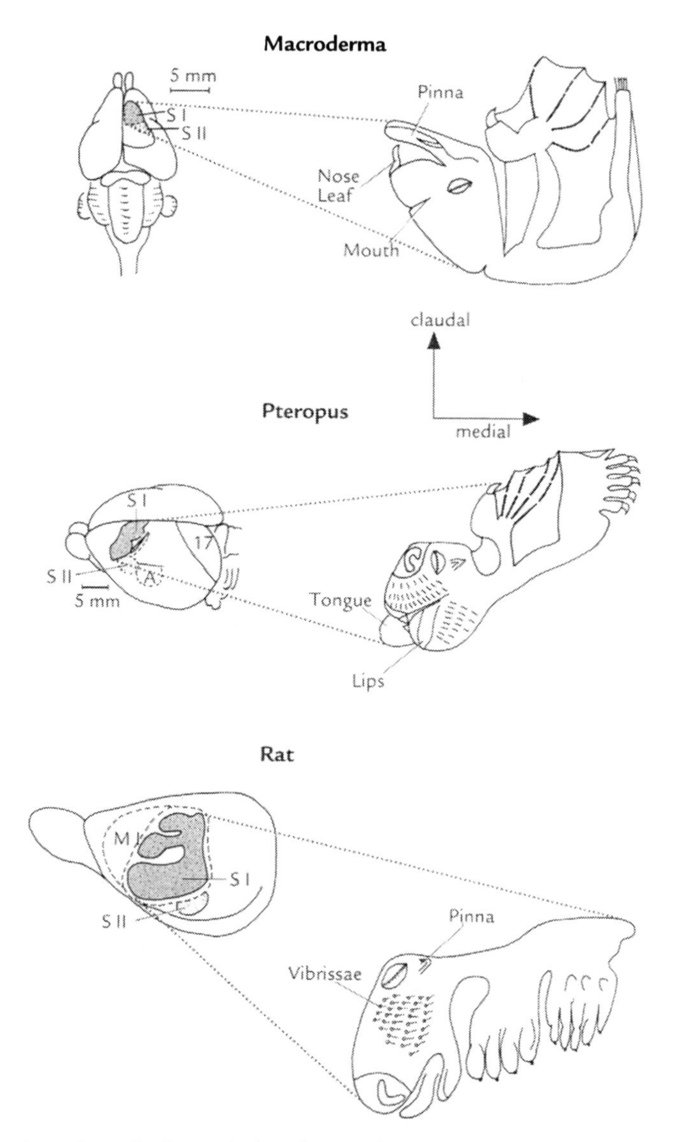

Figure 5.8 Location of primary (SI) and secondary (SII) somatosensory cortex in the telencephalon (left) and sensory representation of the different body regions in primary somatosensory cortex (right) in an echolocating microchiropteran (*Macroderma gigas*), a nonecholocating megachiropteran (*Pteropus*), and a rodent (rat). In contrast to the rat, where the cortical representation of the extremities is below the ventral surface of the body, the representation of the extremities in the bat is above the dorsal surface. This suggests that bats perceive their limbs as being above their body. There is an expanded representation of the pinnae and nose-leaf in *Macroderma* and of the lips and tongue in the frugivorous species, *Pteropus*. A, auditory cortex; MI, motor cortex; 17, area 17, primary visual cortex. *Macroderma* from Wise et al. (1986); *Pteropus* sp. from Calford and Tweedale (1988); rat (not to scale) from Kaas (1983).

- The pinnae and the nose, which is used for sound emission, have a larger representation in the echolocating bat than in the flying fox.
- In the flying fox, each toe has its own receptive field. In contrast, in the echolocating bat, the entire foot is represented in a single small, undifferentiated receptive field. This difference can be attributed to the fact that echolocating bats use their feet only for hanging from a surface and for spreading their wings, whereas frugivorous flying foxes use all four limbs to crawl around in trees where they search for fruit.
- In the flying fox, there is a large cortical receptive field for the tongue. In echolocating bats, no such field can be distinguished. Again, the reason for this is the difference in diet.
- In both Microchiroptera and Megachiroptera there is an expanded representation of the thumbs, probably because they are the only part of the forelimb that can be used for grasping.

The relatively small size of neocortex and the virtual absence of convolutions in the bat cortex have been interpreted as indicating that the cortex of bats is primitive or plesiomorphic. However, this conclusion should be regarded with caution because it is based strictly on macroscopic, and therefore superficial, criteria. Microscopic studies using modern anatomical tracing methods have shown that in *Antrozous pallidus* (Vespertilionidae), for example, the connections among the different cortical areas are as complex as they are in other mammals. The *corpus callosum*, which connects the two hemispheres, is well developed in flying foxes and phyllostomids. Nevertheless, in many Microchiroptera, the corpus callosum provides only a weak connection between the two hemispheres. Instead, fibers cross from one hemisphere to the other in the hippocampal commissure and in the anterior commissure. This commissural pattern is also considered to be a characteristic of plesiomorphic cortex and is otherwise found only in nonmammals and marsupials.

Pyramidal tract. In parallel to the neocortex, mammals have developed a new motor pathway to the spinal cord, the pyramidal tract. This pathway benefits from the highly developed analytical capability of the cortex and provides a hitherto unattainable degree of fine motor control of the head and limb muscles. The axons of cortical neurons first project to the pons. Some of the fibers terminate in the pontine nuclei while others cross below the olivary complex and continue from there to the spinal cord. One group of pyramidal fibers remain uncrossed and project to the thoracic portion of the spinal cord. In contrast to other mammals, the pyramidal decussation in Chiroptera occurs very early. This is considered to be a characteristic of primitive mammals. It is thought that the pyramidal tract of bats crosses a second time just before it enters the spinal cord. It is through the pyramidal tract that cortical neurons gain direct access to the motor neurons that innervate the head and spinal cord. These neurons are primarily involved in volitional movements.

In Microchiroptera, the pyramidal tracts are not well developed. In humans, the pyramidal tract contains 1,100,000 fibers, in the macaque it contains 554,000 fibers, in the cat 186,000, and in the mouse 32,000. In *Tadarida brasiliensis*, the pyramidal tract at the level of the medulla contains only 8000 fibers. Nevertheless, even among bats there are considerable species differences. It is possible that motor control of the wings is mainly under reflex control, primarily through input from the hypertrophied cervical spinal cord, and that the influence of the cortex via the pyramidal tract is small. The pinnae and larynx, on the other hand, are thought to be controlled via pyramidal fibers.

Paleopallium and archipallium. In mammals, the older parts of the brain are covered over by the neocortex and pushed inward (fig. 5.7). The paleocortex contains mainly olfactory centers, which in bats often lie on the cortical surface. Among the structures derived from the paleocortex is the amygdala. The septum provides a link between the olfactory centers and the thalamus. The archicortex has given rise to three important nuclei of the limbic system, the hippocampus, the dentate gyrus, and the cingulate gyrus. The limbic system is defined as a group of functionally related corticothalamic nuclei including the amygdala and olfactory centers. The limbic system plays an important role in linking learned cognitive behavior with instinctive behavior and emotion.

In the last few years, a number of different types of studies have been performed on the hippocampus and amygdala of laboratory mammals, inspired by the highly organized connections of these structures. These studies have shown that these forebrain structures play an important role in memory, especially spatial memory. In bats, both the hippocampus and the amygdala are unusually large. The nuclei of the amygdala in Microchiroptera are even more highly developed than those of flying foxes. These nuclei reach their highest degree of differentiation in horseshoe bats. The hippocampus is the part of the brain that exhibits the most interspecific variability among bats. Interestingly, the amygdala of echolocating whales is also hypertrophied. This enlargement may be related to the importance of these structures for acoustic imaging, perhaps especially for spatial memory.

The amygdala also has connections with the olfactory and extrapyramidal systems. Thus, ablation of the amygdala leads to a loss of reflex spreading of the wings. The auditory system has access to the amygdala through the medial geniculate nucleus of the diencephalon. The basal ganglia, which include the caudate nucleus, the globus pallidus, and the putamen (fig. 5.7), perform a variety of functions. They are a central component of the extrapyramidal motor system.

Extrapyramidal motor system. Although the name of this motor control system implies that it is separate from the pyramidal tract, this is now known not to be true. The extrapyramidal system consists of a number of interconnected circuits with multiple outputs; one of the most important outputs is, in fact, the pyramidal tract. The extrapyramidal system controls reflex motor actions such as postural adjustment, motor coordination, and muscle tone.

Structures involved in the extrapyramidal system include the basal ganglia as well as specific thalamic and hindbrain nuclei. The extrapyramidal system gains access to the limbic system via the habenula. The most important targets of the extrapyramidal system are the cortical areas that give rise to the pyramidal tracts, which, in turn, directly contact the motor neurons of the spinal cord. The fibers of the pyramidal tract also give rise to collaterals that project to the ventrolateral and centromedian thalamic nuclei, the red nucleus, and the reticular formation. These pathways frequently send recurrent collaterals to the *corpus striatum*, the subdivision of the basal ganglia that consists of the caudate and putamen. Other pathways originate in the substantia nigra and project to the medial and ventrolateral nuclei of the thalamus, and to the superior colliculus which then projects back to the reticular formation. From there, motor commands are transmitted to the spinal cord. Like the pyramidal tract, the extrapyramidal tract in bats is poorly developed.

5.3 AGING AND THE CENTRAL NERVOUS SYSTEM

As a general rule, the life span of an animal is directly proportional to the size of its brain. Bats, however, are a flagrant contradiction to this rule. Their brain is no larger than that of a mouse but, unlike a mouse that lives only 2–3 years, bats can live for more than 20 years. On average, bats have a life span at least three times as long as that of nonflying mammals. This extreme longevity in bats is not related to their ability to hibernate because it is also true for tropical species.

Baudry et al. suggested the hypothesis that aging is related to the calpain content of neurons. Calpain is a calcium-dependent proteolytic enzyme, one function of which is to break down cytoskeletal proteins. There is supposed to be a linear negative correlation between calpain activity and life expectancy and between calpain activity and brain size. Calpain activity has been measured in various regions of the bat brain and found to be five to seven times less than in the mouse. Reptiles with long life spans are also reported to have low levels of calpain activity. It would be productive to pursue this hypothesis, including flying foxes in the analysis. Low levels of calpain activity in Megachiroptera would provide support for the idea that both suborders have a common phylogenetic origin. Another line of reasoning suggests that the longevity of bats is due to their flying lifestyle, which makes them less vulnerable to environmental hazards. This idea is corroborated by the observation that nine species of gliding mammals, including squirrels, lemurs, and marsupials, have life spans 1.7 times longer than would be expected from their body size.

References

Anstead SN, Fischer KE (1992). Mammalian aging, metabolism, and ecology: Evidence from bats and marsupials. J Gerontol Biol Sci 46:B47–B53.

Baron G (1969). Etude comparative de quelques noyaux moteurs encephaliques chez des chiropteres neotropicaux. I. Nucleus alaris et nucleus nervi hypoglossi. Rev Can Biol 28:241–257.

Baron G (1970). Etude comparative de quelques noyaux moteurs encephaliques chez des chiropteres neotropicaux. II. Nucleus nervi facialis et Nucleus motorius nervi trigemini. Rev Can Biol 29:115–128.

Baron G (1970). Etudes comparative de quelques noyaux moteurs encephaliques chez des chiropteres neotropicaux. III. Noyaux oculomoteurs. Rev Can Biol 29:233–251.

Baudry M, Dubrin R, Beasly L, Leon M, Lynch G (1986). Low levels of Calpain activity in chiroptera brain: Implications for mechanisms of aging. Neurobiol Aging 7:255–258.

Calford MB, Tweedale R (1988). Immediate and chronic changes in responses of somatosensory cortex in adult flying fox after digit amputation. Nature 332:446–448.

Hackethal H (1973). Zur vergleichenden Anatomie des Kleinhirns der Chiropteren. Period Biol 75:71–76.

Hafiza B, Sood PP (1979). Histological and histoenzymological studies of medulla oblongata of *Taphozous melanopogon*. Acta Anat 105:439–451.

*Henson OW JR (1970). The central nervous system. In W.A. Wimsatt, ed., Biology of Bats, Vol. II, pp. 58–152. Academic Press, New York.

Humphrey T (1936). The telencephalon of bats. I. The noncortical nuclear masses and certain pertinent fiber connections. J Comp Neurol 65:603–711.

Kaas JH (1983). What, if anything, is SI? Organization of first somatosensory area of cortex. Physiol Rev 63:206–231.

Krubitzer LA, Calford MB (1992). Five topographically organized fields in the somatosensory cortex of the flying fox: Microelectrode maps, myeloarchitecture, and cortical modules. J Comp Neurol 317:1–30.

Kurepina M (1968). Einige Besonderheiten der Struktur des Neocortex bei Fledermäusen im ökologischen Aspekt. J Hirnforschung 10:39–48.

Machin C, Rua C, Taberprierce E, Carrato A (1983). Ultrastructure development of raphe nuclei in the bat (*Myotis myotis*). J Hirnforschung 24:405–415.

Mann G (1963). Phylogeny and cortical evolution in Chiroptera. Evolution 17:589–591.

*McDaniel VR (1976). Brain anatomy. In R.J. Baker et al., eds., Biology of Bats of the New World Family Phyllostomidae, Part I, pp. 147–200. Spec. Publ. Museum Texas Tech, Lubbock.

Quay WB (1970). Peripheral nervous system. In W.A. Wimsatt, ed., Biology of Bats, Vol. II, pp. 153–179. Academic Press, New York.

Quay WB (1970). Pineal organ. In W.A. Wimsatt, ed., Biology of Bats, Vol. II, pp. 311–318. Academic Press, New York.

Rose M (1912). Histologische Lokalisation der Großhirnrinde der kleinen Säugetiere (Rodentia, Insectivora, Chiroptera). J Psychol Neurol 19:389–479.

Schneider R (1957). Morphologische Untersuchung am Gehirn der Chiropteren. Abh Senckenb Natf Ges 495:1–92.

Schober W (1967). Zur Lage der Decussation pyramidum bei den Fledermäusen (Chiroptera). Anat Anz 120:174–180.

Starck D (1982). Vergleichende Anatomie der Wirbeltiere, Vol. 3. Springer-Verlag, Berlin.

Stephan H (1977). Encephalisationsgrad südamerikanischer Fledermäuse und Makromorphologie ihrer Gehirne. Gegenbaurs Morph Jahrb 123:151–179.

Stephan H, Frahm HD, Baron G (1987). Brains of Vespertilionids. Zeitschrift Zool. Systemat. Evolforschung 25:67–80.

Stephan H, Nelson JE (1981). Brains of Australian chiroptera: I. Encephalization and macromorphology. Aust J Zool 29:653–670.

Stephan H, Pirlot P (1970). Volumetric comparisons of brain structures in bats. Z Zool Syst Evolutforsch 8:200–236.

Tamura I (1950). Comparative anatomical studies on the brainstem with special reference to the reticular formation and its relating nuclei of chiroptera. Acta Anat Niigata Ensia 50:65–98.

Wise LZ, Pettigrew JD, Calford MB (1986). Somatosensory cortical representation in the Australian ghost bat, *Macroderma gigas*. J Comp Neurol 248:257–262.

Yamamoto S, Shimoda B, Momma R, Sai H (1955). Studies on brain stem. IX. Comparative anatomical study of brain stems of some species of bats. Tohoku J Exp Med 61:339–344.

6

ECHOLOCATION

6.1 THE DISCOVERY AND BASIC PRINCIPLES OF ECHOLOCATION

THE DISCOVERY OF ECHOLOCATION

HUMANS ARE NOT the only organism that depends on vision to perceive the world; most animals are highly dependent on vision. Even animals that hunt prey at night use their eyes to find their way around. The earliest demonstration of this fact was through the experiments of Lazzaro Spallanzani, the Bishop of Padua, and an active experimental scientist in the late eighteenth century. Spallanzani brought owls into his study and observed that these nocturnal birds of prey refused to fly if all of the candles in the room were extinguished. When he repeated the same experiment using bats, these small mammals flew confidently around the bishop's study, even in total darkness, managing to avoid the wires that Spallanzani had hung from the ceiling. On the ends of the wires were attached small bells, but the bats did not hit the wires and cause the bells to ring even after he had blinded the bats with a red-hot needle. However, the bats became irritated when Spallanzani placed brass tubes in their ear canals. When these tubes were closed, the bats hit the wires as they flew and rang the bells. As soon as the tubes were opened again, these same bats regained their ability to fly in the darkness. Because the bats did not produce any sounds that were audible to Spallanzani as they flew through the darkness, he was not able to discover the secret of their orientation. Somewhat puzzled, he ended his experiments in 1794 with the following entry in his journal, "thus, blinded bats are able to use their ears when they hunt insects . . . this discovery is incredible."

This puzzle was not solved until 150 years later, when an American biology student and a Dutch professor independently discovered the secret of echolocation. In 1938, when Donald Griffin was visiting the laboratory of the physicist G. W. Pierce at Harvard, he discovered that the world's first ultrasound microphone transformed apparently silent bats into a hoarde of extremely noisy animals. It suddenly became obvious that the bats were using their mouths to emit short ultrasound signals, that is, sounds whose frequency lies above about 20 kHz, the upper limit of human hearing. In 1943, the zoologist Sven Dijkgraf, who had very sensitive hearing, noticed the audible component, or "*Ticklaut*" emitted by the bat, *Myotis emarginatus*. When he fitted the bats with small muzzles that prevented them from emitting these sounds, they became just as helpless and disoriented as Spallanzani's bats with the blocked ear canals. These observations provided the answer to the puzzle: bats emit high-frequency sounds through their mouth or nose and perceive their nocturnal environment by listening to the echoes from the objects that surround them.

The term *echolocation* does not fully describe the acoustic imaging system used by bats. Echolocating animals can determine not only the location of an echo source, but they can also perceive its size, form, and surface texture. It would therefore be more accurate to speak of "echo perception" or "echo imaging."

GENERAL PRINCIPLES

The echo imaging system consists of two main components, a transmitter and a receiver. The transmitter is the larynx, which produces the echolocation sounds. The receiver is the two ears and their associated neural systems (fig. 6.1). Echolocation, like electroreception in electric fish, is considered an active orientation system because the carrier signal for information is produced by the animal itself. Through this system, bats have succeeded in becoming independent of sunlight as a medium for perceiving their world.

DISADVANTAGES OF ECHOLOCATION

Although independence from sunlight has some advantages, these are offset by certain disadvantages compared to visual perception, as described below.

Energy cost. Acoustic imaging is energy intensive. For example, if a horseshoe bat (*Rhinolophus rouxi*) flys out at 6:00 PM to forage for insects and returns to its roost in the cave at 5:00 AM, over this 11-h period it will emit approximately 400,000 echolocation calls. Each echolocation call corresponds to an energy flux density of 6×10^{-6} J/m^2. During the night, this amounts to about 2.4 J/m^2.

Duty cycle. In contrast to vision, which provides a continuous flow of images of the outside world, echolocation provides a series of discrete "stroboscopic" images of the surroundings, with each image corresponding to an emitted echolocation call. The horseshoe bat mentioned above, even with its 400,000 echolocation calls per night, remains "in the dark" for 6.5 h—the time between echoes. For the horseshoe bat, the duty cycle (the ratio of the time during which a signal is present to the entire duration of the time from one call onset to the next) takes up 40–50% of the nightly activity period. In most species, however, the duty cycle is only 4–20%.

Limited sound field. Compared with the visual field of a mammal, the echo imaging field of a bat is restricted due to the narrow range of the emitted sound, and it is concentrated in the direction of flight. The only way in which the bat can gain a broad image of its surroundings is by emitting multiple, successive calls in different directions.

Limited range. The range over which echolocation operates is quite restricted. Sound is attenuated in air due to two processes: (1) geometric attenua-

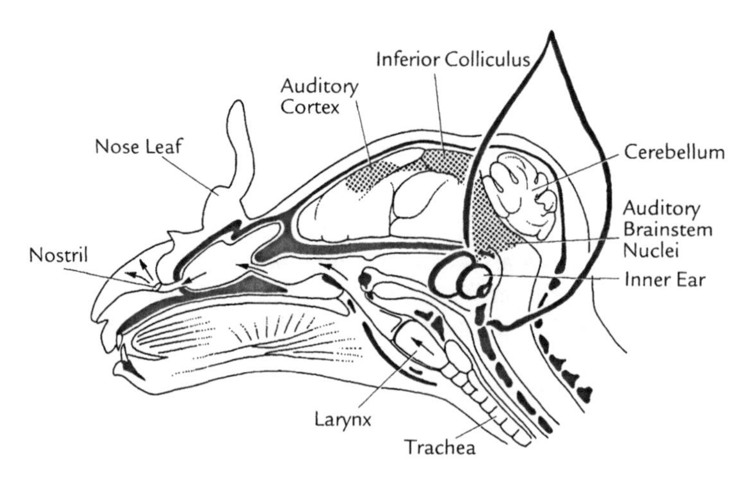

Figure 6.1 Schematic section through the head of a horseshoe bat, showing the sender (the larynx) and the receiver (the ear and the auditory pathways of the brain, shaded). The arrows indicate the path taken by sound as it passes from the larynx to the nostrils.

tion—the sound pressure of the emitted signal, which can be described as a circular wave, decreases as the square of the distance from the transmitter; and (2) absorption—sound absorption in air is greatest for high frequencies and also increases with humidity and temperature. Thus, low frequencies have a longer range (fig. 6.2). In echolocation, the path traveled by sound is doubled because it must first travel from the bat to the reflecting object and then back again from the object to the bat's ears. Because of these physical constraints, echolocation can only operate over a limited range, usually less than 20 m, and no more than 50–60 m.

Resolution. In optics, there is a general rule that the wavelength of an information-carrying signal must be no more than half the size of the object to be resolved. This same rule can be roughly applied to acoustic imaging. Because the frequency of an echolocation call is inversely proportional to its wavelength, high frequencies can achieve a better resolution than low frequencies. Consequently, the bat must find a compromise between the long range provided by low frequencies and the high resolution provided by high frequencies.

EVOLUTION

Why are bats the only land-dwelling mammals that have developed an echolocation system? Why is it that even though most small mammals such as rodents and insectivores have good hearing in the ultrasound range and are capable of producing ultrasound vocalizations, none use ultrasound as an aid to orientation? There are approximately 900 species of bats, of which all small bats (Microchiroptera) echolocate. This means that about 750 species have an echo-imaging

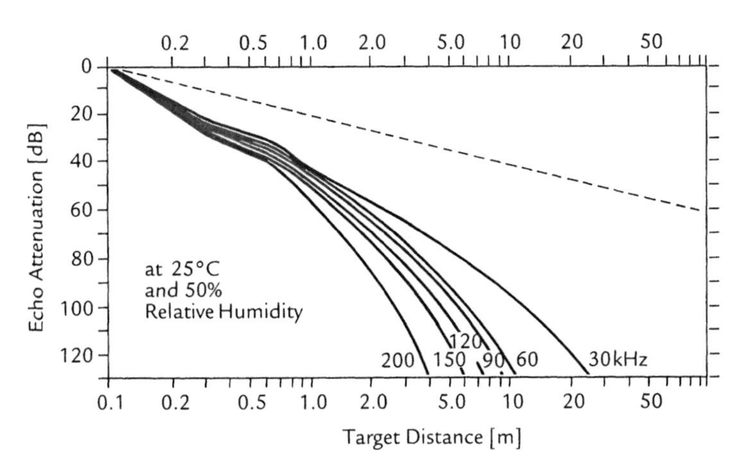

Figure 6.2 Attenuation of sound pressure in air as a function of sound frequency (kHz). This attenuation is due to a combination of geometric attenuation and atmospheric absorption. Dotted line = geometric attenuation of the sound impinging on an object. From Lawrence and Simmons (1980).

system. Flying foxes (Megachiroptera) that feed on fruit juice, nectar, and pollen, with only a few exceptions, lack an echo-imaging system. These facts have given rise to two very different hypotheses regarding the origin of echolocation:

1. Echolocation evolved due to animals' need to find their way around in caves. This hypothesis is supported by the fact that echolocation is not only found in cave-dwelling Microchiroptera, but also in the cave-dwelling megachiropteran *Rousettus*, in the only cave-dwelling birds, the cave swiftlet (*Collocallia*), native to Asia, and the oilbird (*Steatornis*), native to South America.
2. Echolocation evolved to aid in capturing insect prey. As bats began to fly, they experienced an increasing need to identify and pursue flying insects in the dark. The low light levels at night were no longer sufficient for conducting their nightly foraging activities. Therefore, to occupy an ecological niche in which they could exploit the supply of night-flying insects, bats had to develop a system for echolocation. According to this hypothesis, the selective pressure due to the bats' foraging style was higher than that due to their use of caves and hollow trees as roosting sites.

Modern technology, especially medicine and materials science, has shown that acoustic imaging does not present any insurmountable technical problem. However, an engineer is free to use any available technology in designing a sound transmitter and receiver, but a bat's system for acoustic imaging is constrained by the features of the mammalian body. The transmitter is necessarily the larynx, and the receiver is the ear. Nevertheless, the larynx and ears of bats are not drastically different from those of humans and other nonecholocating mammals.

6.2 ECHOLOCATION CALLS AND THEIR PRODUCTION

TYPES OF ECHOLOCATION CALLS

The echolocation calls of bats, like the vocalizations of other mammals, are produced by the larynx. Unlike communication calls, echolocation calls are very short in duration, generally lasting only a few milliseconds. Echolocation calls consist of up to three different types of *signal elements*, which can be present singly or in combination (fig. 6.3):

- *Downward FM*. The most common echolocation signal is a downward frequency-modulated (FM) pulse (left column in fig. 6.3). This type of signal starts at a high frequency and sweeps downward to progressively lower frequencies.
- *CF*. A constant-frequency (CF) signal is a pure tone or a signal that is only slightly modulated in frequency. CF signals typically last for 10–100 ms and are commonly used as search signals (middle column in fig. 6.3). The echolocation calls of horseshoe bats and hipposiderids always contain a CF component.
- *Upward FM*. Sometimes the CF component of the echolocation call is preceded by an upward FM component (right column in fig. 6.3). So far, upward CF signals have only been described in association with CF signals.

Echolocation calls frequently have a complex harmonic structure, meaning they contain a number of different frequencies that are multiples of a fundamental frequency. For example, 40 kHz, 60 kHz, and 80 kHz are 2nd, 3rd, and 4th harmonics of a fundamental frequency of 20 kHz. The fundamental frequency is also commonly referred to as the first harmonic. The largest amount of energy in the echolocation call is usually in the 2nd or 3rd harmonic, not in the fundamental frequency (fig. 6.3).

Flying foxes of the genus *Rousettus* are the only Megachiroptera that produce echolocation calls. *Rousettus* is also the only flying fox that spends its days in caves rather than roosting in trees. Unlike other bats that produce echolocation signals with their larynx, *Rousettus* produces them with its tongue. Each time the tongue is raised from the floor of the mouth, it produces a short double click—one click for each of the two lips. The clicks are only a few milliseconds long and contain a broad frequency spectrum from about 15 kHz to 150 kHz. Using these clicks, *Rousettus* can navigate even in the total darkness of the caves where it roosts. Whereas most bats emit echolocation signals even in broad daylight, flying foxes of the genus *Rousettus* only produce echolocation signals in the dark and otherwise use their large eyes for orientation.

The fact that *Rousettus*, a megachiropteran, uses a completely different method to generate echolocation signals than do the microchiropterans is sometimes used to support the argument the two suborders have different phylogenetic origins.

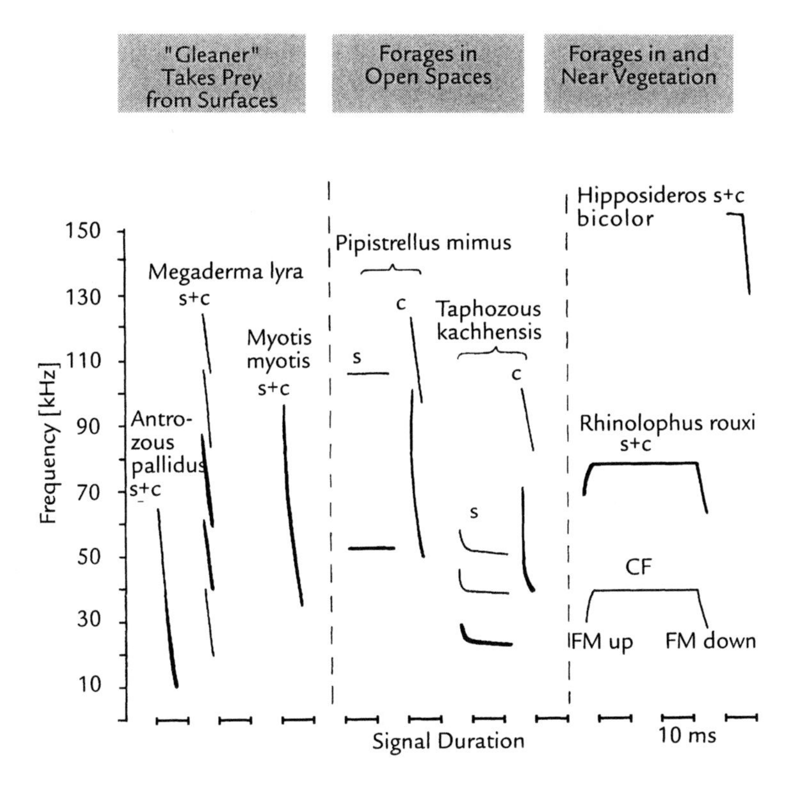

Figure 6.3 Types of echolocation signals used by bats, illustrated as sonagrams (plots of frequency as a function of time). For signals with multiple harmonics, the most intense harmonic is indicated by a thick line. The structure of the echolocation calls is correlated with foraging habits, indicated by the three columns. s, search call, used when the bat is searching for prey; c, pursuit calls, used when the bat is pursuing prey; s+c, calls that are used for both search and pursuit of prey. FM (down), downward frequency-modulated component; FM (up), upward frequency-modulated component; CF, constant-frequency signal component.

ANATOMY OF THE LARYNX

As amphibians left the water and developed lungs, the vertebrate larynx evolved, not as an organ for producing sound, but rather for the purpose of protecting the respiratory tract by preventing particles of food from entering. The larynx is located precisely in that part of the throat where the esophagus and trachea join together. This vulnerable area in which very different functions cross paths is protected by the epiglottis and by the elevation of the larynx during swallowing. In echolocating bats and whales, the larynx extends farther cranial in the pharyngeal cavity than it does in other mammals, allowing breathing to occur through the nose even while swallowing occurs.

The larynx (fig. 6.4) is suspended from the broad arc of the hyoid bone and the muscular areas in the throat and at the base of the tongue. As it extends upward from the trachea, the larynx forms a conical-shaped membranous tube that contains three cartilaginous structures. Two of these are moved by the laryngeal muscles during vocalization. The laryngeal cavity can be closed off horizontally into an upper ventricle and a subglottal space by pressing the vocal cords against one another. Thus, the glottis is closed. The base of the supporting structure of the larynx forms a thick ring, the *cricoid cartilage*. In bats, the cricoid is ossified and open ventrally. On both sides of the cricoid cartilage are protrusions that support the *thyroid cartilage*, which can be tilted. Positioned on a cricothyroid joint, the thyroid cartilage can be pulled down like the visor of a helmet (fig. 6.4b).

Inside the membranous laryngeal cavity, on the upper, dorsal surface of the cricoid cartilage, sits a pair of movable structures, the *arytenoid cartilages* (4 in fig. 6.4c). Each arytenoid has a long, ventrally directed process that secures one of the vocal folds. The paired vocal folds narrow the diameter of the larynx to form a slit through which air can pass. Through movements of the arytenoid cartilage, the diameter of this aperture can be varied from wide open to completely closed. During respiratory movements when no vocalization takes place, the vocal aperture remains continuously open. On inspiration, the pair of arytenoid cartilages are rotated outward along their long axes through contraction of the crycoarytenoid muscles. This causes the vocal folds to be pulled outward, and the vocal aperture to open to its widest extent (fig. 6.4).

SOUND PRODUCTION

The energy for sound production comes from the stream of expired air, and hence from the muscles that power respiration. When vocalization occurs, the amount of time spent in expiration increases relative to the time spent in inspiration. In humans, the ratio of expiration to inspiration is about 5:4 when no vocalization occurs, but it increases to 6–7:1 during speech and 10–50:1 during singing. About 20 ms before a sound is produced, the two arytenoid cartilages, which move about on the cricoid cartilage, are pressed together through the contraction of the oblique and transverse arytenoid muscles. This closing of the glottis completely blocks the airway.

In bats, the two arytenoid cartilages are fused together at their upper ends so that the whole structure is stiff. The *lateral cricoarytenoid muscles* act to rotate the arytenoid cartilages along their long axes toward the midline (fig. 6.4b). In addition, the paired *thyroarytenoid muscles*, which extend from the thyroid cartilage to encircle the ventral part of the arytenoid cartilage, press the vocal folds together. These muscular contractions act together to close the glottis.

Sound intensity. During expiration, air pressure builds up beneath the closed glottis, resulting in a *subglottic pressure* and an associated increase in the volume of expired air. In a species of bat weighing 18–26 g, the air volume needed for vocalization is 0.3–0.6 ml, compared to 0.05–0.1 ml for expiration in the absence of vocalization. In a human speaking at a normal conversational level, subglottic

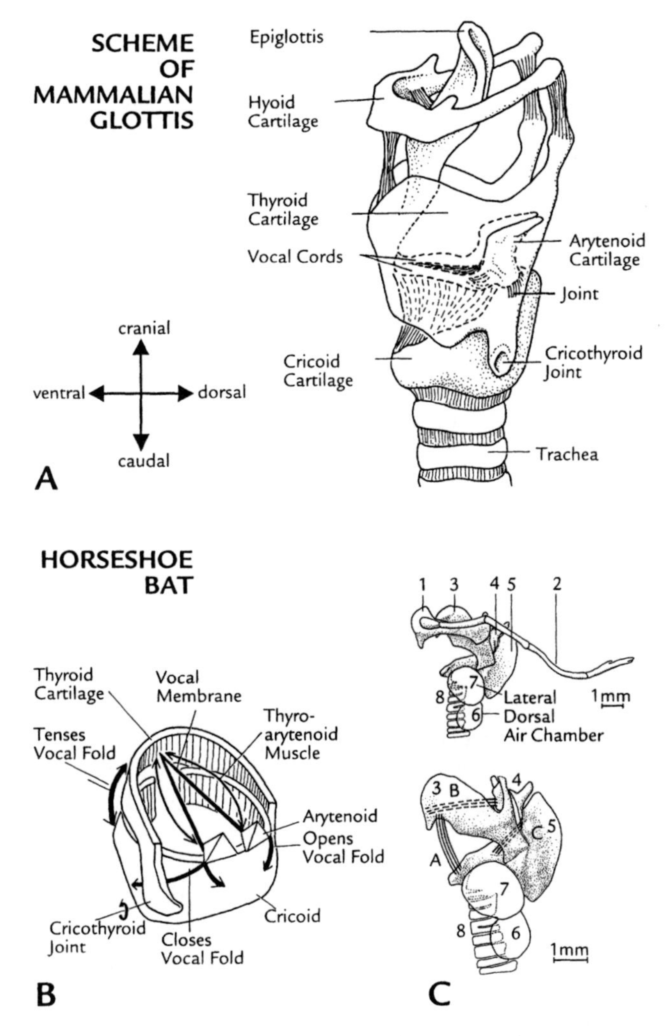

Figure 6.4 The larynx of bats. (A) Side view of the human larynx for comparison. (B) Functional model. The black arrows indicate the most important muscular contractions. "Opens vocal fold" = posterior cricoarytenoid muscle; "closes vocal fold" = lateral cricoarytenoid muscle; "tenses vocal fold" = cricothyroid muscle. The thyroarytenoid muscle assists in closing the glottis. (C) Cartilaginous elements of the larynx in *Rhinolophus ferrumequinum*. At the top, these are shown in their natural positions. The lower drawing illustrates the most important actions of the muscles on the cartilages: A, cricothyroid muscles; B, thyroarytenoid muscles; C, cricoarytenoid muscles; 1, hyoid bone; 2, anterior process of hyoid; 3, thyroid cartilage; 4, arytenoid cartilage; 5, cricoid cartilage; 6, dorsal membranous air chamber; 7, lateral membranous air chamber; 8, tracheal cartilages.

pressure is equivalent to about 8 cm H_2O. The echolocation signals emitted by bats are extremely loud, reaching sound pressures of 110 dB SPL. Accordingly, in the echolocating bat, *Eptesicus fuscus*, subglottic pressures of 30–45 cm H_2O have been measured. These high subglottic pressures are thought to result from two specializations of the bat's larynx. First, the lips of the vocal cords are thick rolls that form a tight seal when pressed against one another. Second, the membranous walls between the thyroid cartilage and the cricoid cartilage are unusually thick in bats. During the buildup of subglottic pressure, they become stretched, thus storing energy to be released during vocalization. This mechanism allows a high subglottic pressure to be maintained even during a series of two to three calls.

Sound envelope. The sound envelope describes the intensity of a sound over time, from its onset to its end. As long as subglottic pressure is maintained and the glottis is only opened as a narrow slit, sound is produced. As soon as the muscles responsible for closing the glottis relax, it opens fully. The resulting loss of resistance causes subglottic pressure to drop rapidly, and vocalization ceases. Bats frequently emit multiple sound pulses over the course of one expiratory cycle (fig. 6.5). Subglottic pressure remains high over the entire expiratory phase. The opening and closing of the glottis determine when a sound pulse begins and when it ends. The arytenoid muscles thus control the envelope of the bat's vocalizations.

Powering sound emission. The subglottal pressure required for vocalization is powered by muscles of the abdominal wall, which start to contract just before the emission of echolocation sounds. These muscles produce a strong intraabdominal pressure, which compresses the thoracic air volume. The generation of abdominal pressure is enhanced by a resilient collagenous plate that covers one-third of each flank and is unique to echolocating Microchiroptera. This "aponeurosis" acts like a diaphragm. When the bat stops vocalizing, the contractions of the abdominal wall cease.

In flying bats, the respiratory cycle, and hence sound emission, is synchronized in a one-to-one ratio with the rhythm of the wingbeats. Sound is emitted when the muscles that power the downstroke contract and exert pressure on the rib cage. This happens during the final phase of the upstroke, and during the transition between the upstroke and the downstroke (see fig. 1.7). Thus, during flight, the powerful thoracic muscles responsible for the downstroke facilitate sound production to such an extent that most of the power required for sound emission can be considered to arise as a by-product of the muscular contractions that power the bat's flight.

The sound pressure levels of echolocation sounds range from 80 dB SPL (sound pressure level) in "whispering" bats such as the false vampire that inspect potential prey at close range to 110 dB SPL in horseshoe bats and vespertilionids that search for flying insects in open spaces. The pressure level for a single echolocation sound corresponds to an energy flux level on the order of 10^{-9} to 10^{-5} J/m^2 (see box 6.1). Assuming a sound emission rate of 1/s, the average power output

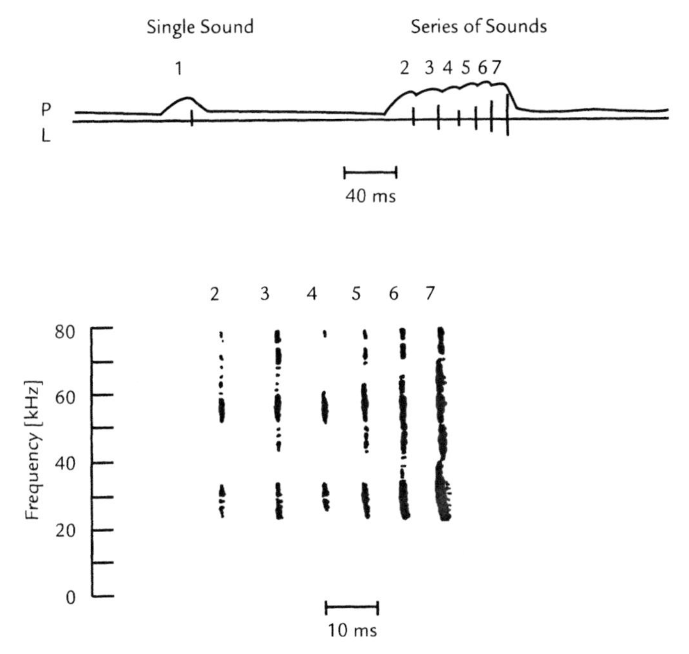

Figure 6.5 Subglottic pressure during production of a single sound pulse (1) and a series of sound pulses (2–7). *Upper diagram*: P, trace showing subglottic pressure; L, emitted echolocation pulses. *Lower diagram*: Sonagram of sound pulses 2–7, illustrated above. From Suthers and Fattu (1982).

Box 6.1 Sound Pressure (dB SPL) and Energy

The unit dB SPL (decibels sound pressure level) is a logarithmic measure of the relative sound pressure of an acoustic stimulus. The decibel scale is constructed relative to a reference point of 20 μPa = 0 dB SPL. This value corresponds to the average threshold for human hearing (i.e., the minimal sound pressure that can be perceived under optimal conditions). An increase or decrease of 20 dB corresponds to a change of one power of ten in the sound pressure. Humans and many mammals have a hearing range that extends from 0 dB SPL to about 140 dB SPL, which is the threshold of pain. This range corresponds to over 7 logarithmic units. The sound pressure level of normal conversation is about 40–50 dB SPL.

A measure of the energy output of a sound is its energy flux density, E:

$$E = P^2_{rms} * T$$

where P_{rms} = root mean square pressure of a signal (N/m²), and T = signal duration (Pa²/s). In energy units:

$$E = [1/\rho\,(c)]\; P^2_{rms} T$$

where ρ = density of the medium (kg/m³); c = velocity of sound (m/s), and T is in J/m².

would be 1 μW in a "whispering" bat and 330 μW in a horseshoe bat. These values are similar to the range of power output in other calling animals such as crickets and frogs.

Ultrasound. It is still not entirely clear which elements of the larynx act to create high-frequency vibrations for ultrasonic sound production. Structures that may possibly act as sound generators are the *vocal membranes*, found only in the larynx of echolocating bats. These vocal membranes are only a few micrometers thick and contain no muscle fibers. They are located along the rims of the vocal cords. Other structures that may function as vibrating membranes are the thin ventricular folds, or plicae ventriculares, and the aryepiglottic fold, or plica aryepiglottica, of the laryngeal ventricle. Experiments in *Eptesicus fuscus* have shown that cuts in the vocal membrane render the bat practically mute, whereas cuts in the ventricular folds have little effect. This suggests that the vocal membrane is an important structure for sound production.

When a bat vocalizes, the glottis is partially opened to form a small slit. The expired air rushes through this slit at speeds up to 100 m/s, into the ventricle, or upper chamber of the larynx. The speed of this stream of air is controlled by the subglottic pressure. As the air passes over the vocal membranes, it causes them to vibrate, and this is probably the origin of the sound. In *Eptesicus fuscus*, the cricothyroid muscle, which stretches the vocal membranes, also plays a role in sound production. When the thyroid cartilage is tilted beyond a certain angle, a horizontal component of its movement causes the glottis to open enough to form a small slit.

Modifications. The emitted echolocation sound contains not just one frequency of oscillation; it also contains integer multiples, or harmonics, of the fundamental frequency. For example, the echolocation call of *Pteronotus parnellii* is composed of four harmonics at 30.5 kHz, 61.0 kHz, 91.5 kHz, and 122 kHz. The geometry and volume of the throat, mouth, and nasal cavities determine which of the harmonics are loud and which are low in intensity. The cavities through which the sound travels act as *passive resonators*, intensifying some frequency components and attenuating others. In bats that emit echolocation calls through the mouth, the influence of the supraglottic air passages on the composition of the signal is small. The sound spectrum of the emitted sound is approximately the same as that produced by the larynx.

The situation is different for bats such as rhinolophids and hipposiderids that emit sound through their nostrils. In these species, the epiglottis fits precisely in the nasolaryngeal opening to create an airway from the larynx to the nasal cavity that can be isolated from the oral cavity. This nasolaryngeal tract contains additional air chambers (fig. 6.6) that could be viewed as resonators. Experiments on the filling of the air chambers in *Rhinolophus hildebrandti* have shown that the *tracheal chambers* (fig. 6.6) do function as resonators: they attenuate the first harmonic of the echolocation call (20 kHz; fig. 6.6). The role of the *nasal chambers* is not fully understood. They appear to attenuate the first as well as the fourth har-

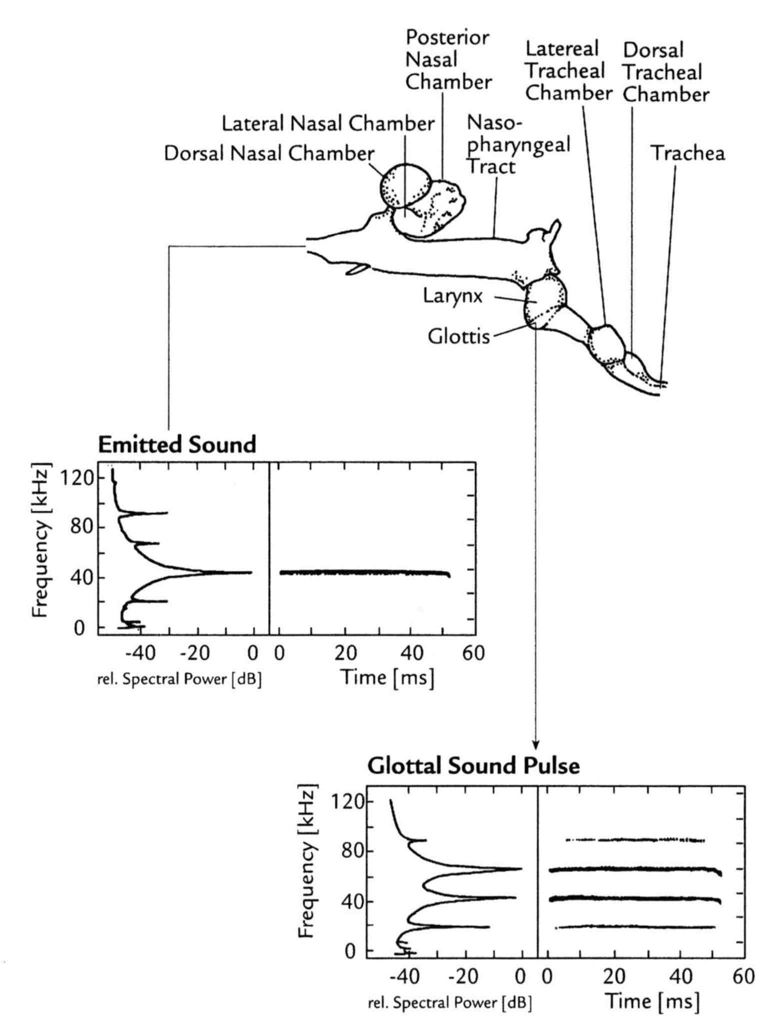

Figure 6.6 The nasolaryngeal tract in the horseshoe bat, *Rhinolophus hildebrandti*. The comparison of the spectrum (left panels) and sonagram (right panels) of the signal produced in the larynx (glottal sound pulse) with that emitted through the nostrils (emitted sound) demonstrates how the harmonic structure of the echolocation signal has been transformed by the nasopharyngeal tract and its air chambers. From Suthers et al. (1988) and Hartley and Suthers (1990).

monics. In any case, the acoustic properties of the nasolaryngeal tract alter the properties of the echolocation call in such a way that the "classic" echolocation call of horseshoe bats has an intense second harmonic and a highly attenuated first harmonic. If a horseshoe bat is forced to vocalize through the mouth, the most intense harmonic is the first, not the second.

Signal frequency. The frequency of vocalization depends on the amount of tension on the vocal cords. There are two fundamental mechanisms whereby mammals can alter the frequency of vocalization:

- Contraction of the vocalis muscle, which originates on the rear surface of the thyroid cartilage and runs within the vocal folds to the arytenoid cartilage. Contraction of this muscle produces tension on the vocal folds, thus influencing their frequency of vibration.
- Tilting of the thyroid cartilage. Tilting of this structure causes the distance between the thyroid cartilage and the arytenoid cartilage to lengthen. Because the vocal folds are stretched between the thyroid cartilage and the arytenoid cartilage, this movement also produces tension on the vocal cords (fig. 6.4).

Because bats lack the vocalis muscle, the only way they can control the tension of the nonmuscular vocal membrane, and hence the frequency of vocalization, is through tilting the thyroid cartilage. The *cricothyroid muscles*, which extend from the thyroid cartilage to the cricoid cartilage and serve to tilt the thyroid cartilage (fig. 6.4b), are appropriately large. It has been demonstrated experimentally in horseshoe bats that contraction of the cricothyroid muscles causes the emitted frequency of vocalization to increase. The cricothyroid muscles remain contracted until the end of the echolocation call. *Eptesicus fuscus* emits downward frequency-modulated signals with a first harmonic in the range between 33–18 kHz. The cricothyroid muscle contracts about 12 ms before the onset of vocalization and relaxes during vocalization. Thus, the decreasing tension on the vocal membrane results in a downward frequency modulation. If the cricothyroid muscle is completely denervated, the emitted frequency decreases so that it lies around 7.9 kHz.

NEURAL CONTROL OF VOCALIZATION

The various types of echolocation calls shown in figure 6.3 all result from precisely timed activity in the cricothyroid muscles and the inner laryngeal muscles that open and close the glottis. Pye has developed a model to illustrate the mechanisms that produce the different types of calls (fig. 6.7). Assuming that the cricothyroid muscles produce a rhythmic pattern of contraction and relaxation, a periodic, trapezoidal pattern of frequency modulation results. By opening and closing the glottis at appropriate times, the bat can select specific portions of this pattern for emission. Theoretically, the different types of signals could be produced by sliding an amplitude envelope over the periodic pattern on frequency modulations so that it produces a temporal window in which sound is emitted (fig. 6.7). If this is the method actually used by bats, the production of vocalizations is reduced to the problem of fine temporal control by the motor neurons that innervate the arytenoid muscles and thus control the onset and offset of sound, and the outer laryngeal muscles that control sound frequency.

In all mammals, the laryngeal muscles are innervated by branches of the vagus nerve (cranial nerve X). The *superior laryngeal nerve* innervates the

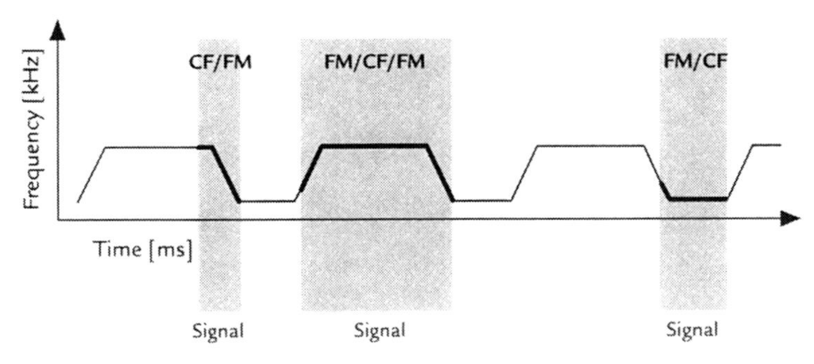

Figure 6.7 Model of sound production developed by Pye. A simulated continuous rhythmic pattern of contraction and relaxation of the muscles that regulate vocalization frequency (trapezoidal curve) results in a corresponding change in vocalization frequency. Depending on when and for how long the glottis is opened and closed, portions of the trapezoidal waveforms can be isolated and used to create sounds with different patterns (thick lines and shaded portions of the curve). All of the different types of echolocation calls used by bats can be produced through the appropriate temporal pattern of activity in the muscles that control the frequency of vocalization and the opening and closing of the glottis (compare with fig. 6.3). CF, constant-frequency component; FM, frequency-modulated component.

cricothyroid muscles and therefore controls frequency modulations. The *inferior laryngeal nerve* supplies the arytenoid muscles and therefore controls the amplitude envelope. The cell bodies of the motor neurons that give rise to both laryngeal nerves lie in the brainstem, in the *nucleus ambiguus* (fig. 6.8). The frequency and the amplitude envelope of vocalization are controlled by two separate populations of neurons. The neurons of the *superior laryngeal nerve* that are responsible for regulating frequency lie in the ventral division of the nucleus ambiguus (shaded area in fig. 6.8). The neurons of the *inferior laryngeal nerve* that are responsible for regulating the amplitude envelope are located in the caudal division of the nucleus ambiguus. Recordings from the fibers of the inferior laryngeal nerve and the cells of the nucleus ambiguus in *Rhinolophus* show that neural activity is precisely correlated with the onset, duration, and offset of emitted echolocation pulses. The activity of the superior laryngeal nerve is also correlated with vocalization. Horseshoe bats can precisely maintain the pure tone component of their echolocation signal within 30 Hz of a set frequency that varies somewhat from one individual to another, but is within a range of 4 kHz (79–83 kHz in fig. 6.8c). Summated activity in the superior laryngeal nerve is linearly proportional to the frequency of the emitted pure tone. The nucleus ambiguus also contains neurons whose activity varies as a function of the frequency of the emitted tone (fig. 6.8).

Vocalization is initiated by activity in the forebrain. It has been shown that in *Pteronotus parnellii*, electrical stimulation of a small region within the rostral cin-

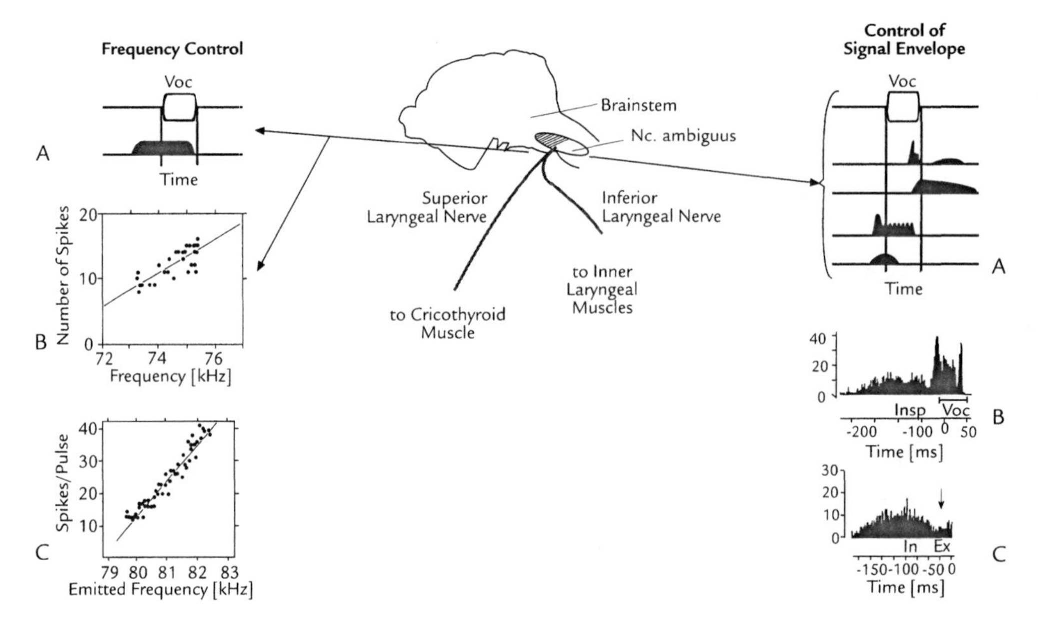

Figure 6.8 Neural control of the frequency and amplitude envelope of vocalization in horseshoe bats through the nucleus ambiguus of the hindbrain and the two branches of the laryngeal nerve that originate in this nucleus. In the center is a schematic parasagittal section through the brain. The shaded area indicates the division of the nucleus ambiguus that contains neurons responsible for regulating the frequency of vocalization. The superior laryngeal nerve innervates the cricothyroid muscle and controls the emitted frequency. The inferior laryngeal nerve originates in the caudal part of the nucleus ambiguus and innervates the muscles that open and close the glottis, thereby controlling the amplitude envelope of vocalization. *Left panel*: Neural control of frequency. (A) the black bar on the lower trace shows the temporal correlation of the activity of a neuron in the nucleus ambiguus with the onset and end of an echolocation pulse (Voc; upper trace). (B) correlation between activity (spike counts) in a neuron in the rostral division of the nucleus ambiguus (shaded region in center diagram) and the frequency of the emitted vocalization. (C) correlation between activity in the superior laryngeal nerve and the frequency of the emitted vocalization. *Right panel*: Neural control of amplitude envelope. (A) timing of the activity of different classes of neurons in the caudal division of the nucleus ambiguus (unshaded area of the nucleus ambiguus in center diagram) in relation to the onset and offset of an echolocation pulse (Voc). (B) summated activity in the inferior laryngeal nerve during inspiration (Insp) and emission of an echolocation pulse (Voc). (C) summated activity in the inferior laryngeal nerve during expiration with no vocalization (Ex). From Rübsamen and Schuller (1981), Rübsamen and Betz (1986), Schuller and Rübsamen (1981), and Schweizer et al. (1981).

gulate cortex elicits vocalizations. This cortical area is tonotopically organized. A stimulus in the anteroventral part elicits a signal at 58 kHz and, as the stimulus is moved progressively farther in the posterodorsal direction, the emitted signal frequency increases up to a maximum of 60 kHz. This cingulate area has connections with the periaqueductal gray in the midbrain, a region that also elicits vocalization when stimulated electrically.

DIRECTIONAL BEAMING

A point source of sound sends out energy in the form of a spherical wave. As a first approximation, the larynx can be considered as a point source of sound. However, sound is not emitted directly from the larynx, but rather is funneled through the open mouth or the nostrils. Thus, most of the sound energy is concentrated in the direction in which the bat is flying, creating a cone-shaped directional beam. The energy contained in this beam falls off rapidly on either side. The higher the frequency, the narrower the beam. One measure of the *diameter* of the sound beam is the angle (relative to the direction of flight) at which the sound pressure has decreased by half—that is, by 6 dB.

In bats that emit calls through the mouth, such as *Myotis grisescens*, the width of the sound beam for a signal at 75 kHz is 40° on either side in the horizontal plane. In the vertical plane, the beam extends upward 25° and downward 55°. At higher signal frequencies, the sound beam becomes narrower, and side bands, called secondary beams, may appear. In bats that emit calls through the nostrils, the sound beam is usually broader in the vertical plane than in the horizontal and is slightly tilted in the downward direction. In *Rhinolophus ferrumequinum*, the sound beam for a signal at 84 kHz extends out 25° on either side in the horizontal plane. In the vertical plane, it extends upward 25° and downward 70°.

In bats that emit calls through their nose, each nostril acts as a point source of sound and, therefore, as the point of origin of a spherical wave. Thus, the two spherical waves interact with one another to produce interference patterns, the exact nature of which depends on the distance between the two nostrils. If the distance between the two nostrils is half of the wavelength of the emitted frequency, the sound energy at points straight ahead will be intensified, but the energy on either side will be attenuated, resulting in a narrowing of the sound beam. Experiments in the phyllostomid bat *Carollia perspicillata* have shown that interference between the spherical waves originating in the two nostrils cause the azimuthal sound beam to narrow from 120° to 50–60° (fig. 6.9). The vertical, lancet-like portion of the nose leaf has a similar effect on the spread of sound in the vertical dimension. If the nose-leaf is removed, the sound beam broadens in the vertical plane from 50–60° to 120°. Thus, the nose leaves help determine the form of the sound beam and are clearly important components of the system for production of echolocation signals.

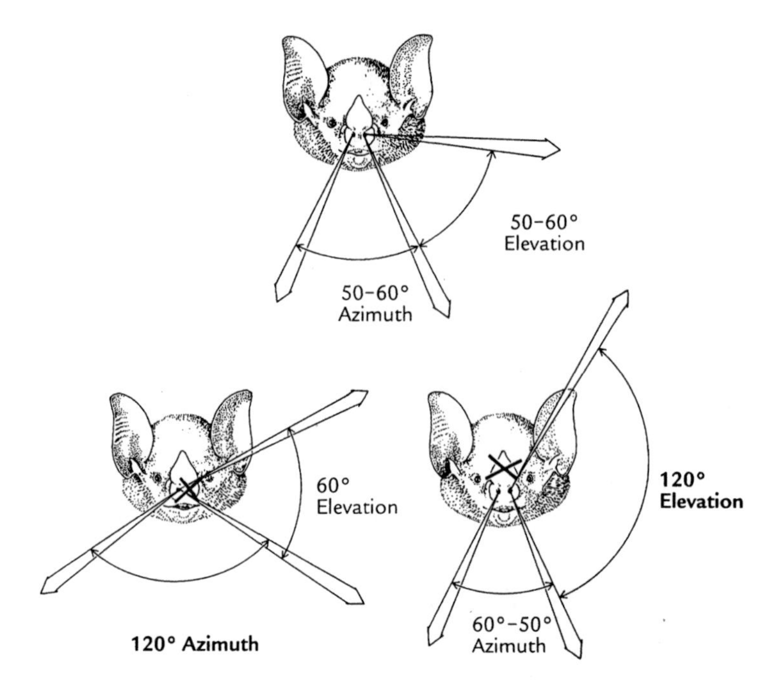

Figure 6.9 The width of the sound beam and its dependence on the nose-leaf in *Carollia perspicillata* (Phyllostomidae). *Top*: Under normal conditions the sound beam is approximately 50–60° wide in both horizontal and vertical dimensions. *Lower left*: If one of the two nostrils is blocked (X), the width of the sound beam doubles in the horizontal plane. *Lower right*: If the upper, lancet-shaped portion of the nose-leaf is removed (X), the extent of the sound beam doubles in the vertical plane. From Hartley and Suthers (1987).

6.3 THE AUDITORY SYSTEM

The ear acts as a receiver for echoes (fig. 6.10). It is made up of the outer and middle ear, which transmit sound to the inner ear, the actual receptor organ. Sound-evoked excitation is relayed via the fibers of the auditory nerve to the ascending pathways of the central auditory system in the brain, where echo signals are analyzed in the frequency and time domains and with reference to the emitted echolocation sound.

FUNCTIONAL ANATOMY OF THE EAR

Outer ear. The outer ear of mammals can be divided into the pinna and the ear canal or meatus (fig. 6.10). Sound impinges on the eardrum, or tympanic membrane, which transmits vibrations to the middle ear ossicles. The fact that mammals are the only animals that possess pinnae is indicative of the importance of listening and acoustic communication in their lives.

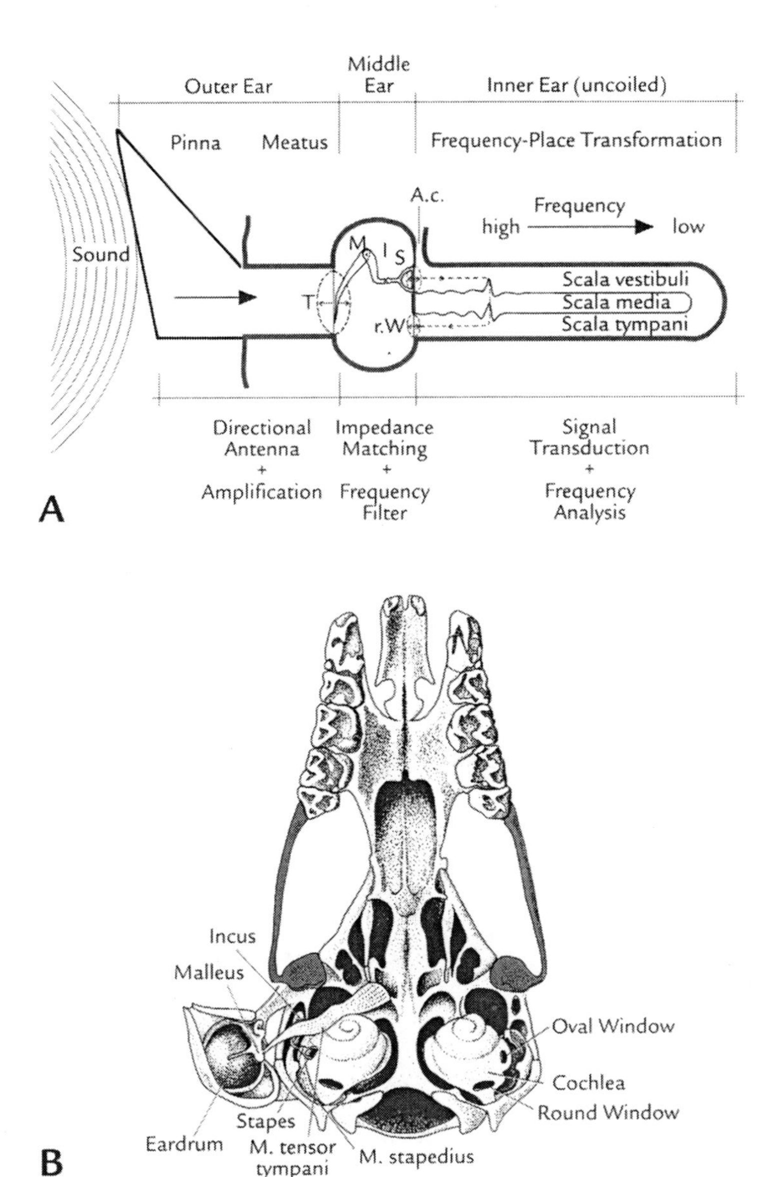

Figure 6.10 The ear. (A) Functional schematic of the mammalian ear. The cochlea, which is normally coiled in a spiral shape, is stretched out here to aid in illustration. Middle ear: T, tympanic membrane (eardrum) with the three ossicles: M, malleus; I, incus; S, stapes. A.c., cochlear aqueduct; r.W., round window. The scala vestibuli and the scala tympani are filled with perilymph, and the scala media with endolymph. The dashed lines and arrows indicate the direction in which sound energy travels through the cochlea. High-frequency traveling waves die out near the basal end of the cochlea, whereas low frequency traveling waves are able to reach the apex of the cochlea. (B) The ear of an echolocating bat *in situ*, viewed from below. The tympanic membrane and malleus are reflected to the left to provide a better view. The middle ear on the right side is removed. Modified from Stanek (1933).

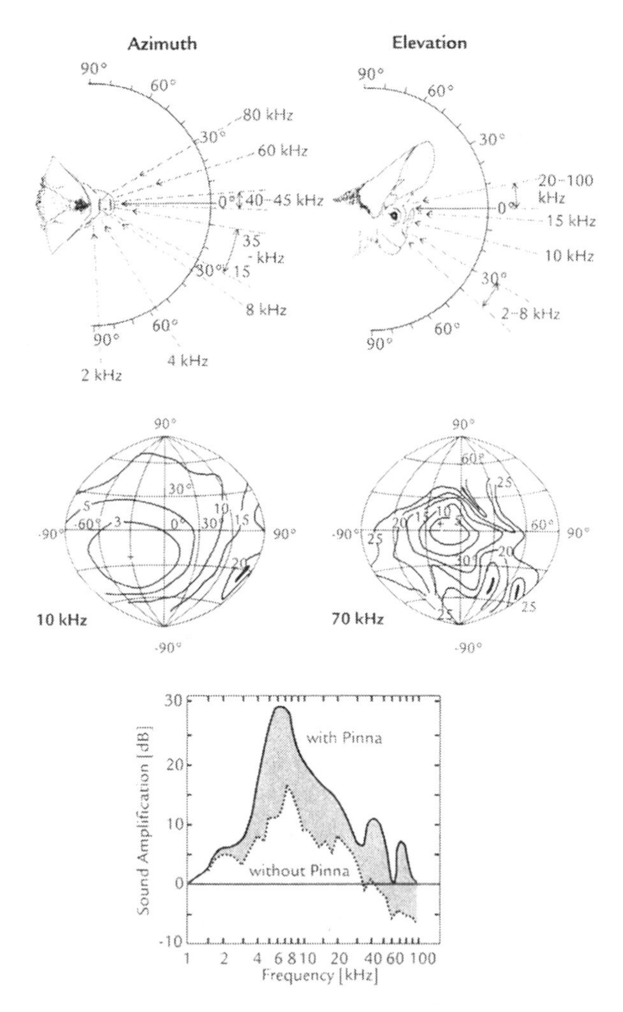

Figure 6.11 The function of the outer ear using *Macroderma gigas* (*Megadermatidae*) as an example. *Top*: The acoustic axis, i.e., the direction of optimal transmission of sound to the eardrum (dashed lines with arrows), changes in both azimuth and elevation as a function of sound frequency (kHz). *Middle*: The directional characteristics of the left pinna, measured at the eardrum, for sound at frequencies of 10 kHz and 70 kHz. Negative angular values indicate directions in the left azimuthal hemifield and in the lower elevational hemifield. Crosses indicate the acoustic axis for each frequency. Isointensity lines indicate the decrease in sound pressure in decibels compared to the sound pressure measured at the acoustic axis. Black areas indicate the regions of minimal sensitivity (nulls). The directional characteristic of the ear is more pronounced for higher frequencies. *Bottom*: Amplification by the outer ear. Solid curve: Amplification by the pinna and meatus as a function of frequency, measured as the difference in sound pressure between that measured at the eardrum and that measured in the free field. Dotted curve: Amplification with pinna removed. Shaded area = amplification due to the effect of the pinna. From Guppy and Coles (1988).

Box 6.2 The Pinna as a Directional Antenna

The most important parameter in determining the directional characteristics of the pinna is the relationship between the size of the pinna opening and the wavelength of sound:

$$a/\lambda \text{ or } ka$$

where $k = 2\pi/\lambda$; the wave number; a = radius of the pinna opening; and λ = wavelength.

The pinna functions as a directional antenna when $ka > 1.25$, that is, when the wavelength is smaller than 4–5 times the radius of the pinna opening. For example, in the big-eared bat, *Plecotus auritus*, the cutoff frequency is 6–8 kHz. At shorter wavelengths, the pinna creates a directional characteristic with a frequency-dependent *acoustic* axis [the direction of maximal sound conduction relative to a reference point of zero at the opening of the meatus (fig. 6.11)] and a frequency-dependent *slope* (the angular sound direction at which sound transmission is reduced by 3 dB compared to the acoustic axis). The higher the frequency, the more pronounced the directional characteristic of the pinna.

The *pinnae*, which can be tilted and rotated, act as movable directional antennae. The pinnae create a characteristic directionality in the auditory field (fig. 6.11). The *directional characteristics* of the pinna are due to sound scattering and therefore depend on the size and geometry of the pinna as well as the wavelength of sound (see box 6.2). At certain frequencies, sound reflections from the cone-shaped interior of the pinna simultaneously cancel each other, resulting in directional nulls. These *directional nulls* enhance the directional gradients within the sound field. The pinna is no longer able to exert its directionality when the wavelength of sound exceeds the dimensions of the pinna.

Acoustically, the pinna can be thought of as a horn with a slanted, elliptical opening. The amplification provided by such a horn depends on its length (l), the area of its opening, and the area of its termination relative to the wavelength of sound. The wavelength should be less than $2\pi l$. The larger the pinna, the better low frequencies will be transmitted to the eardrum. Amplification is further increased due to the resonance properties of the ear canal, which can be thought of in acoustic terms as a tube closed at one end by the tympanic membrane. The wavelength of the resonant frequency of such a tube is equal to four times the length of the meatus.

The different families of bats are characterized by highly variable pinna forms and sizes (fig. 6.12). Presumably the different types of pinnae have correspondingly diverse acoustic properties. The pinnae of most echolocating bats are not conspicuously large because they are tuned to ultrasound frequencies with short wavelengths between 2 and 16 mm. In contrast, species that detect prey on the ground (gleaning bats) have gigantic pinnae. In megadermatid bats and in the European long-eared bat, the greatest pinna amplification (ca. 25 dB) is in the range

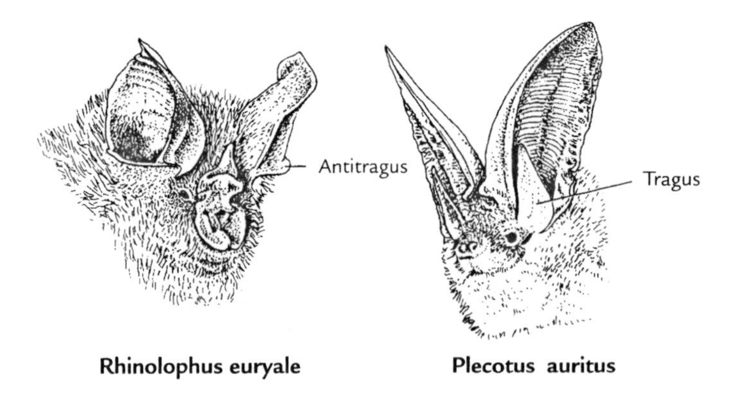

Figure 6.12 Typical pinna forms in echolocating bats. The tragus contributes to localization of a sound source in elevation. The function of the antitragus is unknown.

of 6–10 kHz, below the range of frequencies used for echolocation (fig. 6.11). This is presumably an adaptation for the detection of prey that make rustling noises while moving about.

The acoustics of the outer ear are more complicated than is generally described because sound takes paths of different lengths depending on the conditions imposed by the structure of the pinna. Most species of bats also have a stiff fold of skin just in front of the opening to the pinna, the *tragus* (fig. 6.12). Exceptions to this rule are the molossids, in which the tragus is poorly developed, and the horseshoe bats and hipposiderids, where it is completely absent. Species that lack a tragus have instead a horizontal fold of skin along the opening of the ear canal, called the *antitragus* (fig. 6.12). The tragus is involved in vertical localization of sound.

Middle ear. The middle ear of mammals consists of the air-filled tympanic cavity that is bounded on one side by the eardrum, or tympanum, and on the other by the oval window. Three small bones, or ossicles, span the length of the tympanic cavity. These are the malleus, the incus, and the stapes (figs. 6.10 and 6.13). Sound energy sets the tympanum in motion, and this motion is transmitted via the lever action of the ossicles to the oval window and thence to the inner ear. This chain of transmission has three properties that ensure that the amplitude and speed of vibration is increased so that when sound waves pass from the "low impedance" medium of the air to the "high impedance" medium of the perilymph in the inner ear, little energy is lost. This process is referred to as *impedance matching*. Sound energy from the relatively large area of the tympanic membrane converges on the small area of the oval window, thus increasing the sound pressure. The larger the area of the tympanum relative to the oval window, the greater the resulting amplification. The ratio of the area of the tympanum to that of the oval window is 16:1 in horseshoe bats and 53:1 in *Tadaridae*. For comparison, the ratio in the cat is 35:1. The lever arm of the malleus is longer than that of the incus. This causes

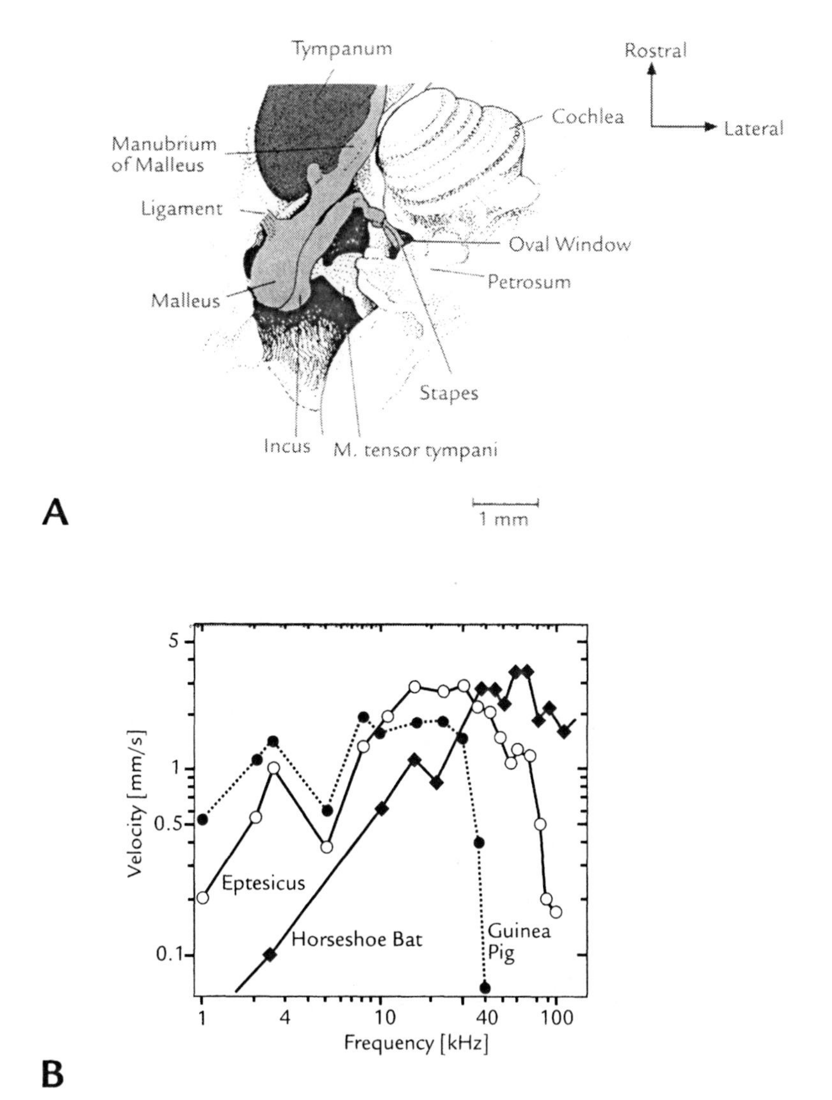

Figure 6.13 The middle ear of the bat. (A) The middle ear of *Trachops cirrhosus* (Phyllostomidae) *in situ.* (B) The frequency-filtering function of the middle ear in the bats *Eptesicus pumilis* and *Rhinolophus ferrumequinum* (horseshoe bat) and, for comparison, in the guinea pig. The measurements represent the speed of displacement of the tympanic membrane in response to acoustic stimuli at 100 dB SPL as a function of frequency. From Bruns et al. (1989), Manley et al. (1972), and Wilson and Bruns (1983).

the stapes to move more slowly, but with greater force. The lever ratio in bats is 3–5:1, but in cats it is only 1.15:1. A third factor can exert an influence when the conical form of the tympanum causes the amplitude of the movements of the incus to be smaller than those of the tympanum itself.

The values of the different amplification factors listed here can only be approximations because the amount of amplification, like many other parameters, depends on sound frequency. Nevertheless, measurements have shown that the impedance matching provided by the middle ear allows approximately two-thirds of the sound energy received by the outer ear to be transmitted to the inner ear.

The middle ear also functions as a *frequency filter*. The smaller the mass and the stiffer the ossicles that transmit the sound, the higher the resonance frequency. Indeed, echolocating bats that listen to high-frequency sounds do have a middle ear with low mass and a stiff ossicular chain. The ossicles are thin and linked together by deep interlocking grooves that increase the stiffness of the system. The head of the stapes pushes against the edge of the oval window, preventing it from becoming distended. Instead, it moves around the outer axis like a tilting window.

There are other, additional specializations that make the middle ear of bats especially well suited for transmission of high frequencies. The eardrum is small, 1.2–11 mm^2, compared to 15 mm^2 in the guinea pig. It is also exceptionally thin (2–11 μm). The volume of the middle ear cavity is also small, thereby reducing the amount of air that is set in vibration. The result of all of these adaptations is that the ears of bats transmit high frequencies much more efficiently than low frequencies (fig. 6.13). The displacement of the tympanum due to a sound at an amplitude of 100 dB SPL and a frequency of 2.5 kHz is 0.0656 μm. A sound of the same amplitude, but at a frequency of 100 kHz, causes a displacement of only 0.00029 μm.

All mammals possess two small *middle ear muscles*, the tensor tympani and the stapedius (figs. 6.10 and 6.13). When these muscles contract, they cause the ossicular chain of the middle ear to stiffen, thereby attenuating the transmission of low frequencies. The muscles start to contract as soon as sound emission begins and are thought to attenuate the reception of the bat's own vocalization. The results of experiments designed to test this hypothesis in bats are contradictory; therefore, the function of the middle ear muscles in echolocating bats remains unclear.

The inner ear. The inner ear of mammals is a micromechanical wonder that not only transforms movements on the order of nanometers into neural activity, but also resolves frequencies with an accuracy of one-tenth of 1%. The function of this highly complicated sound receiver and frequency analyzer can be illustrated by a simplified model (figs. 6.10 and 6.14). The first component of the model is a long tube with a wedge-shaped cross section, the scala media. The tube is closed at both ends, and filled with a fluid, the endolymph, which is rich in K^+ ions. This first tube is covered by a second tube which curves around at the apical end forming the helicotrema and continues on the bottom side back to the basal end. This second tube contains another fluid, the perilymph, which is rich in Na^+ ions. The upper perilymph-filled space, called the scala vestibuli, and the

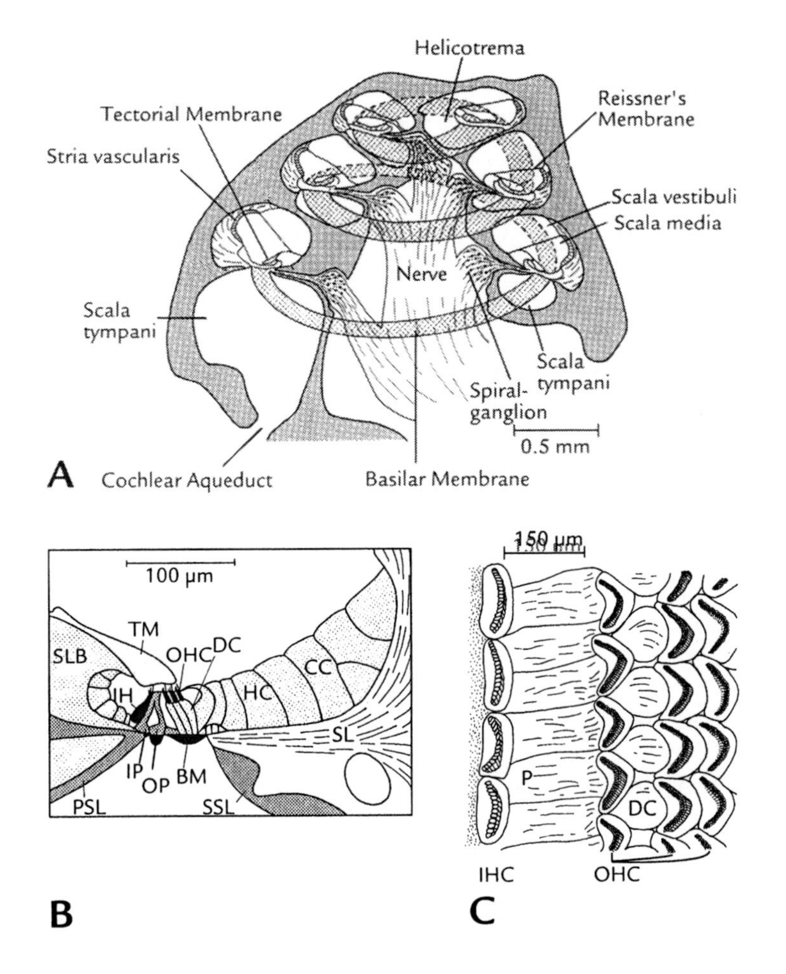

Figure 6.14 The inner ear of a bat. (A) Schematic cross-section through the cochlea of *Trachops cirrhosus*, Phyllostomidae. (B) Schematic cross-section through the receptor organ (scala media) in *Hipposideros speoris*. TM, tectorial membrane; BM, basilar membrane; PSL, primary spiral lamina; SSL, secondary spiral lamina; SLB, spiral limbus; ICH, inner hair cells; OHC, outer hair cells; DC, Deiter's supporting cells; HC, Hensen's cells; CC, Claudius' cells; IP, inner pillar cell; OP, outer pillar cell; SL, spiral ligament. (C) Schematic view of the hair cells in the cochlea of the horseshoe bat (*Rhinolophus ferrumequinum*) as seen from above. Abbreviations as for panel B. From Bruns et al. (1989) and Dannhof and Bruns (1991).

lower perilymph-filled space, called the scala tympani, surround the highly sensitive scala media, forming a sort of sandwich. Both of the perilymph-filled spaces are separated from the middle ear by elastic membranes, the oval and round windows. The footplate of the stapes contacts the membrane that covers the oval window. When the stapes moves and presses against the oval window, the pressure in the scala vestibuli increases. The momentary pressure difference between the scala vestibuli and scala tympani causes a traveling wave to arise in the scala media. The wave will ultimately be dissipated through a compensatory movement of the round window, the membrane that closes off the end of the scala tympani.

The three fluid-filled tubes of the inner ear are embedded in a bony capsule. Because of space constraints, this whole structure, the *cochlea*, is wound around in a spiral, like a snail shell. In bats, the cochlea usually consists of 2.5–3 turns. The most basal turn has by far the largest diameter (figs. 6.10, 6.13, and 6.14).

The scala media is covered over by a thin, ion-impermeable membrane, called Reissner's membrane. The floor of the scala media is formed by the basilar membrane. As can be seen from table 6.1, the basilar membrane is especially long in bats (7–15 mm). The elastic basilar membrane supports the sensory epithelium, which is made up of hair cells (fig. 6.14b). From the base to the apex of the cochlea, the hair cells are arranged like fence posts in closely packed rows. There is one inner row and three outer rows of hair cells. *Trachops cirrhosus* (Phyllostomidae), for example, has 1590 inner hair cells (110/mm basilar membrane length) and about 5720 outer hair cells (395/mm). The apical membranes of the hair cells, together with those of several types of supporting cells, form a stiff plate, the cuticular plate (fig. 6.14c). Only the short mobile hairs of the sensory cells, the *stereocilia*, extend above the cuticular plate (fig. 6.14b). Each hair cell possesses a bundle of stereocilia. The relatively short stereocilia of the outer hair cells become progressively longer going from the base to the apex of the cochlea. In the horseshoe bat *Rhinolophus ferrumequinum*, for example, the outer hair cell stereocilia at the base of the cochlea measure 0.8 μm, but at the apex they are 1.5 μm long, nearly double the basal length. In contrast, the inner hair cell stereocilia are 4.3 μm long at the base of the cochlea and hardly increase in length even at the apex.

Displacement, or shearing, of the hair cell cilia gives rise to a receptor potential. If a hair cell bundle is displaced in the direction of its longest cilium, the membrane potential becomes depolarized. A displacement in the opposite direction leads to hyperpolarization. The inner ear of mammals contains a jellylike membrane that extends from the inner concavity of Reissner's membrane to the interior of the scala media and covers the stereocilia tips of the outer hair cells. Vibration of the basilar membrane against the tectorial membrane (fig. 6.14b) causes shearing of the hair cell stereocilia and, consequently, oscillations in the receptor potential that follow the rhythm of basilar membrane movement. If an electrode is placed on the round window, it is possible to record the *cochlear microphonic potential*, which represents the summated receptor potential generated by the hair cells (fig. 6.15).

Table 6.1 Length of the basilar membrane (BM) and density of innervation (n/mm) by neurons of the spiral ganglion (SG neurons) in different species of bats (data from the house mouse are shown for comparison)

	BM length (mm)	BM length/bw (mm/g)	SG neurons Avg. no.	N/mm
Hipposideros bicolor	8.8	1.51	13,500	1,530
Hipposideros speoris	9.9	1.32	15,000	1,510
Rhinolophus ferrumequinum	16.1	0.89	16,000	990
Pteronotus parnellii	14.3	1.19	30,000	3,500
Rhinopoma hardwickei	11.8	0.81	18,000	1,590
Megaderma lyra	9.9	0.28	17,500	1,770
Myotis lucifugus	6.9	0.19	55,000	7,970
Eptesicus serotinus	8.9	?	?	?
Nyctalus noctula	11.8	0.44	?	?
Trachops cirrhosus	14.5	0.33	40,500	2,790
Taphozous kachhensis	12.1	0.25	23,000	1,900
Molossus ater	14.6	0.39	31,800	2,180
House mouse	6.8	?	10,500	1,544

From Bruns et al. (1989).

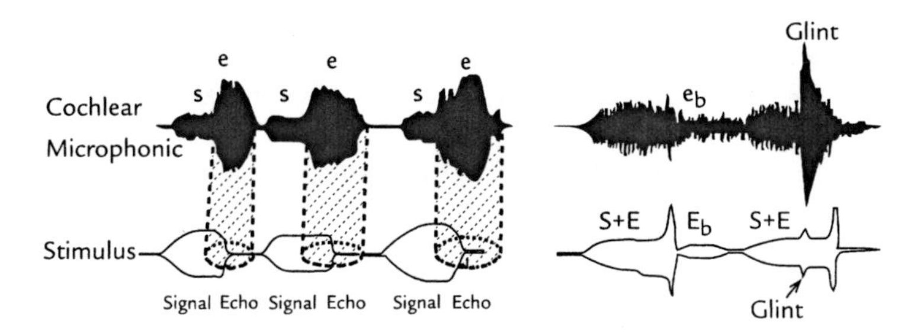

Figure 6.15 Summated inner ear response to echolocation pulses and echoes in the mustached bat, *Pteronotus parnellii* (Mormoopidae), recorded as the cochlear microphonic (black trace). The cochlear microphonic primarily reflects the activity of the outer hair cells. Left panel: Cochlear microphonic in a nonflying animal. The echolocation pulses (stimulus) elicit smaller microphonic potentials (s) than do the overlapping echoes (e). Right: Cochlear microphonic potentials recorded from a flying mustached bat in pursuit of a moth. The echolocation pulses and the echoes from the prey overlap (S + E), and the echoes from background objects (E_b) are visible during the intervals between vocalizations. Because the emitted pulse and the echo have different frequencies due to the Doppler effect, interference patterns result. These interference patterns can be seen as amplitude modulations in the cochlear microphonic. If the echolocation pulse is reflected from an insect with beating wings, the echo contains a glint that causes an especially large response in the cochlear microphonic. From Henson et al. (1982, 1987).

After acoustic information has been converted to receptor potentials in the hair cells, this information is transmitted to the brain by the fibers of the auditory nerve. Paradoxically, more than 90% of the auditory nerve fibers are devoted to innervating the single row of inner hair cells, even though the outer hair cells are three times as numerous. In horseshoe bats, for example, the inner hair cells are innervated by 16,000 nerve fibers. This means that 9–24 fibers contact a single inner hair cell, whereas each fiber that crosses to innervate the outer hair cells contacts multiple hair cells in all three rows. Thus, the single row of inner hair cells is largely responsible for transmitting information about the acoustic environment to the brain.

This pattern of innervation raises the question of what the outer hair cells might be good for. Recent research on isolated outer hair cells has resolved this enigma by showing that outer hair cells not only act as mechano-electrical transducers as do the inner hair cells, transforming mechanical deflections of their stereocilia into a receptor potential, but they also act as electro-mechanical transducers which respond to their own receptor potential by changing their mechanical properties. In outer hair cells, the receptor potential induces an intricate, submembranous system of molecular motors to contract. In isolated outer hair cells, these molecular changes result in a shortening of the hair cell body. In the cochlea, however, the outer hair cells are tightly anchored to the basilar membrane and therefore do not move. Instead, they exert mechanical force on the basilar membrane. The force of the outer hair cells reduces the viscous damping of fluid on the basilar membrane. This release from damping increases the amplitude and narrows the extent of the traveling wave envelope (see below). Thus, the outer hair cells act as amplifiers and increase the selectivity of the cochlear frequency filters. At threshold, the peak amplitude of the traveling wave is commonly about 0.3 nm. Without the outer hair cell motors it would be 100-fold less.

The inner ear is not only a sense organ that transduces sound energy into neural activity, but it also analyzes the stimulus to determine its frequency composition. The inner ear is able to exploit the transduction process to perform a simultaneous frequency analysis due to the complex hydromechanical properties of the cochlea. The sound-induced movements transmitted via the stapes to the scala vestibuli cause a momentary pressure difference between the scala vestibuli and the scala tympani, resulting in a "traveling wave" along the basilar membrane (fig. 6.10). The basilar membrane moves up and down at the same frequency as that of the stimulus. This oscillation progresses along the length of the basilar membrane, from one turn of the cochlea to the next, all the while decreasing its speed. At some point the basilar membrane excursion reaches a maximum and quickly diminishes thereafter due to mechanical damping. The traveling wave always begins at the stiffest part of the basilar membrane, at the basal end next to the oval window. When the stimulus is a high-frequency sound, the traveling wave moves only a short distance along the basilar membrane. The lower the frequency, the farther the traveling wave moves within the cochlea, and the more distal the point at which the excursion of the basilar membrane reaches its maximum. Thus, high frequencies activate only the most basal hair cells, and lower frequencies activate the apical hair cells most strongly. In this way, the frequency of a stimulus is trans-

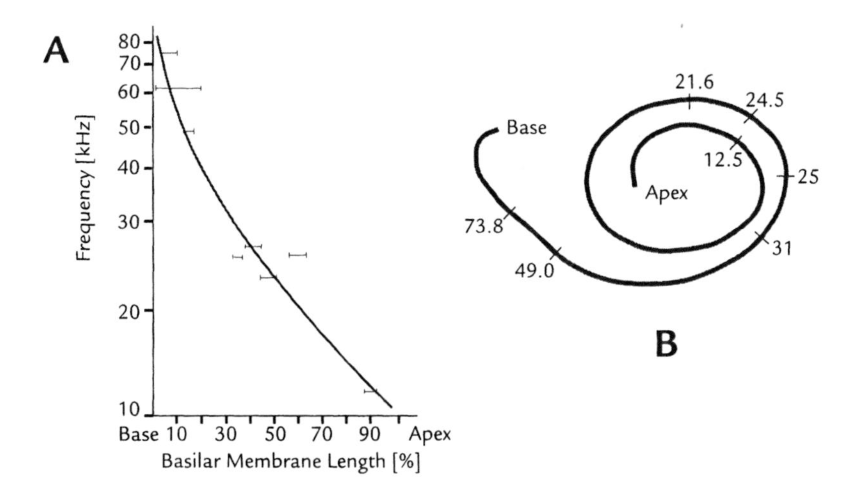

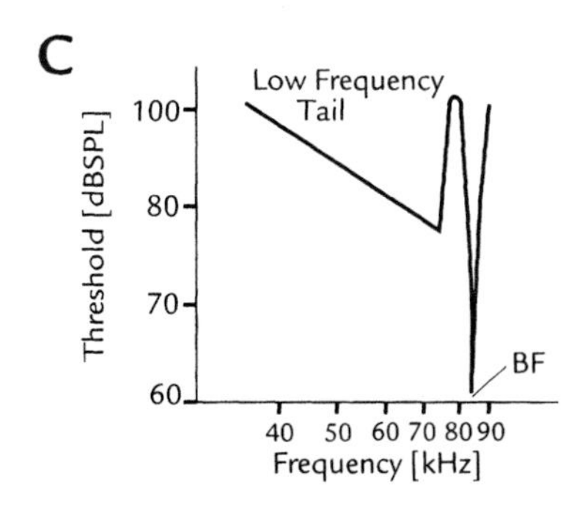

Figure 6.16 Frequency representation in the inner ear. (A) The frequency representation on the basilar membrane of *Tadarida brasiliensis* (*Molossidae*). (B) The sites where selected frequencies are represented (numerals in kHz), shown on a horizontal projection of the basilar membrane. (C) Tuning curve of an auditory nerve fiber in *Rhinolophus ferrumequinum*. BF, best frequency. From Neuweiler and Vater (1977) and Siefer (unpublished).

formed into a specific place of activation on the basilar membrane. According to its position on the basilar membrane, each hair cell is maximally activated by a specific frequency, its characteristic frequency or best frequency (BF) (box 6.3). The result is that each hair cell and the auditory nerve fibers associated with it are responsible for transmitting information about this particular frequency (fig. 6.16).

Several factors contribute to the transformation of sound frequency to a place of maximal excitation on the basilar membrane. The first factor is the mechanical properties of the basilar membrane. At its basal end, the basilar membrane is narrow and thick, making it very stiff. The stiffness of the basilar membrane is proportional to the third power of its thickness and inversely proportional to the fourth power of its width. Over its course toward the apex of the cochlea, the basilar membrane becomes progressively wider and thinner, making it more elastic.

Box 6.3 Best Frequency and "Tuning Curves"

In mechanical terms, the basilar membrane functions as a bank of low-pass filters set to progressively lower frequencies going from the base to the apex of the cochlea. Each auditory nerve fiber, in turn, is tuned to a characteristic frequency, its best frequency (BF), according to its position along the length of the cochlea. The best frequency is defined as the frequency at which the minimal amount of sound energy is required to activate the nerve fiber. As can be seen from the tuning curve of an auditory nerve fiber in *Rhinolophus ferrumequinum*, the threshold for frequencies above and below the BF increases steeply (fig. 6.16). Thus, auditory nerve fibers behave as finely tuned frequency filters. At frequencies below BF, there is a region of insensitivity and maximal threshold, followed by a region of sensitivity for loud, low-frequency sounds. In figure 6.16, this region is found in the range between 75 and 40 kHz. This "low frequency tail" is a typical feature of mammalian auditory tuning curves and reflects the behavior of the traveling wave on the basilar membrane. Because low-frequency traveling waves always traverse the regions where high frequencies are represented, they can excite these regions provided the sound pressure level is high enough.

High-frequency traveling waves oscillate maximally at the stiff basal end of the basilar membrane and then rapidly die out. Lower frequency traveling waves oscillate maximally at the more elastic portions of the basilar membrane. The stiffness gradient along the basilar membrane thus determines the distribution of the oscillatory maxima for the various frequencies. It is the general rule that the stiffness gradient is such that each frequency range of one octave (e.g., 120–240 Hz, 240–480 Hz, 480–960 Hz, etc.) occupies an equal distance along the basilar membrane (fig. 6.16). For example, the bat *Taphozous kachhensis* hears over a range from 85 kHz to <1 kHz. Its basilar membrane is 14.5 mm in length and contains a representation of at least 8 octaves (85.0–42.5 kHz, 42.5–21.5 kHz, 21.5–10.26 kHz, 10.26–5.31 kHz, 5.31–2.66 kHz, 2.66–1.33 kHz, 1.33–0.66 kHz, and 0.66–0.33 kHz) so that each octave occupies a length of 1.8 mm on the basilar membrane. Generally, the space occupied by the lowest and highest audible frequencies is rather compressed. There are some remarkable exceptions to the usual logarithmic frequency representation in the inner ear (see fig. 6.29).

As indicated previously, frequency filtering due to the mechanical gradients of the basilar membrane is greatly enhanced by the outer hair cell amplifiers. The interaction between passive filtering through basilar membrane mechanics and active, localized undamping by the outer hair cells confers on the cochlear frequency filters a high degree of selectivity and a nonlinear pattern of sensitivity.

Even though the contribution of outer hair cells to cochlear frequency analysis is now largely understood, the mechanisms by which local undamping is produced at the site of the envelope peak of the traveling wave are not yet known. Currently it is hypothesized that mechanical gradients in the tectorial membrane might also assist in producing highly selective frequency filtering. Finding conclusive answers is not

Box 6.4 Common Abbreviations Used to Designate Nuclei of the Ascending Auditory Pathway

IC	Inferior colliculus
AVCN	Anteroventral cochlear nucleus
DCN	Dorsal cochlear nucleus
PVCN	Posteroventral cochlear nucleus
MNTB	Medial nucleus of the trapezoid body
DNLL	Dorsal nucleus of the lateral lemniscus
INLL	Intermediate nucleus of the lateral lemniscus
VNLL	Ventral nucleus of the lateral lemniscus
LSO	Lateral superior olive
MSO	Medial superior olive
NCAT	Nucleus of the central acoustic tract

likely to be easy because even the least invasive experimental procedures upset the delicate micromechanical balance of the cochlea and alter the results.

AUDITORY PATHWAYS IN THE CENTRAL NERVOUS SYSTEM

Once the initial frequency analysis has been performed by the cochlea, the resulting information is transmitted to the ascending pathways of the brain. The central auditory system of bats is made up of the same nuclei as that of other mammals, although some of these nuclei are relatively larger than they are in other mammals. At every level of the auditory pathway, the output of the bank of spectral filters in the inner ear is organized into a tonotopic, or cochleotopic, map. The tonotopic organization is based on a laminar arrangement of neurons in which the cells within a given lamina all have a similar best frequency. Thus, each lamina is responsible for processing information from one specific frequency band. In the midbrain, for example, the isofrequency laminae are stacked from dorsal to ventral. Low frequencies are represented dorsally and high frequencies ventrally (fig. 6.17). The tonotopic organization within the brain is related to the frequencies analyzed in the same way a piano keyboard is to the notes used in a musical composition.

The ascending auditory system contains both monaural and binaural pathways that process sound information in parallel to one another. All of these pathways terminate in the *inferior colliculus*, the midbrain auditory center (fig. 6.18). In echolocating bats, the inferior colliculus is so large that it pushes the hindbrain and cerebellum aside, and it is visible on the surface of the brain. In the inferior colliculus, all of the auditory information previously separated into different parallel pathways comes together again. From the inferior colliculus, information is transmitted to the *medial geniculate body* of the thalamus, and from there to the *auditory cortex*.

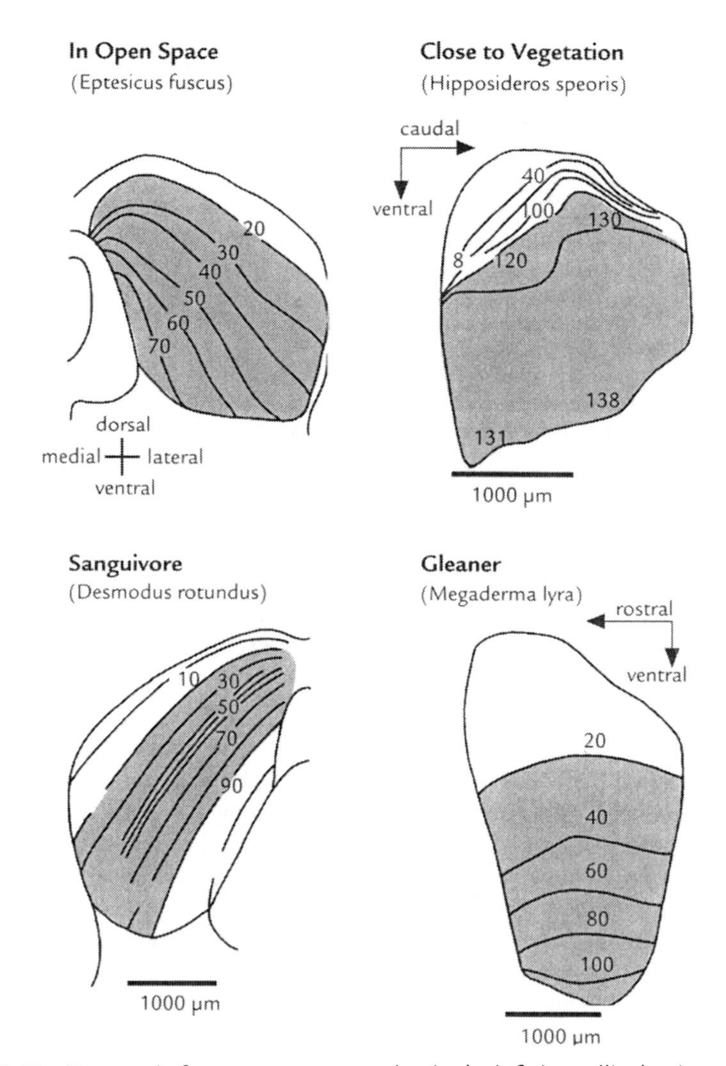

Figure 6.17 Tonotopic frequency representation in the inferior colliculus (transverse section) in bats with different foraging styles. Gray shading indicates frequency range of the bat's echolocation calls. The numerals indicate best frequencies in kHz, the lines indicate isofrequency contours. *Eptesicus fuscus* (Vespertilionidae) emits FM calls in the range 90–20 kHz and forages for insects in open spaces. All the frequency bands within the ultrasound range are rather similar in the size of their representation. *Hipposideros speoris* emits CF-FM calls (CF frequency: 130–134 kHz) and forages for flying insects in the vicinity of vegetation. In this bat, the CF frequency range occupies a greatly expanded representation, the "auditory fovea." *Megaderma lyra* emits short, broad-band clicks (100–20 kHz) and hunts mainly ground-dwelling animals. This bat localizes its prey by passively listening to the rustling sounds they make (10–25 kHz). The true vampire, *Desmodus rotundus*, emits FM sounds (90–40 kHz) and feeds on the blood of mammals (sangiuvore). In these latter two species, there is an expanded representation of low frequencies that correspond to the range of the rustling noises made by their prey. These low frequencies are not contained in the echolocation call. From Pollak and Casseday (1989), Rübsamen et al. (1988), and Schmidt et al. (1991).

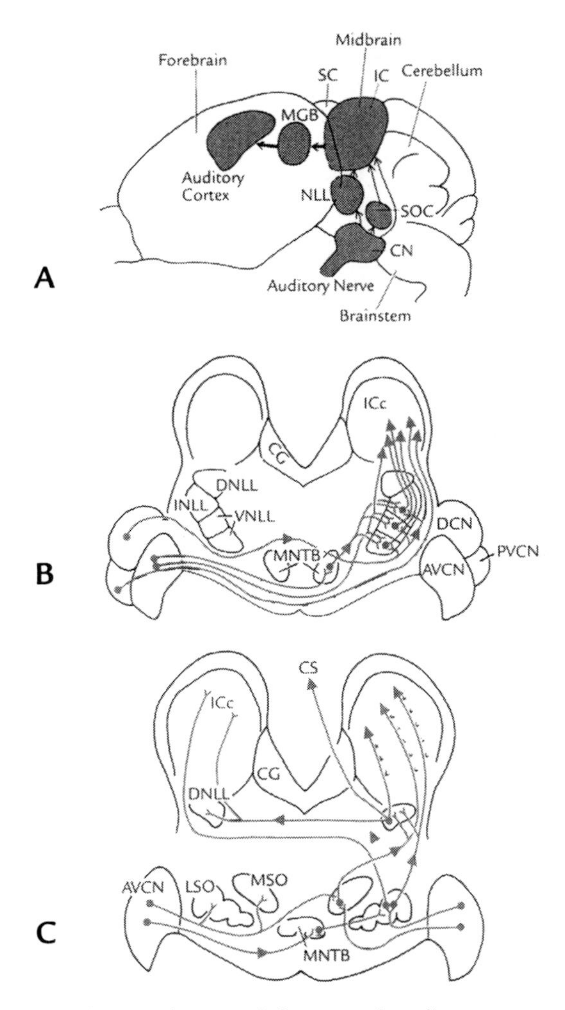

Figure 6.18 The ascending pathways of the central auditory system in bats. (A) Schematic diagram of the ascending auditory pathways. IC, inferior colliculus; SC, superior colliculus (both colliculi are part of the midbrain); MGB, medial geniculate body (thalamus); CN, cochlear nucleus; NLL, nuclei of the lateral lemniscus; SOC, superior olivary complex (CN, NLL, and SOC are auditory brainstem nuclei). (B) The monaural pathways. This system of pathways crosses the midline and ascends on the contralateral side. Here, the pathway crosses from the left cochlear nucleus to the right ICc (central nucleus of the inferior colliculus). In addition to these direct projections, there are also indirect projections from the cochlear nucleus to the ICc, which potentially form delay lines. These indirect projections ascend via the medial nucleus of the trapezoid body (MNTB), intermediate (INLL) and ventral (VNLL) nuclei of the lateral lemniscus. AVCN, anteroventral cochlear nucleus; DCN, dorsal cochlear nucleus; PVCN, posteroventral cochlear nucleus; ICc, central nucleus of the inferior colliculus; CG, central gray. (C) The binaural pathways. LSO, lateral superior olive; DNLL, dorsal nucleus of lateral lemniscus; MSO, medial superior olive. From Pollak and Casseday (1989).

The 15,000–35,000 neurons of the spiral ganglion, located within the turns of the cochlea, receive information from the inner hair cells. Each ganglion cell is responsible for information within a specific frequency band, according to the position it innervates on the basilar membrane, and the frequency-to-place transformation that takes place in the inner ear. The ganglion cell axons assemble to form the auditory nerve. In bats, the auditory nerve is very short. Each frequency-specific auditory nerve fiber splits into three branches, which terminate in the three different subdivisions of the *cochlear nucleus*, located in the hindbrain. These three subdivisions are the anteroventral cochlear nucleus (AVCN), the posteroventral cochlear nucleus (PVCN), and the dorsal cochlear nucleus (DCN). Thus, each subdivision of the cochlear nucleus receives a full set of auditory information. The three subdivisions of the cochlear nucleus give rise to the monaural and binaural pathways of the brainstem.

Monaural pathways. All three subdivisions of the cochlear nucleus give rise to monaural pathways that cross to the contralateral side and project either directly to the inferior colliculus (IC) or to the nuclei of the lateral lemniscus (fig. 6.18b). In echolocating bats the AVCN is unusually large. It contains four distinct types of neurons that can be identified on the basis of their dendritic morphology and the shape of their cell bodies. These four neuronal types are called stellate cells, globular bushy cells, spherical bushy cells, and multipolar cells. The *stellate cells* are the main source of the monaural pathways. Thus, auditory information reaches the IC with only two intervening synapses: the inner hair cell and the AVCN neuron. Auditory information also reaches the IC after one or more delays introduced by the intermediate (INLL) and ventral (VNLL) nuclei of the lateral lemniscus. In echolocating bats the INLL and VNLL are hypertrophied. Both of these nuclei also receive their inputs in graded proportions from the contralateral AVCN and PVCN and project directly to the ipsilateral IC. The INLL and VNLL also receive indirect, delayed, input from the contralateral cochlear nucleus via the medial nucleus of the trapezoid body (MNTB; fig. 6.18). In at least two echolocating bat species, the spherical cells of VNLL are organized in columns. Such a structure has not been found in nonecholocating mammals and therefore may be specific for echolocation. The columnar subdivision of VNLL may function in the analysis of echo timing.

The convergence of multiple, time-delayed inputs from the monaural pathways provides strong evidence that the auditory pathway contains *delay lines*. These delay lines could be the basis for discrimination of sequential auditory events—for example, the time that elapses between the emission of an echolocation call and the return of an echo. Neurons that receive both a direct and a delayed input could act as "coincidence detectors" that would act as a logical "AND" gate and fire only when both inputs reach the cell simultaneously. In a delay chain made up of such neurons, each individual cell would fire in response to a characteristic delay, determined by its position within the chain.

The PVCN, like the AVCN, projects directly to the contralateral IC; it also projects indirectly, with a time delay, via the INLL and VNLL. At present, little is known about the function of this pathway. The PVCN is also the input to the efferent pathways that project back to the inner ear.

The output of the DCN projects directly and without time delays to the contralateral IC. The neural circuitry within the DCN itself is quite complex. It contains bipolar, fusiform cells that receive input from the auditory nerve on the dendrites of one end and input from AVCN neurons on the dendrites of the other end. This circuitry could also be considered as a system of delay lines. In most mammals the DCN is large and organized in layers. This type of organization is thought to indicate a high degree of development. In echolocating bats, however, the DCN is the smallest of the three subdivisions of the cochlear nucleus, and in many species no layering is present. In horseshoe bats, for example, the frequency range that corresponds to the constant frequency part of the echolocation signal is represented in a nonlaminated part of the DCN, and lower frequencies are represented in a thin laminated portion. DCN neurons not only project to the IC, they also project back to the neighboring AVCN and PVCN. Many researchers consider the function of the DCN to be concerned with attention, suggesting that it alerts the auditory system to novel or alarming stimuli.

Binaural pathways. The binaural pathways also originate in the AVCN bilaterally, and include the nuclei of the superior olivary complex and the dorsal nucleus of the lateral lemniscus (DNLL) (fig. 6.18c). In echolocating bats, the lateral superior olive is especially large. This nucleus receives inputs from two sources: one input is from the *spherical bushy cells* of the ipsilateral AVCN and the other input is from the *globular cells* of the contralateral AVCN via the MNTB. The output of the MNTB is glycinergic, thus converting the contralateral input to the LSO into an inhibitory pathway.

Thus, all LSO neurons are excited by ipsilateral input (E_i) and inhibited by contralateral input (I_c). These E_i/I_c neurons are well suited to encode information about the location-dependent interaural intensity differences that are present for high-frequency sounds. Because of the tonotopic organization of LSO, the overall pattern of activity provides a good representation of the difference spectrum between the two ears, another good indicator of sound location. LSO neurons provide excitatory input to the contralateral IC, causing the binaural properties of IC neurons to be reversed, so that they are E_c/I_i. The LSO also projects to the ipsilateral IC. This pathway is thought to be glycinergic and inhibitory. If the inputs from the ipsilateral and contralateral LSO converged on the same neurons, a second comparison of the spectral and intensity differences processed in the LSO might be performed. The LSO also projects to the ipsilateral and contralateral DNLL which, in turn, project bilaterally to the inferior colliculus.

The second binaural pathway ascends via the medial superior olive (MSO) (fig. 6.18c). The MSO receives input directly from the spherical bushy cells of the AVCN on both sides. Many, but not all, MSO cells are excited by input from either side (E/E cells). It is generally assumed that the MSO encodes information about interaural time differences, although this remains to be proven experimentally. The distance between the two ears is very small in most bats, and consequently the interaural time differences experienced rarely exceed 60 μs. For this reason, it has been postulated that bats do not have an MSO and are unable to use binaural time differences for sound localization.

It is true that there are some bat species in which it is difficult to recognize an MSO based on histological criteria. On the other hand, the superior olivary complex does contain one or more cell groups that receive input from the AVCN bilaterally and that project to the ipsilateral IC. The results of recent neuroanatomical and neurophysiological experiments in *Pteronotus parnellii* have called into question the generally accepted concepts about the connections and function of MSO. The first surprise was that most MSO neurons were monaural and could be excited only by input to the contralateral ear. Neuroanatomical experiments support this finding. Although the MSO receives input from two sources, both of these are on the same side. One input projects directly from AVCN, while the other input is delayed by one synapse and projects via MNTB. Like the LSO, the MSO in *Pteronotus* receives one direct excitatory input and one time-delayed inhibitory input. The LSO receives these inputs from the left and right ears, but the MSO receives its two inputs from one ear. It may be that MSO in this species of bat (or perhaps in many species of echolocating bats) is connectionally different from MSO in other mammals. Alternatively, it may be that the region of the bat superior olive that has been termed MSO is actually not homologous to the MSO of other mammals.

Recently, inhibitory inputs of different types have been described in the MSO of nonecholocating mammals. Thus, there might be a continuum of patterns of input to MSO ranging from the dog in which binaural excitatory patterns dominate to an echolocating bat such as *Pteronotus parnellii* in which mixed excitatory and inhibitory inputs from the contralateral ear predominate. However, it is still not clear what different patterns of auditory behavior are reflected in the different patterns of input to the MSO.

In *Molossus ater*, MSO neurons have been described that are capable of encoding interaural time differences of as little as 10 μs. It is debatable, however, whether these time differences can be used to localize sounds. Given these contradictory findings, the question of whether bats possess an MSO that can encode interaural time differences remains. In any case, the structure of the MSO appears to be more phylogenetically variable than that of the larger and more differentiated LSO. The MSO projects to the ipsilateral IC and to the ipsilateral DNLL. Like the monaural pathways, both binaural pathways are connected to the IC through a set of potential delay lines.

Inferior colliculus. All the monaural and binaural pathways converge at the IC. However, little is known about the specific functions of this enormous nucleus. The most obvious organizational principle in the IC is the tonotopic map (fig. 6.17). The inferior colliculus may perform an important function in adjusting the time scale of neural response to auditory stimuli. In auditory-guided behaviors that occur as fixed action patterns, such as orienting movements, the rapid pace of auditory input has to be linked to the much slower generation and execution of motor patterns. Such an adjustment of time scales to link sensory input and motor preprocessing may be a general function of the inferior colliculus, as suggested by Casseday and Covey (1996). In echolocating bats, the inferior colliculus is to date the most peripheral nucleus that contains neurons specifically tailored for echo analysis. Recently, in two different species of bats, some neurons in the inferior colliculus have been found to respond vigorously to a pair of auditory signals but poorly to single sounds. Many of these "combination-sensitive neurons" are facilitated by (1) characteristic "best delays" between the first and the second signal, which correspond to naturally occurring echo travel times of a few ms to 30 ms (see fig. 6.22b) and, (2) a second signal that is fainter than the first one. Because of these characteristics, combination-sensitive neurons are thought to selectively process echoes and encode target-range information based on the neurons' best delays. Many other functions have been attributed to the inferior colliculus, including the representation of auditory space in the form of a neural map. However, no unequivocal evidence for this or other functions has been provided so far.

Medial geniculate body. The inferior colliculus projects to the ipsilateral medial geniculate body, which, in turn, is made up of a number of different subdivisions. Each subdivision receives input from the entire frequency range. The different subdivisions of the medial geniculate body innervate different regions of auditory cortex. Each cortical region also projects back to the subdivision of the medial geniculate that provides its input. Presumably these reciprocal connections make it possible for the cortex to control its own excitatory input.

Central acoustic tract. In addition to the pathways of the "classical" auditory system described above, bats, and presumably all other mammals, possess a rapid ascending pathway that bypasses the inferior colliculus. This pathway originates in a cell group of the superior olivary complex, now known as the nucleus of the central acoustic tract (NCAT), but formerly also called the anterolateral periolivary nucleus. The NCAT receives input from the contralateral AVCN (fig. 6.19). The ascending fibers of the central acoustic tract divide into two branches. One branch terminates in the deep layers of the superior colliculus, which are thought to control head and pinna movements. The other branch terminates in the suprageniculate nucleus of the diencephalon which, in turn, projects to the auditory cortex and the frontal cortex. The frontal cortex projects back to the auditory cortex and the deep superior colliculus. This connectional pattern suggests that the central acoustic tract plays a role in the reflex movement of the head and ears toward a sound source. It is also thought that the frontal cortex may be involved in spatial memory.

Efferent pathways. No sensory organ can escape some form of control by the brain. The auditory system is no exception; it contains a number of efferent pathways that project from the central nervous system to the periphery. The most thoroughly studied efferent auditory pathway is the olivocochlear system (fig. 6.20).

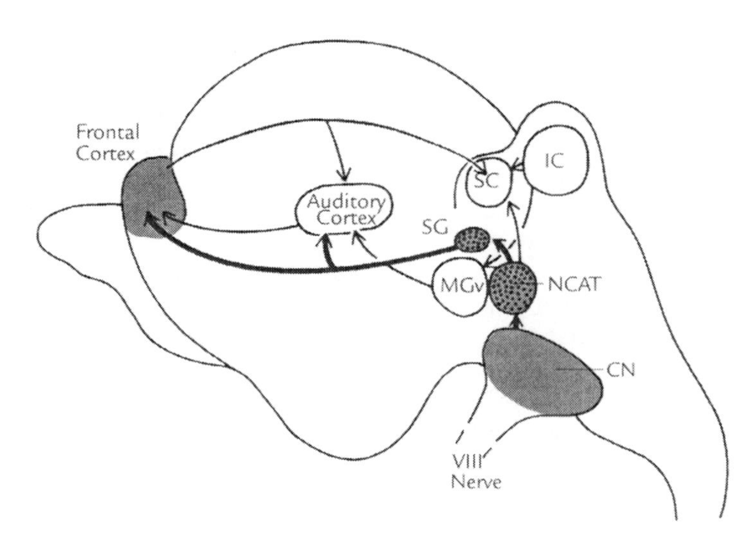

Figure 6.19 The central acoustic tract in *Pteronotus parnellii* (Mormoopidae). The primary pathway, shown as a thick line, projects from the nucleus of the central acoustic tract (NCAT) to the suprageniculate nucleus and thence to the auditory cortex and frontal cortex. VIII, auditory nerve; IC, inferior colliculus; SC, superior colliculus; MGv, ventral division of the medial geniculate; CN, cochlear nuclei. From Pollak and Casseday (1989).

A medially located group of small nuclei within the superior olivary complex innervate the outer hair cells. These pathways use the neurotransmitter acetylcholine. A laterally located group of cells innervate the afferent fibers that come from the inner hair cells. The neurons in this pathway may use some neuropeptides such as enkephalins and dynorphins as transmitters. One hypothesis about the function of the efferent system is that in the presence of a background noise, it improves the ability to hear a narrow-band signal such as a pure tone by suppressing the response to the noise. In addition, it is thought to protect the ear from overstimulation by very loud stimuli. Efferent nerve fibers induce changes in the length and mechanical properties of outer hair cells. In the mustached bat, noise-induced activation of the efferents to the outer hair cells results in damping of cochlear vibrations produced by a pure-tone stimulus. To date, horseshoe bats and hipposiderids are the only mammals in which no efferents to the outer hair cells have been found. This lack of efferents to the outer hair cells may be related to the highly specific, narrowly tuned and damped cochlear filter in these species, which is capable of rejecting broad-band noise that would otherwise degrade detection of the signal.

The nuclei of the ascending central auditory pathways are also subject to efferent influence. The auditory cortex, for example, innervates the auditory nuclei of the thalamus as well as the dorsal part of the inferior colliculus. The nuclei of the lateral lemniscus project back to the superior olive, and the nuclei of the superior

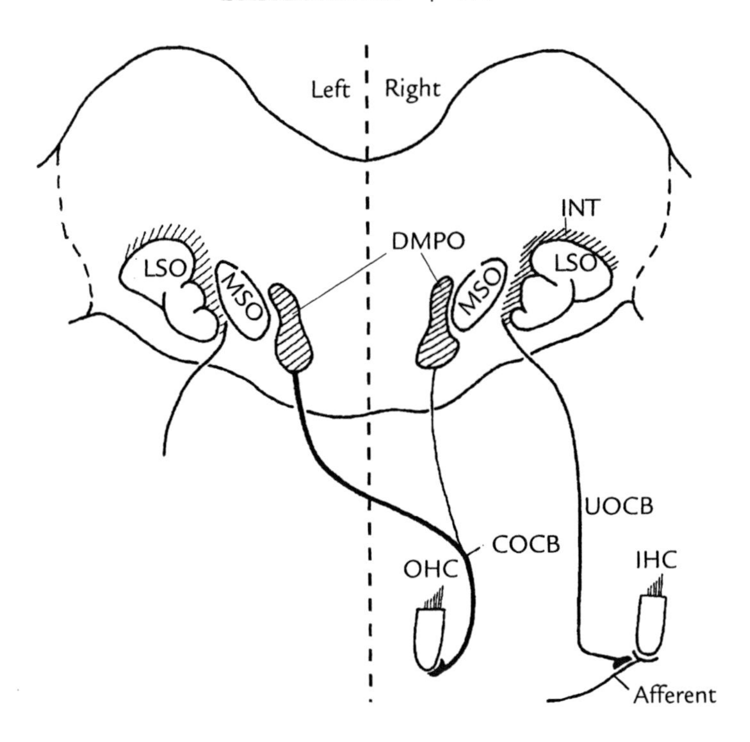

Figure 6.20 The efferent, olivocochlear system in *Pteronotus parnellii* (Mormoopidae). The efferent pathways originate in the brainstem. The crossed olivocochlear bundle (COCB) originates bilaterally in the dorsomedial periolivary nuclei (DMPO); the uncrossed olivocochlear bundle (UOCB) originates in the ipsilateral intermediate periolivary nucleus (INT), indicated by the shaded area surrounding the dorsal part of the lateral superior olive (LSO). IHC, inner hair cell; OHC, outer hair cell; MSO, medial superior olive. From Bishop and Henson (1987).

olive project back to the cochlear nuclei. As already mentioned, the DCN projects back to the ventral cochlear nucleus.

In addition, the entire auditory system receives more or less dense noradrenergic innervation. This diffuse, modulatory system probably originates in the locus coeruleus, located in the hindbrain. The locus coeruleus modulates the attentional state of the brain. It has been shown that in *Pteronotus parnellii,* noradrenalin increases the precision of temporal coding in the cochlear nucleus. It is therefore possible that the noradrenergic system improves echolocation ability through regulation of attention.

(The auditory cortex will be considered in section 6.4, "Echolocation Performance.")

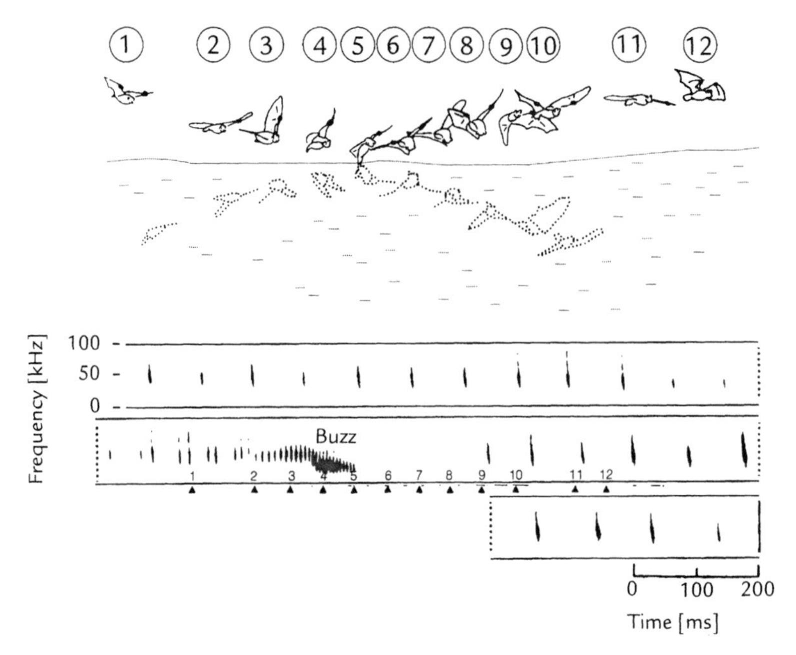

Figure 6.21 Echolocation sequence of a foraging bat, *Myotis capaccinii* (Vespertilionidae), as it captures an insect flying over the surface of the water. Drawings 2–11 are taken from photos snapped at 45-ms intervals; drawings 1 and 2 and drawings 11 and 12 are separated by 90 ms. At position 6, the bat captured an insect using its feet and tail membrane; at position 8, the bat used its mouth to retrieve the insect from the pouch formed by its tail membrane. The lower panels are sonagrams to show the accompanying sequence of echolocation sounds. The times corresponding to positions 1–12 are marked with the corresponding numbers. Just before the bat made contact with the prey (3–5), it emitted a rapid series of pulses, with a progressively decreasing frequency range, the final buzz. During capture, while the bat's head was buried in the tail membrane (6–8), no echolocation sounds were recorded. From E. Kalko (unpublished).

6.4 ECHOLOCATION PERFORMANCE

Echolocation is used primarily for the detection and localization of prey (fig. 6.21). A flying bat foraging for insects emits on average 4–12 echolocation calls per second, separated by irregular time intervals. These "search signals" are often more than 10 ms in duration. Because the pauses between the signals are considerably longer than the signals themselves, no echolocation information is available during approximately four-fifths of the flight time (duty cycle). As soon as the bat detects a target, it begins to emit signals at a higher repetition rate, with progressively decreasing intervals between signals. When the bat is pursuing prey, the signal repetition rate increases to 40–50/s. Just before the bat captures its prey, it emits a "final buzz," consisting of a sequence of 10–25 short pulses separated by minimal intervals. During the final buzz, the duty cycle increases to about 90%. The entire sequence of detection, pursuit, and final buzz usually lasts less than 1–2 s (fig. 6.21).

The detection of prey is not a significant problem for a bat flying in an open space. As long as there is no outside (human) interference, most echoes that the bat hears probably come from flying insects. First the bat must localize its intended prey by determining the direction and distance of the object giving rise to the echo.

RANGE DETERMINATION

Theoretically, echolocating bats could use any one of three methods to measure the distance of an object: (1) measure the time between the emitted signal and the return of the echo; (2) use difference in intensity between the emitted signal and the returning echo to measure target range; (3) for broad-band echolocation signals, use the relative increase in low-frequency energy within the echo spectrum as an indicator of target distance, as low frequencies are less attenuated in air than are high frequencies.

Bats certainly use the first of these strategies to determine target range. In behavioral studies using a positive reinforcement paradigm, bats are able to resolve range differences of about 12 mm for two identical objects. The actual physical targets can be replaced by simulated echoes presented from two loudspeakers located at the same distance from the bat, with one signal delayed relative to the other. In this case, the bat chooses the signal with the shortest delay as being the closer target, for which it was reinforced in the previous experiment. Under these conditions, bats can resolve delay differences on the order of 60 µs. This time difference corresponds approximately to the distance that bats are able to resolve using real targets.

The measurement of time delay is performed by the central auditory system. To achieve stopwatch precision on the order of several microseconds, precise starting and ending signals are required. Short, broad-band signals are better suited for this purpose than long, narrow-band signals or constant-frequency pure tones.

Suga and O'Neill were the first to demonstrate how such a neural stopwatch mechanism operates in *Pteronotus parnellii*, a bat that emits a complex echolocation signal made up of four harmonics. The vocalization consists of a long constant-frequency tone at about 60 kHz (second harmonic) and is followed by a short frequency-modulated signal. The bat uses the short FM component for echo delay measurement and range determination. Suga and O'Neill discovered an area in the auditory cortex of *Pteronotus* in which neurons respond preferentially to the FM portion of the echolocation call (fig. 6.22). However, these neurons respond only if three specific conditions are met:

- They do not respond to single sounds, but require a pair of sounds corresponding to the emitted pulse and a returning echo. Such neurons are called "combination-sensitive neurons."

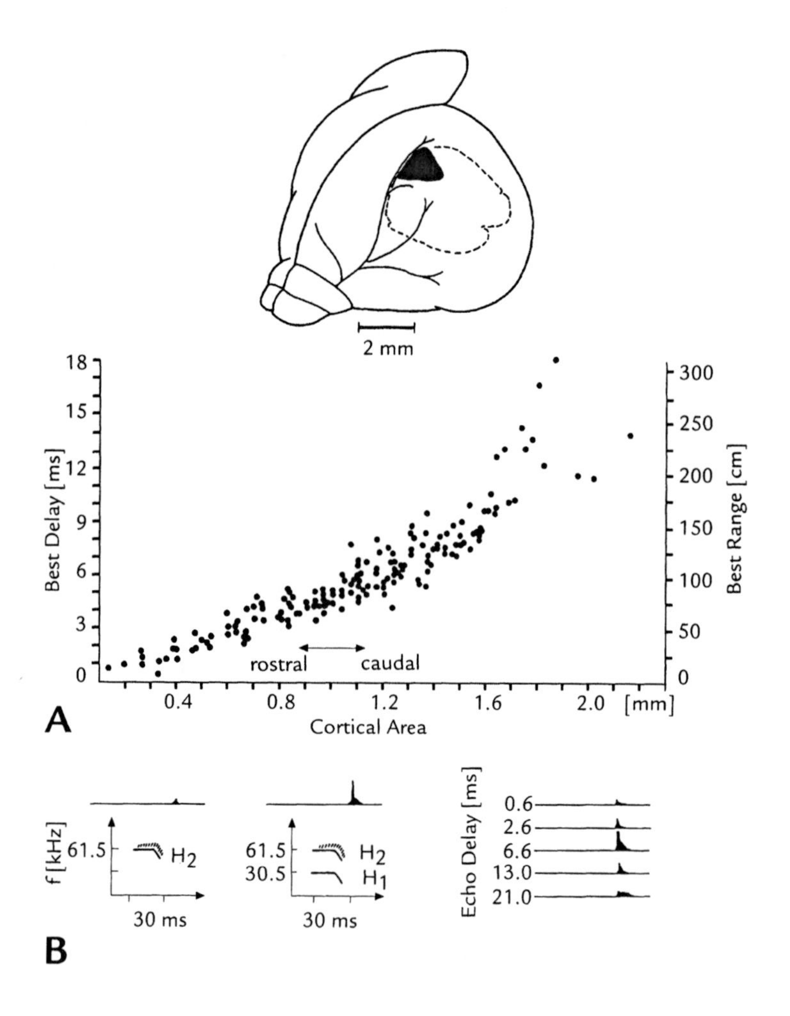

Figure 6.22 (A) Range-finding neurons in the auditory cortex of *Pteronotus parnellii* (Mormoopidae). The auditory cortex (dotted line) occupies a large part of the lateral forebrain. Within the auditory cortex is an area (shown in black) where neurons respond selectively to pairs of sounds (pulse and echo) separated by a specific delay. This range-finding area of the cortex is organized so that neurons tuned to short delays are located rostrally and neurons tuned to long delays are located caudally. In the graph, the left vertical axis represents the time delay between the first and second sound of a pair; the right vertical axis represents the corresponding target distance for each delay. The cortical area contains a map of the space around the bat out to a distance of about 3 m. (B) Responses of neurons to various experimental conditions. The histograms on the right show the responses of a neuron that responded best to a delay of 6.6 ms. The graph in the middle shows the responses of a neuron to a combination of the high-intensity second harmonic (H_2 = 61.5 kHz) and the low-intensity first harmonic (H_1 = 30.5 kHz). The echo, shown by the hatched signal, contains only the second harmonic. The graph on the left shows that when the first harmonic is omitted from the first signal, the neuron responds poorly. From Suga and O'Neill (1979).

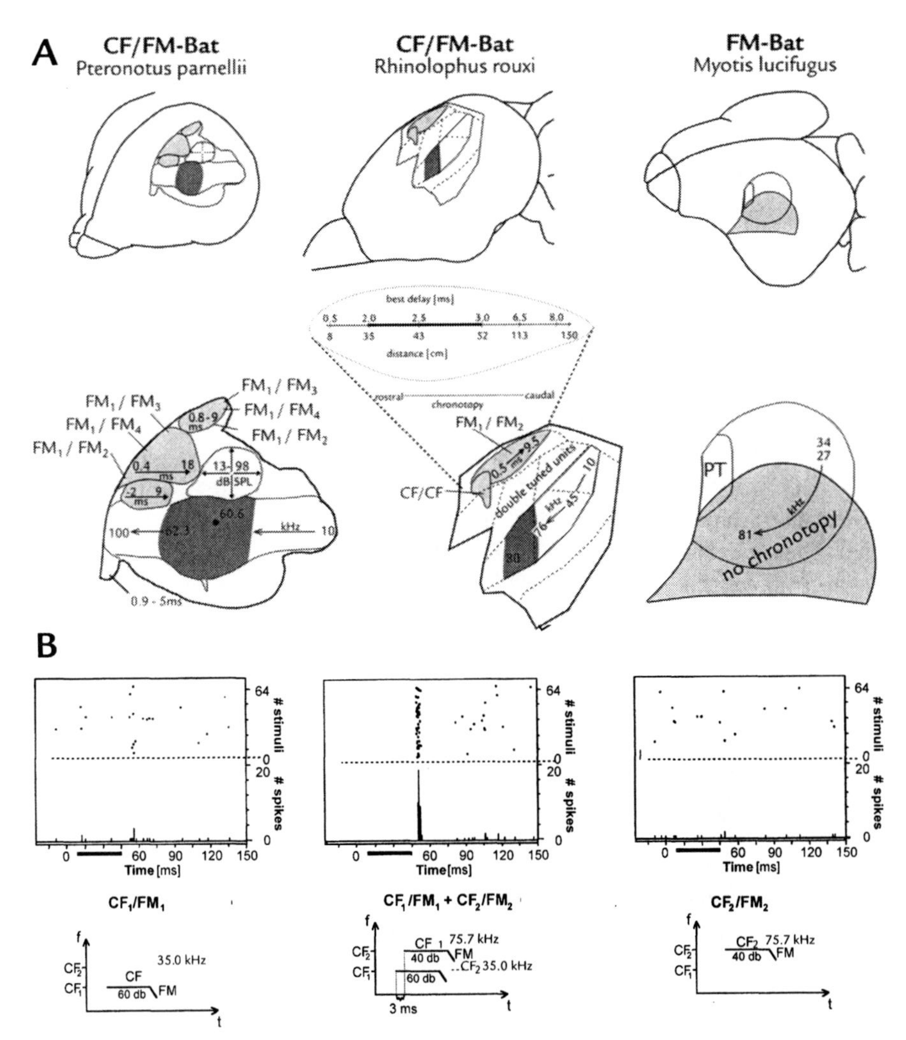

Figure 6.23 Functional differentiation of auditory cortex in an FM bat (*Myotis lucifugus*), an Old World CF/FM bat (*Rhinolophus rouxi*), and a New World CF/FM bat (*Pteronotus parnellii*). (A) Upper row shows site of auditory cortex areas within the forebrain. Lines within the auditory cortex delineate cytologically and functionally identified areas. Lightly shaded areas contain range-finding neurons; darkly shaded tonotopically arranged areas represent the narrow frequency range of the individual CF echoes. Lower row: The auditory cortices. Note that in the FM bat range-finding neurons are not chronotopically organized as in the FM/FM areas in CF/FM bats (shown for *Rhinolophus rouxi*). In *Pteronotus* the FM/FM area is represented several times with specific combinations of FM_1 and higher harmonics. FM: neurons only respond to a combination stimulus consisting of the first harmonic (FM_1) with one of the higher harmonics ($FM_{2,3,4}$). (B) Cortical neuron in *Rhinolophus rouxi* which only responds to the combination of the first harmonic with a delayed (3 ms) second harmonic (FM_1/FM_2, middle graph). The neuron does not respond to the first or second harmonic alone (left and right graph). Upper halves of the graphs depict a scatter histogram of spikes to 64 repeated stimuli; the lower halves represent the sum of activities as a poststimulus-time histogram.

- The first sound must contain the frequency of the first harmonic, about 30 kHz, just as the emitted pulse does (fig. 6.23). Without the first harmonic, the neurons respond poorly or not at all (fig. 6.23).
- The second, "echo" signal, must occur within a specific time following the first, "pulse" signal. If it occurs earlier or later, the neuron fails to respond (fig. 6.23). Thus, such a neuron will only respond to echoes returning from targets at a particular distance. Such neurons are called "range-finding" neurons and belong to the class of combination sensitive neurons (see p. XX).

A range-finding neuron responds maximally when there is a specific delay between the first and second sound. Neurons are organized within the cortical area according to their best delays (fig. 6.22). The rostral part of the cortical area contains neurons that respond best when the delay between the first and second signal is between 0.6 and 2.0 ms, corresponding to a target range of 10–35 cm. Neurons with progressively longer best delays are found more caudally. Neurons near the caudal border of the area respond to delays up to 18 ms, corresponding to a target distance of about 3 m. Thus, the target ranges represented in this cortical area correspond to those that are important for prey capture.

For a bat flying toward a target, the cortical activity evoked by the echoes starts in the caudal part of the cortical range-finding area and progresses rostrally until the bat reaches the target. A prerequisite for range discrimination is the existence of neurons that precisely encode arrival times of signals. In the inferior colliculus and more peripheral nuclei, a type of neuron has been described that responds only to the onset of an auditory stimulus, with a single spike. The latency of the spike (i.e., the time from onset of the stimulus until the occurrence of the spike) does not change with stimulus intensity and varies only within about 250 μs. These "constant latency neurons" may be inputs to neuronal circuits for time analysis.

"Coincidence detectors" are thought to be the basic elements of neural circuits subserving delay sensitivity. A coincidence detector will respond only when it is simultaneously excited by two different inputs; it therefore corresponds to a logical "AND" gate. Synchronous excitation of a coincidence detector by an emitted echolocation sound and a subsequent echo may be achieved in different ways. These include:

- *Delay lines.* Arrival of excitation elicited by the first signal of a pair (the echolocation sound) is delayed by some time interval. When this time interval corresponds to the delay between the emission of the echolocation sound and the subsequent return of its echo, the coincidence detector will be activated. The "best delay" of the coincidence detector is determined by the magnitude of the time interval introduced by the delay line transmitting the response to the first signal. Delay lines may be formed by addition of extra synapses, with each synapse introducing a delay in transmission of at least 0.4 ms. The time required for conduction of the nerve impulse also depends on the length and thickness of nerve fibers, with long, thin fibers conducting more slowly than short, thick ones. Delay lines of this type would be well suited to produce "best delays" of up to 4 ms.

- *Paradoxical latency shift.* Neural response latency commonly shortens with increases in stimulus intensity. However, the auditory cortex of *Myotis lucifugus*, an "FM-bat," has been shown to contain a class of neurons that behave differently. For these neurons, response latency becomes longer as stimulus intensity is increased. As a result of this paradoxical latency shift, excitation evoked by a faint echo might arrive at the same time as excitation evoked by an earlier but louder echolocation signal.
- *Inhibition.* There is ample evidence that inhibition is also involved in creating neural delays. In this case, the echolocation sound first elicits inhibition, which persists for a certain amount of time. As long as the inhibition prevails, excitation by subsequent signals (echoes) will be blocked. As a result, the shortest delay to which the neuron will respond is set by the time at which inhibition fades away. Inhibition is well suited to produce relatively long delays of 4 ms to 10+ ms. However, this mechanism does not explain why neurons do not respond to any echo that occurs after the end of the inhibitory period initiated by the emission of the echolocation sound. This might be due to "time windows" of neural facilitation for processing of echoes, also triggered by the onset of sound emission.

Neural time windows. Time windows may be set by the bat's vocalization. When the bat hears its own emitted echolocation sound, this opens a "time window" of increased sensitivity for subsequent echoes in certain populations of neurons. If a sound arrives after the time window has closed, the neurons do not respond. Behavioral experiments have shown that the bat's ability to measure and discriminate range differences is disrupted when loud artificial sounds are presented within a period up to 40 ms after the emission of an echolocation pulse. However, the artificial sounds interfere with echolocation only if they resemble the bat's own species-specific echolocation call. The artificial sounds are probably effective in disrupting performance during the period of the open time window because the animal mistakes them for echoes. This finding suggests that the window of increased sensitivity to sound is open only for signals that resemble echolocation pulses.

There are types of neurons in the inferior colliculus that only respond to simulated echoes when the bat has actually emitted an echolocation signal 30–40 ms before the simulated echo. The events responsible for opening the time window in these neurons might include an efferent copy of the motor command for vocalization or the bat's hearing of its own emitted signal. Psychoacoustic studies indicate that there may be differences in auditory efficiency between silent bats and ones that are actively echolocating. Therefore, a realistic picture of how the specific features of echoes are analyzed in the central nervous system will only be gained when it becomes possible to record from the brain of freely flying, actively vocalizing bats.

The common belief that any animal with ultrasonic hearing is capable of echolocation is false. Many animals that do not echolocate, including mice, rats, and cats, have excellent hearing in the ultrasound range. On the other hand, the lowest-pitched echolocation signals (Molossidae: *Otomops martiensseni,* 15 kHz; *Tadarida brasiliensis*, as low as 10 kHz) are well within our audible range. Echolocation does not depend on hearing within a specific frequency range; instead, it depends on hearing faint signals which must be preceded by one loud signal with a similar structure, the emitted echolocation call. The vocally triggered neural time window of heightened sensitivity is thus the key to understanding the specific adaptations of the auditory system for echo imaging.

The difference in sound level between the emitted call and the echo may be as high as five orders of magnitude. This enormous difference in sound level raises the question of how the bat is able to hear an echo at all when it arrives just after a very loud pulse. Bats have evolved a number of strategies for dealing with this problem. First, as the bat approaches an object, it progressively decreases the intensity of its vocalizations so that the sound level of the returning echoes remains approximately constant (e.g., *Pteronotus parnellii*). Second, the bat's hearing threshold increases by 30 dB when the delay of an artificial echo is decreased from 6 ms to 1 ms, corresponding to a change in the virtual target distance from 110 cm to 17 cm (e.g., *Eptesicus fuscus*). Third, neural responses to the bat's own vocalization are attenuated by 20–25 dB somewhere at or below the level of the nuclei of the lateral lemniscus (e.g., *Myotis grisescens*). Fourth, many neurons in the central auditory system have upper thresholds and do not respond to sounds louder than 40–70 dB. Experiments on the role of the middle ear muscles in attenuating the bat's own vocalizations have yielded contradictory results, so the question of whether they are involved in diminished ability to hear vocalizations remains unresolved. It is not clear whether these mechanisms for matching the perceived intensity of the pulse and echo are valid only in certain cases, or whether they can be generalized to echo processing in all species.

In the mustached bat, Suga and his colleagues not only described a chronotopically organized FM/FM area, but also cortical maps representing different echo parameters (fig. 6.22). The FM/FM map is repeated in two smaller maps, one of which represents delays from 1–9 ms and the other of which represents delays of 1–5 ms. In each area, the combination of the first harmonic with each of the three higher harmonics is represented in a separate cluster. Other cortical maps are thought to represent relative velocity between the bat and a target, subtended target angle, and target location. Parts of the "echo maps" are overlapped by populations of neurons that respond to communication calls. To date, the auditory cortex of the mustached bat is the most highly differentiated auditory brain structure of any mammal.

SOUND LOCALIZATION

The location of a target is defined in terms of its direction and its range. Bats are able to determine the vertical and horizontal coordinates of a target with an accuracy of 2–5°. Nevertheless, the region from which echoes can originate is restricted to an angle of approximately 50°, which corresponds to the sound beam in front of the bat. For directional hearing, bats use the same mechanisms as other mammals.

Directional characteristic of the pinna. Although the inner and middle ear have no directional sensitivity, the funnel shape of the outer ear creates a highly directional and frequency-specific axis (fig. 6.11). In addition, the pinnae act as a barrier for sound waves that impinge on their back side. Most neurons of the central auditory system reflect the directional characteristics transmitted by the outer ear to the eardrum (fig. 6.11). The shorter the wavelength (i.e., the higher the frequency of sound), the greater the directional specificity of the pinna.

In both bats and humans, the form, structure, and size of the pinnae determine the relative intensity of sounds across the audible spectrum, and the resulting spectral characteristics depend on the direction from which the sound originates. When sound waves reflected from the surfaces of the pinna interact with sound waves that enter directly, interference patterns arise. The resulting spectral patterns are produced by attenuation and amplification of specific frequencies due to the directional characteristics of the pinna and the interference patterns. Thanks to these variations in timbre, a human who is deaf in one ear can still localize sound. In humans, the fine structure of the pinna plays a highly individual role in the perception of music. Each individual perceives the spatial characteristics of music differently. We learn unconsciously the characteristic spectral patterns that result when sounds originate from in front, from behind, from above, or from below. It seems reasonable to suppose that bats also use information resulting from their pinna structure to determine the spatial characteristics of sounds.

Vertical localization. Localization of sounds in the vertical dimension is not as important for a land-dwelling animal as it is for a bat that has to pursue flying prey through the open air. Behavioral experiments on bats that emit a broad-band echolocation signal have shown that the tragus, a skin flap that sticks up across the lower part of the external ear canal, is important for determining the vertical angle, or *elevation*, of a sound source (fig. 6.12). It is likely that interference patterns arise due to interaction of sound waves reflected from the tragus with those reflected from the wall of the pinna and that the resulting pattern differs systematically according to the vertical direction of the sound source.

Horseshoe bats and hipposiderids that use narrow-band echolocation signals lack a tragus (fig. 6.12). Instead, they move their pinnae in a stereotyped alternating pattern correlated with the emission of their echolocation calls, so that as one ear moves forward and downward, the other ear moves upward and back. Behavioral tests on horseshoe bats have shown that when the pinnae are immobilized, they perform poorly on sound localization in the vertical dimension. Thus, the vertical ear movements in these species probably perform the same function as the interference patterns created by the tragus in species that use broad-band echolocation calls.

Horizontal localization. The cues used to determine the horizontal angle, or *azimuth*, of a sound source, are the differences in the sound that reaches the left and right ears (fig. 6.24). The binaural pathways of the ascending auditory system encode information about interaural differences in excitation. E/I neurons in the bat's

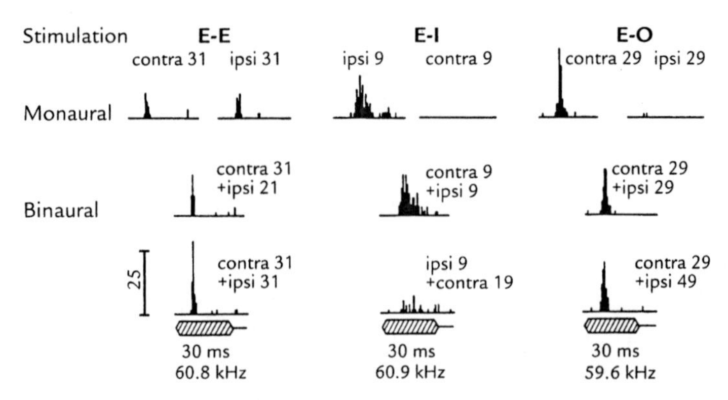

Figure 6.24 Responses of neurons in the superior olivary complex to binaural combinations in *Pteronotus parnellii* (Mormoopidae). Responses of three types of neurons are shown. In the left column are responses of E/E neurons. These neurons are excited by sound at both the contralateral ear (contra) and the ipsilateral ear (ipsi), and are found in the MSO. The middle column shows responses of E/I neurons. These neurons are excited by an ipsilateral sound and inhibited by a contralateral sound. E/I neurons are well suited to encode interaural intensity differences and are found in the LSO. The right column shows responses of E/O neurons. These neurons respond to contralateral sound but are unaffected by sound at the ipsilateral ear. E/O neurons are commonly found in the MSO. The upper row of histograms shows responses to monaural stimulation (numerals give sound level in dB SPL). The middle and bottom rows illustrate responses to different binaural combinations of tones having identical frequency but different sound levels at the two ears. The stimulus is indicated by the hatched symbol. From Pollak and Casseday (1989).

large LSO respond to interaural intensity differences. Many neurons are sensitive to very small differences in the sound level at the two ears, and hence are sensitive to small deviations of a sound source from straight ahead. Because the distance between the two ears is only a few centimeters in most species of bats, the range of interaural time differences experienced is at the most 50–70 μs. Neurons have been found in *Molossus ater* that are sensitive to interaural time differences of only 15–20 μs. These findings show that echolocating bats may also use interaural time differences as a cue to determine the direction of a sound source. Humans are capable of resolving interaural time differences of as little as 5 μs, so there is no reason a bat would not be able to use this time cue as well.

To date there has been no convincing demonstration in bats or any other mammals of a neural map of auditory space such as has been found in the owl. This failure to find a map might be due to the fact that auditory space is represented relative to the position of the mobile pinna. It would seem reasonable to look for neurons that integrate information about the pinna position with information about the location of a sound source in space. The primary function of echolocation is to guide the animal toward a target. It should be possible to achieve this based on very precise information about what is straight ahead and about small deviations of the target from the midline. For the remainder of auditory space, it might be sufficient for the bat to receive signals from "alerting" neurons that indicate in a general way the spatial sector in which an insect or an object is present.

OBJECT DISCRIMINATION

The range and direction of a target define its position in space, but they do not provide any information about what it is. Bats commonly pursue stones that are tossed up in the air just as they would insects, so they seem to pay little attention to the nature of objects that move through the air. However, if one continues to throw stones in the air, the bat quickly learns to distinguish the stones from insects and ignores them. Behavioral tests have shown that horseshoe bats are even capable of distinguishing between different species of insects based on their echoes. A number of other behavioral experiments have shown that bats can discriminate two- and three-dimensional forms and objects of different sizes.

Object size. The smallest object that a bat can detect is less than 1 mm in diameter. If a swarm of fruit flies or mosquitoes are released in the laboratory, bats will immediately begin pursuing them. Bats can detect these small insects with diameters of 1–4 mm from a distance of 35 cm, and quickly capture them. The absolute detection threshold for objects has been determined in obstacle avoidance tests. Different species of bats can detect and avoid wires with a diameter of 0.06–0.1 mm. The wires are 20–280 times smaller in diameter than the wavelengths of the sounds that the bats use to detect them. To date it has not been possible to measure the amount of sound energy reflected from such small objects. Spectral analyses of the sound reflected from objects of different materials, forms, and sizes have shown complex interactions in the reflected waveforms. For example, "creeping waves" that channel energy around the object have been observed. This and similar effects may help explain how bats are able to perceive extremely small objects based on their echoes.

Bats' ability to discriminate the size of objects has been tested using equilateral triangles as targets. The triangles are perceived as different in size when their area differs by 17% or more. This difference in area corresponds to an echo sound level difference of about 1–3 dB. Thus, differences in the sound level of echoes may be important for discriminating target size.

Material and form. After a relatively long training period, echolocating bats can learn to discriminate differences in the form or the material of targets. For example, bats can learn to discriminate among circular, square, and triangular targets with equal areas and between a square with rounded corners and one with angular corners. Bats can discriminate among three-dimensional targets of equal volume such as a cylinder, a cube, and a pyramid. They can also discriminate among three-dimensional objects that are identical in form but made from wood,

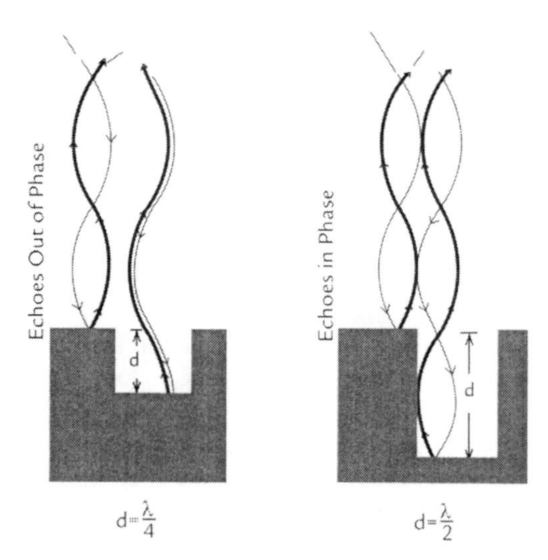

Figure 6.25 The principle of how interference patterns are created through reflection from a two-front target consisting of a raised surface and a recessed surface separated by a distance *d*. When $d = \lambda/4$, echoes from upper and recessed front are 180° out of phase, and the resulting interference causes cancellation of the waveform so that the echoes are nullified. When $d = \lambda/2$, echoes are additive (summation).

plexiglass, or plastic. Unfortunately, there is no satisfactory explanation of what echo characteristics bats use to identify the form or material of an object.

Surface structure and "echo colors." Under natural conditions no bat would attempt to land on a smooth wall. Presumably echoes provide information about the texture of surfaces. Behavioral discrimination experiments have shown that bats are able to discriminate between plates in which 8-mm-deep holes have been drilled and plates with holes 7.0–7.2 mm deep. However, if randomly structured objects with different textures are used as targets, bats' performance is not nearly as impressive. The false vampire, *Megaderma lyra*, could distinguish sandpaper plates covered with grains having an average diameter of 0.4 mm from plates covered with larger diameter grains only when the size difference was at least 2 mm.

When broadband signals (120 kHz–20 kHz) of the type used by *Megaderma lyra* were reflected from the plates with the drill holes or sandpaper covering and the echoes measured, the echoes were found to contain peaks and nulls at different frequencies, depending on the depth of the textured surface of the target (fig. 6.25). When there is interference between sound waves of identical frequency and amplitude, addition of the energy in the waves is maximal when they are in phase; cancellation of the energy in the waves is maximal when they are 180° out of phase. Such a cancellation within a spectrum is called a notch. A notch occurs when the distance between the high and low points on a target is λ/4 or an uneven multiple of this relationship. The frequency of the notch (f_{notch}) is then:

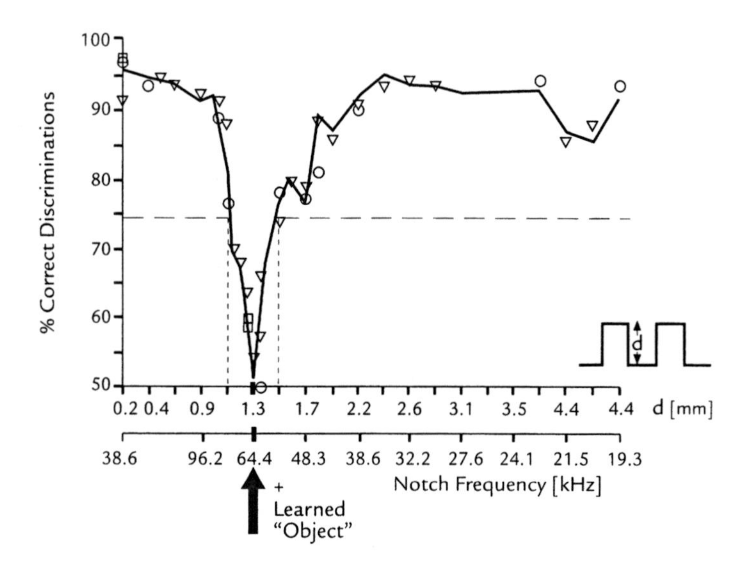

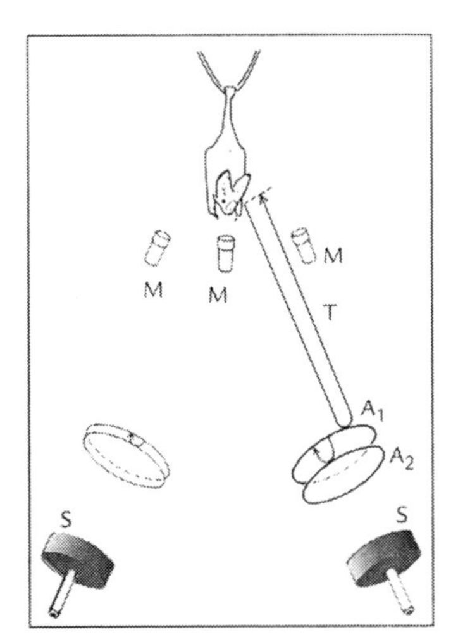

Figure 6.26 Discrimination of electronically simulated two-front echoes by the echolocating false vampire bat, *Megaderma lyra*. Lower panel: Experimental setup. The bat's echolocation calls are picked up by the microphones (M) and played back over the loudspeakers (S) as two interfering echoes separated by a time delay of t_i = 7.7 μs. The bat perceives a two-front target located at a distance *T*; the two simulated surfaces A_1 and A_2 are separated from one another by a distance of 1.3 mm, corresponding to the delay of 7.7 μs. Bats learn to discriminate this virtual target from targets with surfaces A_1 and A_2 separated by different distances. The upper panel shows the percentage of correct discriminations between a learned target and another virtual two-front target (distance *d* of the two fronts: ordinate). Arrow: Frequency notch caused by interference. From Schmidt (1992).

$$f_{\text{notch}} = \frac{1c}{4\lambda}, \frac{3c}{4\lambda}, \frac{5c}{4\lambda 1}, \ldots; c = \text{velocity of sound}$$

The interfering waveforms are additive when the difference in depth between the target surfaces is $\lambda/2$ or an even multiple thereof (fig. 6.25).

The positions of the peaks and nulls within the spectrum of a broad-band echo, or the "echo color," provide a measure of the surface texture of a target. The smoother the target, the higher the frequency of the maximal notch. A textured object with a depth of 2.4 mm between the high and low points on the surface produces an echo with a notch at 35 kHz; if the depth is 0.5 mm, the notch is at 155 kHz. Thus, the echo no longer contains all the frequencies that were present in the emitted signal. In an analogy with sunlight that contains all wavelengths in the visible spectrum, an emitted sound may be white, but the returning echo is no longer "white"—hence the term "echo colors." The interference pattern or echo colors reflects the structure of the target giving rise to the echo.

To test the discriminability of different patterns of echo colors, the false vampire, *Megaderma lyra*, was presented with signals from virtual two-front targets in which the first front represented the raised surface and the second front the recessed surface (fig. 6.26). The virtual two-front targets were created by allowing the bat to emit its usual echolocation call, recording the call and playing it back (the first front) together with a delayed version of the same signal (the second front). The animals were trained on a virtual two-front echo signal with a delay of 7.77 μs. This delay between the first and second fronts corresponds to a depth of 1.3 mm. Interference between the two signals in the virtual echo results in a spectral notch at 64.4 kHz. The false vampire could learn to discriminate this virtual echo signal from others provided there was a difference of at least 1 μs in the delay which results in a 7–9 kHz difference in the position of the spectral notch. These conditions would correspond to a difference in depth of only 0.2 mm in a real object. The animals could make the discrimination not only by means of echolocation, but also by passive listening to the virtual echoes. This indicates that the discrimination of spectral colors is not peculiar to echolocation, but is instead a property of hearing in general.

Although the mechanism used by the bats to make this discrimination might well have been the spectral colors or spectral composition of the signal as postulated here, they could also theoretically have used the arrival times of the two echo wavefronts as a cue. If this were the case, it would mean that the bats were capable of resolving time differences of 1 μs. To date, no such temporal precision has been found in the central auditory system. The only requirement for the spectral analysis is a frequency resolution on the order of 10%, a level of precision that is easily achieved in the auditory systems of most mammals.

Echo colors depend not only on the structure of a target, but also on the angle from which the echolocation signal impinges upon the target. Thus, to discriminate among different species of insects on the basis of echo colors, the bat must be capable of quite sophisticated feats of generalization and memory. On the other hand, changes in echo color could provide an excellent means of detecting living

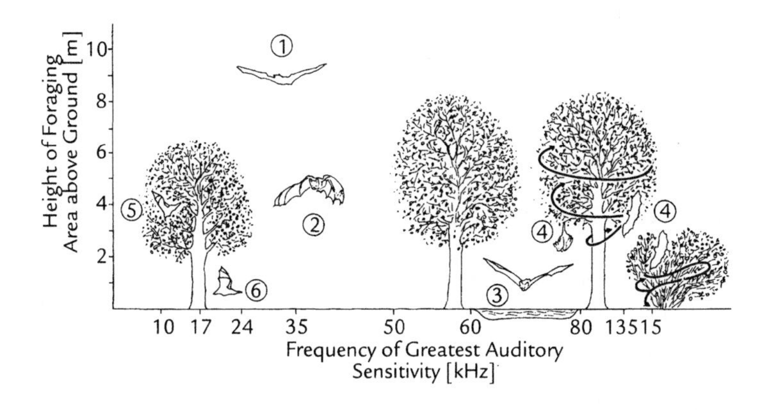

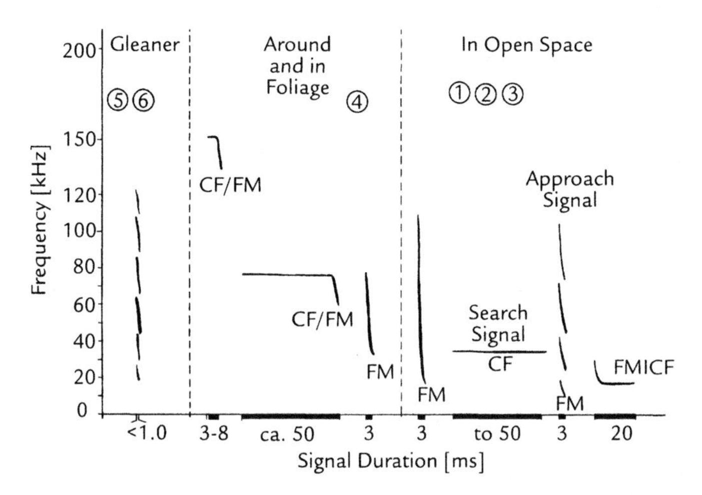

Figure 6.27 Frequency of maximal hearing sensitivity and echolocation calls are correlated with the preferred foraging biotopes of different species of echolocating bats. Top: Typical foraging biotopes (1–6) for different species of bats and frequency of maximal hearing sensitivity (abcissa). The frequency of maximal sensitivity is lowest for bats that hunt farthest above the treetops or above the ground (1). It is also low in bats that glean prey from substrates (5 and 6). Bottom: Sonagrams of typical echolocation calls used in different biotopes (circled numbers refer to the corresponding foraging areas in the upper graph) by the various species of bats. When searching for prey, bats that hunt in open spaces (foraging areas 1–3) emit long, almost pure-tone signals, either alone (CF) or following a frequency-modulated component (FM/CF). Bats that capture prey from the ground (foraging areas 5 and 6; e.g., *Megaderma lyra*), always use short, broad-band signals. In and around the foliage (foraging area 4): signals from *Hipposideros bicolor* (left), *Rhinolophus rouxi* (middle), *Myotis nattereri* (right). In open space: signals from *Myotis myotis* (left), *Rhinopoma hardwickei* (middle, search and approach signals), *Tadarida macrotis* (right).

prey against a background. When an insect appears against a background, this entire structure produces a complex, highly colored echo. Experiments have shown that the echo color changes as soon as an insect changes its position by 0.2 mm. Thus, small movements and the resulting changes in echo colors produced by living insect prey may render it much more easily recognizable by the bat than a similar stationary object.

6.5 ADAPTATION TO DIFFERENT BIOTOPES

Bats do not forage everywhere prey are found. Instead, they restrict their foraging to certain preferred areas. The flight style, wing shape, and echolocation system of different bat species are adapted specifically for the biotopes in which they forage. Figure 6.27 shows six different foraging areas shared by a number of sympatric bat species. Most species of bats do not restrict their foraging to their preferred area. During seasons when prey are not plentiful, bats will forage in other biotopes and employ different foraging strategies. The fishing bat *Noctilio leporinus*, for example, occasionally feeds on flying insects. *Myotis myotis* forages for insects in the open air as well as on the forest floor.

Echo imaging is especially well suited for detecting and capturing insects in open space because there the echoes generally originate from a single object, the potential prey. Foraging within the leaves or branches of trees generally offers a larger selection of insects but is complicated by the fact that echoes from the tiny prey may be masked by echoes from the large mass of background vegetation.

6.5.1 ECHOLOCATION IN OPEN SPACES

When searching for prey, all bats that forage for insects in the space above the treetops or lower down but at a distance of more than 2 m from vegetation use relatively long (6–60 ms), narrow-band or CF signals (fig. 6.26). As soon as the bat detects a target and starts its pursuit, the echolocation call changes to a short (1–5 ms) FM signal (pursuit signal in fig. 6.27). Such broad-band sounds are well suited for target range determination.

When searching for prey, however, there are two reasons a narrow band signal is better: the echo of a narrow-band signal is particularly well suited for encoding information about the wingbeat patterns of flying insects, and the range over which the bat can scan is probably increased due to the fact that most of the energy in the signal falls in the frequency range at which hearing is most sensitive.

Species of bats that forage in open spaces do indeed have the most sensitive hearing for the frequencies of the CF part of their own echolocation calls. For example, the frequency at which hearing is most sensitive is 18 kHz in *Tadarida aegyptiaca*, 25 kHz in *Taphozous melanopogon*, 35 kHz in *Rhinopoma hardwickei*, and 50 kHz in *Pipistrellus mimus*. The higher the frequency of maximal sensitivity in these species, the lower they fly, and the more restricted their foraging area (fig. 6.27). This correlation is related to the range over which the echolocation signal can reach. Because higher frequencies are more attenuated in air, fast-flying species that detect and hunt prey over long distances above the treetops must use the lowest possible frequencies in their echolocation calls when searching for prey. It is likely that the wider range of these calls, probably up to 50 m, is offset by a decreased ability to detect small insects.

FORAGING IN AND NEAR VEGETATION

Bats that forage in open spaces need only detect and localize single punctate targets; in contrast, bats that forage in thick vegetation or near the ground must contend with "echo clutter" from background objects. These bats must somehow detect the echoes of small moths and beetles from among a multitude of other competing echoes. Many species of bats seem to have mastered this situation without any obvious adaptations in their echolocation systems. *Myotis myotis*, for example, flies in the forest beneath deciduous and evergreen trees, feeding on ground-dwelling arthropods such as beetles, spiders, and millipedes. In laboratory experiments, *Myotis* emits ordinary FM signals while foraging for insects on the floor, even emitting a typical final buzz during the capture phase. However, if the prey is located on the floor within a few centimeters of a wall, the bat ceases to emit echolocation calls, runs toward the wall, and scans the wall with its long whiskers until it touches the prey.

Although *Myotis myotis* ceases to echolocate when background echoes are too loud, many other bat species have undergone adaptations of their auditory and echolocation systems that make them better suited to function under conditions of echo clutter. These adaptations are of three types: prey detection through analysis of changes in echo colors, specializations for the detection of wingbeating insects, and passive localization.

Detection based on changes in echo colors. When a broad-band signal is reflected from an insect sitting on a surface, the result is a complex "colored" echo, the spectrum of which is determined by both the insect and the background. The spectral color of the echo changes perceptibly when the distance between the insect and the background changes—for instance, when the insect makes small movements. These changes in echo color could alert a bat to the presence of a living object, and thus a potential prey, in the area being inspected. Many observations and documentary films have shown that bats that glean insects from leaves and walls often spend long periods of time hovering in front of their prey, darting to capture it as soon as it moves. All species that prefer to glean prey from surfaces emit short, broad-band "white" signals that pick up any changes in echo color. These echo signals are so low in intensity that they can only be used within close range of the prey.

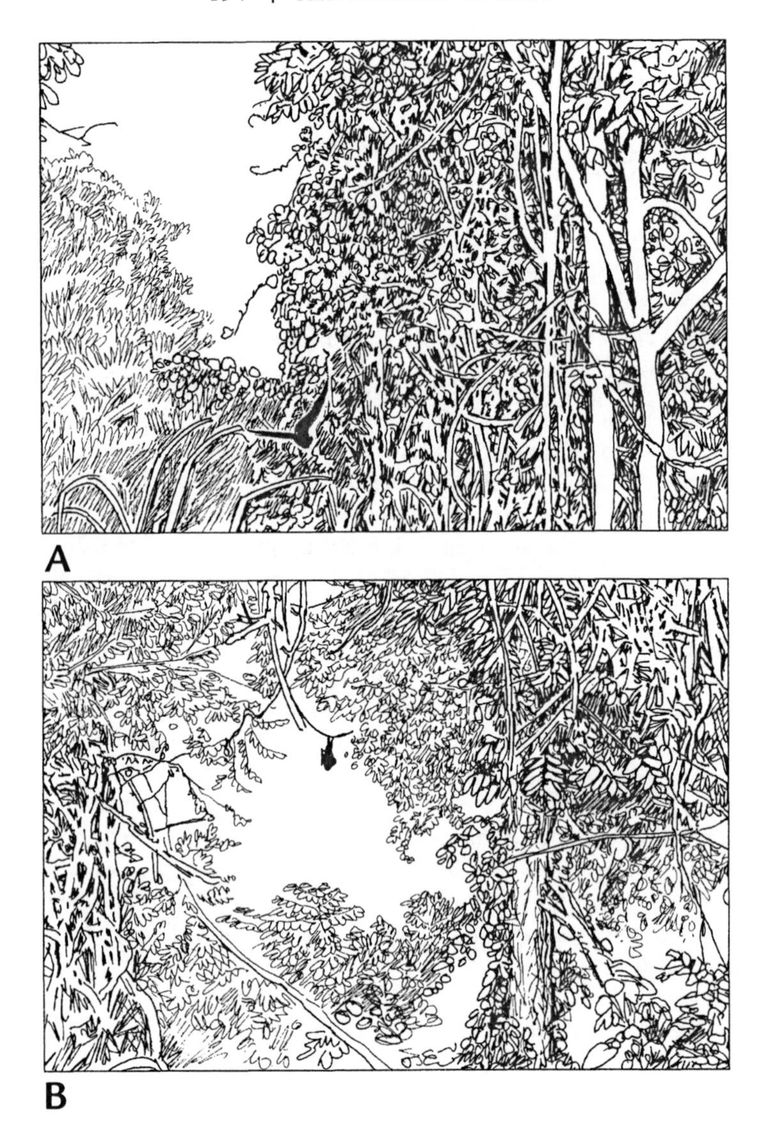

Figure 6.28 Foraging strategy of *Rhinolophus* in the jungle of Sri Lanka. (A) When the bats first leave the cave in the evening, they forage on the wing, near vegetation. (B) The bat spends most of the night hanging from a twig, continually rotating its body and emitting echolocation calls in all directions. If a passing insect is detected, the bat flies out, captures it, and returns to its listening post.

Specialization for detection of wingbeating insects. Horseshoe bats (Rhinolophidae) prefer to forage for flying prey in thick vegetation. During the first 1–2 h of the evening, they fly around bushes and trees (fig. 6.28a). The rest of the night they spend lying in wait for prey. A given bat returns night after night to the same small foraging area where it hangs from an especially favorably placed twig (fig. 6.28b) and rotates its body continually about the axis of its legs, searching the area with its echolocation calls. As soon as the horseshoe bat detects a flying insect within a distance of about 5 m, it takes off from the twig, captures the insect, and returns with its prey to the roost. On average, a *Rhinolophus rouxi* makes about 30 such short foraging flights every hour.

For prey detection, *Rhinolophus rouxi* uses a signal that seems counterintuitive. The echolocation signal is a combined FM/CF/FM signal (see fig. 6.3), and consists mainly of a loud, high-frequency pure tone (ca. 72–85 kHz), 40–60 ms in duration. For a signal of such long duration, echoes from flying prey would interfere with echoes from the vegetation. However, no echo colors would result because the emitted echolocation signals are "monochromatic" pure tones. An additional complication is that echoes from objects closer than 8 m will overlap with the emitted signal. Nevertheless, field observations and laboratory experiments show that in spite of these seemingly adverse conditions, horseshoe bats immediately detect any wingbeating insect in the vicinity. A single real or simulated insect wingbeat is sufficient to cause the bat to fly off and capture its prey. Insects that do not beat their wings are completely ignored, even if they are very close to the bat.

Thus, the echolocation system of horseshoe bats is highly specialized for the detection of insect wingbeats. Comparison of the echoes of FM signals and pure-tone signals reflected off wingbeating insects shows why pure-tone signals are especially well suited to this type of detection task. When a pure-tone signal is reflected off a flying insect, there is a moment during the cycle of the wingbeat when the insect's wing is oriented perpendicular to the impinging signal. The reflected echo will then exhibit a momentary glintlike energy maximum. Such a *glint* is characterized not only by an increase in sound intensity, but by a broadening of the sound spectrum due to Doppler effects resulting from movement of the wing toward or away from the direction of the sound source. The more rapid the wing movements, the larger the Doppler effect, and the broader the spectrum of the resulting sound. The fine structure of the glints depends on the shape of the insect's wings and on the way in which they move. Glints are therefore a potential source of information on which to discriminate different species of flying insects. Glints are prominent in CF signals but are not easily distinguished in a broad-band signal.

The hearing of rhinolophids is highly specialized for the detection of glints. The basilar membrane of the inner ear contains an extremely narrow filter that is tuned to the pure-tone echolocation frequency of the individual (fig. 6.29). The resolution of the filter, expressed as the $Q_{10\,dB}$ value, is 400 or higher. The $Q_{10\,dB}$ value is equal to the best frequency of a tuning curve divided by the bandwidth of the tuning curve at 10 dB above threshold. For example, for a filter with a center frequency of 80 kHz and a Q value of 400, the bandwidth at 10 dB above threshold would be only 200 Hz. In nonecholocating mammals and in nonspecialized bats, Q values seldom exceed 20. The cochlear filter of horseshoe bats is so finely tuned that the bat can easily detect a deviation away from the center frequency of just 50 Hz (0.06%). Thus, horseshoe bats can easily discriminate shifts of up to 2000 Hz from the center frequency of the echolocation signal caused by glints that result from insect wingbeats.

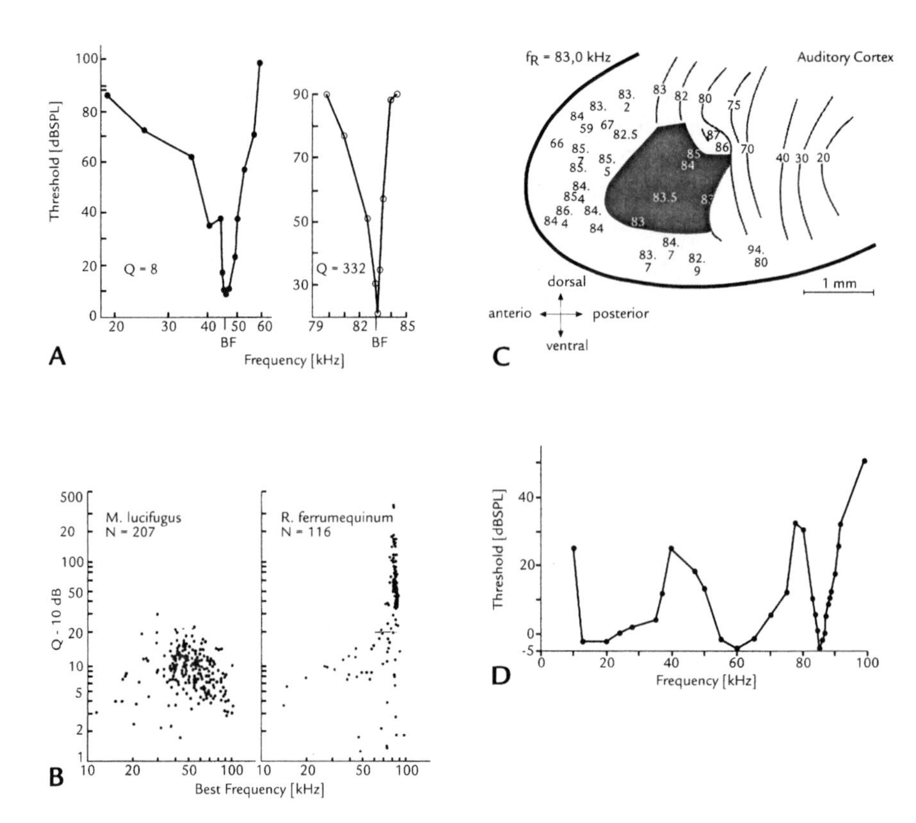

Figure 6.29 Fine tuning to a narrow frequency band in the auditory fovae of horseshoe bats. (A) Tuning curves of a neuron with a best frequency (BF) below the range of the auditory fovea (left) and a neuron with a BF within the range of the auditory fovea (right). Note the different frequency scales in the two graphs. (B) Comparison of Q values from a bat that lacks an auditory fovea (*Myotis lucifugus*) and a bat with an auditory fovea (*Rinolophus ferrumequinum*). Q value = best frequency/bandwidth of the tuning curve at 10 dB above threshold. Q values are a measure of the sharpness of frequency tuning. Neurons with Q values above 20 are found only in the region of the auditory fovea. (C) Tonotopic frequency representation in the auditory cortex of the horseshoe bat *Rhinolophus ferrumequinum*. The numbers indicate BFs of neurons in kHz. The narrow frequency band of the auditory fovea, from 83 to 86 kHz (echolocation frequency f_R = 83.0 kHz), occupies a disproportionately large area of the auditory cortex (black area). (D) Behavioral audiogram of the greater horseshoe bat, *Rhinolophus ferrumequinum*. From Neuweiler and Vater (1977), Suga et al. (1976), Ostwald (1984), and Long and Schnitzler (1975).

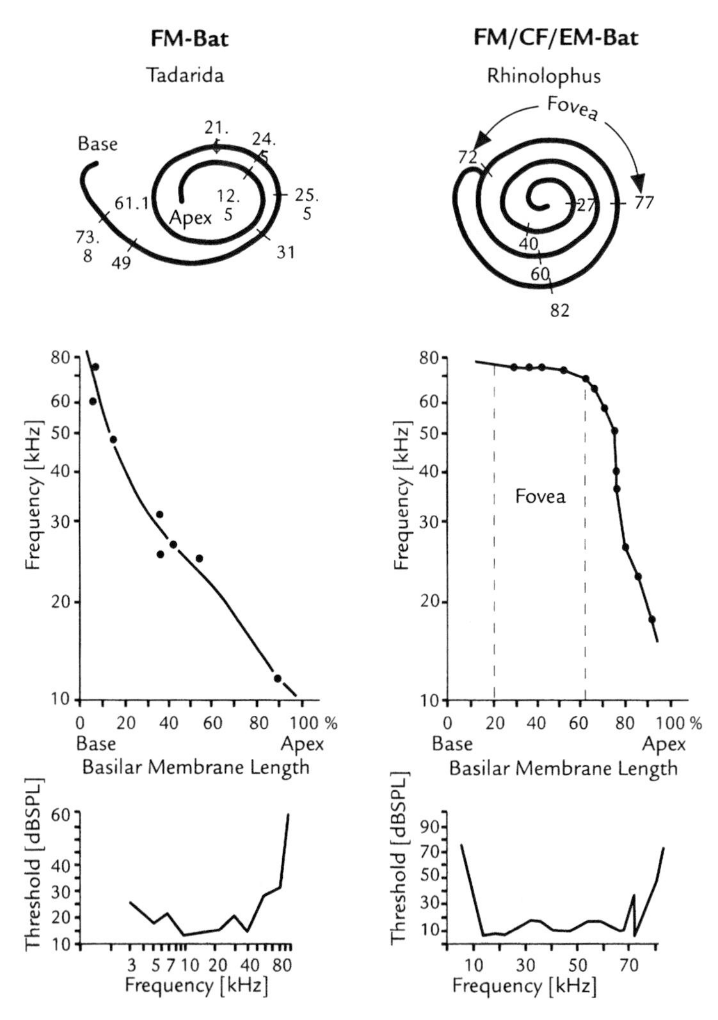

Figure 6.30 Cochlear frequency map and audiogram of a bat that lacks an auditory fovea, *Tadarida brasiliensis* (left column) and a bat with an auditory fovea, *Rhinolophus rouxi* (right column). Top: Horizontal projection of the cochlea with frequency map. Center: Frequency representation from the base to the apex of the cochlea. In *Tadarida*, the frequency progression is approximately logarithmic as it is in most other mammals. Bottom: Audiograms of both species of bats, derived from threshold measurements on neurons in the inferior colliculus. From Sieffert and Vater (unpublished) and Vater et al. (1985).

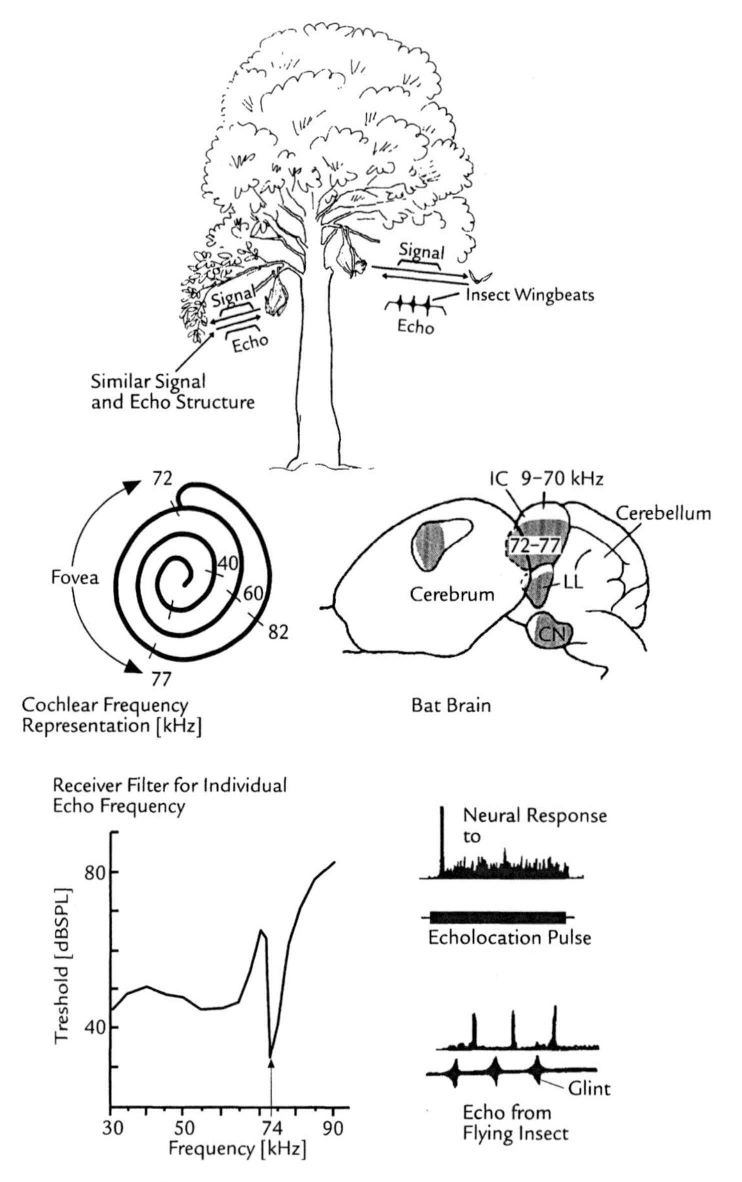

Figure 6.31 Use of the auditory fovea for wingbeat detection in horseshoe bats (*Rhinolophus spp.*). *Top*: Field situation. *Center left*: The auditory fovea in the cochlea (horizontal projection of the cochlea with frequency map in kHz). *Center right*: The expanded representation of the auditory fovea (gray areas, 72–77 kHz) in the ascending auditory system. CN, cochlear nuclei; LL, nuclei of the lateral lemniscus; IC, inferior colliculus; AC, auditory cortex. *Lower left*: Audiogram showing the narrow filter characteristics tuned to the individual echolocation frequency (arrow). *Lower right*: Responses of a filter neuron in the inferior colliculus to the pure-tone frequency of an echolocation call (top) and to an echo from a wingbeating insect (bottom).

Auditory fovea. The cochlear mechanisms that create this narrow filter are still not fully understood. Experiments to map the frequency representation in the inner ear have shown that the filter is based on an *expanded representation* of a very narrow frequency range. A frequency band of only 6 kHz around the individual's pure-tone echo frequency (fovea in fig. 6.30) occupies nearly the entire second turn of the cochlea, corresponding to approximately one-fourth of the entire length of the basilar membrane. Normally, this region would contain the representation of an entire octave, for example, 30–60 kHz. In an analogy with the visual system, this expanded representation of a narrow frequency band has been called the "auditory fovea." As would be expected from the principle of tonotopy within the central auditory system, neurons tuned to the filter frequencies occupy an expanded representation at every level of the auditory pathway, including auditory cortex (figs. 6.18 and 6.28). In the inferior colliculus, one-half to two-thirds of the neural mass is devoted to the analysis of the narrow filter frequency band of the echolocation signal (fig. 6.18).

Many of these "filter neurons" respond to the emitted echolocation signal weakly or with one spike at the beginning of the sound (fig. 6.31). However, as soon as a signal containing glints is presented, the filter neurons respond precisely to each glint and thus encode a faithful representation of the wingbeat pattern of the insect (fig. 6.31). The neural circuitry is such that cells respond not so much to the echolocation signal itself as to rapid changes in the signal, such as those contained in glints. Consequently, the nervous system of horseshoe bats is sensitized to echoes from wingbeating targets.

With the evolution of the auditory fovea and its associated filter properties millions of years ago, rhinolophoids had already developed a sort of "biological radio." Like a radio station, *Rhinolophus* broadcasts a carrier frequency to which the receiver, the auditory fovea, is precisely tuned. Just as in the case of a radio broadcast, the carrier frequency itself is uninteresting. Its only purpose is to match the tuning of the sender (the vocalizing bat) with the tuning of the receiver (the auditory fovea of the bat's ears). What constitutes the meaningful information in a radio signal are the frequency modulations in the carrier signal that correspond to speech or music. For the horseshoe bat, its "music" is in the frequency modulations imposed on the carrier signal by glints from wingbeating insects. Each horseshoe bat broadcasts its own individual carrier frequency. Because the auditory fovea of each bat is tuned to its own emitted frequency (fig. 6.30), the result is a wingbeat detector that is resistant to interference from neighboring senders with slightly different carrier frequencies.

It is not known how each individual horseshoe bat's emitted frequency and foveal frequency become matched with one another. Young horseshoe bats start emitting echolocation calls through their nostrils at about 20 days of age. At this same age, an auditory fovea in the inner ear can first be distinguished physiologically. The auditory fovea is always tuned to the same frequency as that emitted by the individual. During a maturation period of 6–8 weeks, the emitted echolocation frequency progressively increases from about 54 kHz to 78 kHz, and the tuning of the foveal frequency in the inner ear increases in parallel (fig. 6.32). Auditory

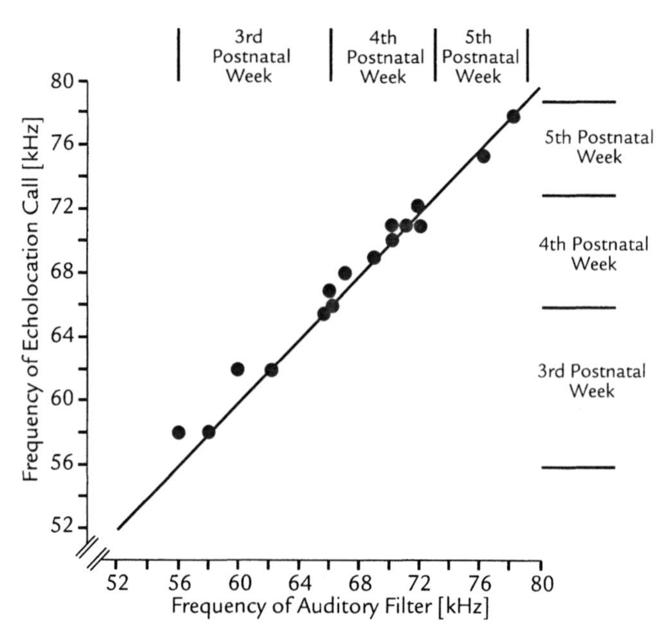

Figure 6.32 Parallel development of the auditory filter frequency and the emitted echolocation frequency in juvenile horseshoe bats (*Rhinolophus rouxi*) from 3 to 5 weeks of age. At every time the frequency tuning of the auditory fovea matches that of the echolocation calls. From Rübsamen (1987).

deprivation experiments in juvenile animals suggest that the emitted frequency and the cochlear tuning develop independently from one another and that hearing the vocalization may only contribute to fine-tuning of the system. In adult horseshoe bats the individual emitted frequency also varies seasonally and is lower by as much as 0.6 kHz during autumn and winter. This might be related to lower body temperature during the cooler season. Because horseshoe bats forage for flying insects during mild winter evenings, one would expect that the frequency of the individual auditory fovea is also lowered. It is not known whether this is indeed the case, nor how such a shift could be achieved.

Among Old World bats, those that have developed specialized echolocation calls and auditory foveas for detecting wingbeating insects are the horseshoe bats and the closely related hipposiderids. Although New World bats that forage in tropical rainforests are faced with exactly the same problem of echo clutter, only one New World species, the mustached bat, *Pteronotus parnellii*, has developed such a specialized echolocation system. When the physiological characteristics of the inner ears of horseshoe bats and mustached bats are compared, it is clear that the auditory filters of the two species are not identical. In both species the basal end of the cochlea exhibits prominent morphological specializations. In horseshoe bats and hipposiderids, the basilar membrane and parts of the supporting structures are massively thickened, whereas in mustached bats the basal scala tympani is greatly enlarged. The fact that these conspicuous features disappear abruptly at the point where the auditory fovea begins suggests that they play a role in the creation of the auditory filter.

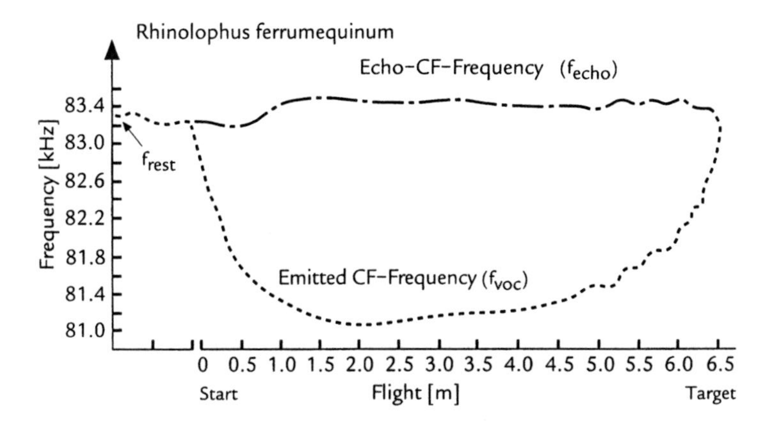

Figure 6.33 Doppler-shift compensation in the horseshoe bat, *Rhinolophus ferrumequinum*. Plots show the frequencies that are emitted (dotted line) and heard (dashed line) as the bat flies from its starting position to a target. During flight, the emitted frequency is lowered in such a way that the returning echo frequency remains constant at a value in the center of the auditory filter region. This behavior compensates for Doppler shifts in the frequency of the echo caused by the bat's flight speed. f_{rest} = frequency emitted by the bat before takeoff. From Schnitzler (1972).

Based on a number of different experiments, Kössl and Vater have developed a working hypothesis. They identified the transition point between the morphologically specialized basal region of the cochlear partition and the beginning of the auditory fovea as the site of a sensitive resonator tuned to the individual echo frequency. When stimulated by pure-tone echoes, the resonator starts ringing and produces a standing wave of vibration between the basal region and the resonator site. In the mustached bat, laser interferometry has shown that the basal part of the cochlear partition does indeed vibrate when the ear is stimulated by the specific echo frequency of 62 kHz. It is thought that the tectorial membrane assists in transferring the energy at 62 kHz from the vibrating basal part of the basilar membrane to the apically situated foveal part where 62 kHz is represented.

Doppler-shift compensation. Logically, one would predict that the entire wingbeat detector system should become nonfunctional as soon as the bat starts to fly. Depending on the bat's flight speed, the complete echolocation signal will be subject to an increase in frequency due to Doppler shift correlated with the bat's flight speed (see box 6.5), affecting the signal emitted through the nostrils as well as the returning echoes that impinge on the "flying" ears. This double shift would be sufficient to move the carrier frequency out of the range of the auditory fovea. However, bats have developed a measure to counteract such a potential mismatch with the foveal frequency, through mechanisms for Doppler-shift compensation. As soon as the frequency of a returning echo exceeds the foveal frequency, the next vocalization is emitted at a lower frequency (fig. 6.33). This feedback loop is capable of compensating for increases in the frequency of the returning echo up

Box 6.5 The Doppler Effect

If a sound source moves toward a listener, more sound waves will occur per unit time than are emitted by the source. The result is that the frequency of the sound heard will be higher than that of the source. If the sound source moves away from the listener, the opposite effect occurs, and the sound heard will be lower in frequency than that of the source. Thus, relative movements between listener and sound source result in characteristic frequency shifts that are called "Doppler effects" after the physicist Christian Doppler. As a first approximation, the Doppler effect can be calculated by the formula $df = fs \cdot 2V_{rel}/c$ (df = frequency shift, fs = source frequency, V_{rel} = relative velocity between source and listener, c = speed of sound).

to 4 kHz above the foveal frequency, but it does not compensate for decreases below this reference frequency. This is a reasonable strategy because the maximal flight speed of horseshoe bats does not produce positive Doppler shifts of more then 4 kHz, and negative Doppler shifts would only occur if the bat were to fly backward. Horseshoe bats only perform Doppler-shift compensation on echoes from objects in the environment, not on echoes from flying insects. In addition, Doppler-shift compensation is only performed on echoes that arrive within 20 ms of the time the pulse was emitted, corresponding to a distance of 3–3.5 m. This control loop uncouples the movement-sensitive echolocation system from the bat's own movement.

Specialization for detection of wingbeats is a practical way to filter out prey-generated echoes from a cluttered background. The advantage of being able to spot flying prey among dense vegetation is offset by the inability to detect stationary prey.

Among the hipposiderids, several species (*H. bicolor*, for example) have unusually large ears that enable them to detect and prey on insects that crawl on the ground as well as those that fly. These species solved the problem of echo clutter before it became an issue by using passive hearing.

Passive hearing: Detection and localization based on prey-generated noises. Bats that glean from the ground cannot use wingbeat characteristics as a means of identifying their prey. As an alternative, these bats detect and localize prey on the basis of the sounds that the prey make when moving. Megadermatids, especially, would have little chance to use echolocation to find the lizards, frogs, birds, mice, and other small animals on which they prey, since these usually move about under the cover of vegetation. The auditory system of *Megaderma* is specialized to detect extremely faint sounds. The enormous pinnae result in a level of sensitivity for rustling sounds in the range of 12–25 kHz that is around –20 db SPL. This means that the auditory sensitivity of this bat is 10 times as high as that of humans (fig. 6.34). All bats that have been investigated to date, including the highly specialized horseshoe bats, have good hearing in this frequency range. Thus, it is possible that sensitivity to rustling noises is a general characteristic of hearing in bats, one that they may have inherited from their insectivorous, non-echolocating ancestors.

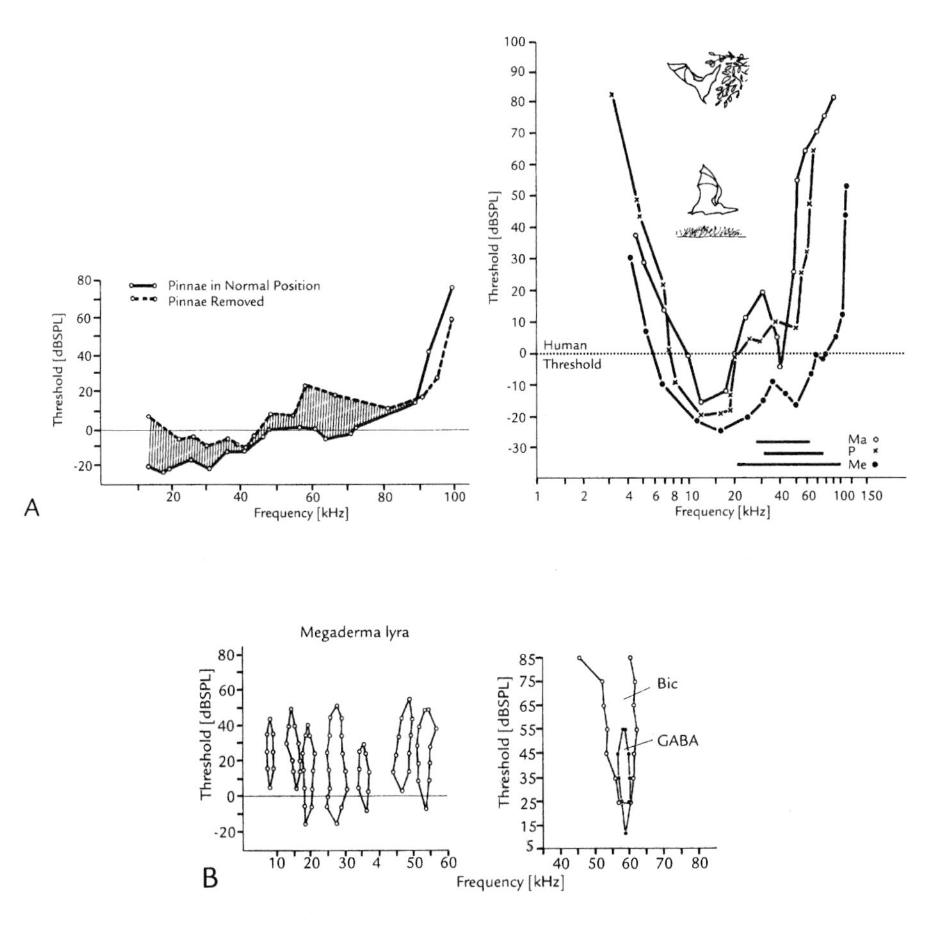

Figure 6.34 Hearing in echolocating bats that preferentially use prey-generated noises to detect their victims. (A) Audiograms of bats that glean prey from the ground or from foliage. Ma, *Macroderma gigas*; P, *Plecotus auritus;* Me, *Megaderma lyra*. Horizontal bars indicate the frequency range of the echolocation calls. The most sensitive hearing range in all three species, up to -27 dB SPL, lies below the range of the echolocation calls, between 15-20 kHz. Below: Spectrogram of fluttering noise of a moth. (B) Neurons with upper thresholds in the inferior colliculus of *Megaderma lyra*. These neurons respond only to low-intensity signals. They are unresponsive to sounds at levels higher than 40–50 dB SPL. The upper thresholds are created through neural inhibition by the neurotransmitter γ-amino butyric acid (GABA). If the GABA antagonist bicuculline (Bic) is applied to the neuron, the upper threshold disappears and the neuron has a "normal" frequency tuning curve. (continued next page)

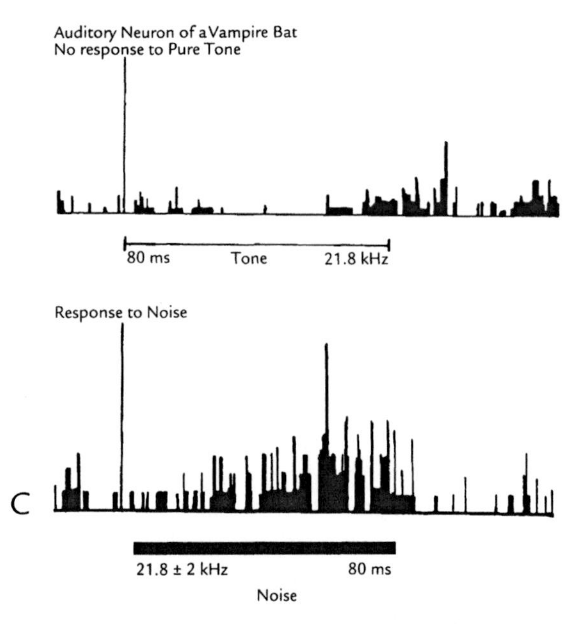

Figure 6.34 (continued) (C) In the true vampire (*Desmodus rotundus*), some neurons in the inferior colliculus are extremely sensitive to low-intensity broad-band signals and are unresponsive to tones or frequency modulated signals. These neurons respond especially strongly to breathing noises of mammals, including humans. From Neuweiler (1990), Schmidt et al. (1991), and Anderson and Racey (1993).

The central auditory system of the false vampire contains an unusually large number of neurons with upper thresholds (fig. 6.34b). These neurons respond only to faint sounds and cease responding as soon as the sound level exceeds 40–50 dB SPL, approximately the level of normal conversation. Additionally, the rostral part of the inferior colliculus in this species contains a special population of neurons that respond only to faint noiselike stimuli. Similar neurons specialized for noise have been found in the true vampire, *Desmodus rotundus* (fig. 6.34c). In the vampire, a bat that feeds on the blood of sleeping mammals, noise-specific neurons respond especially well to low-intensity breathing noises.

As "visual animals," humans like to attribute the complex and highly differentiated tasks of pattern recognition and object identification to the visual system. However, as shown by the above examples of pattern recognition in bats, the mammalian auditory system is capable of similarly complex and differentiated recognition tasks when it is placed under specific evolutionary pressure.

DEFENSIVE MEASURES TAKEN BY PREY

In nature, every successful strategy has a counter-strategy. Thus, many moths have developed effective defensive mechanisms to protect themselves from their high-speed predators. In a field study in India it was found that out of 68 species of moths

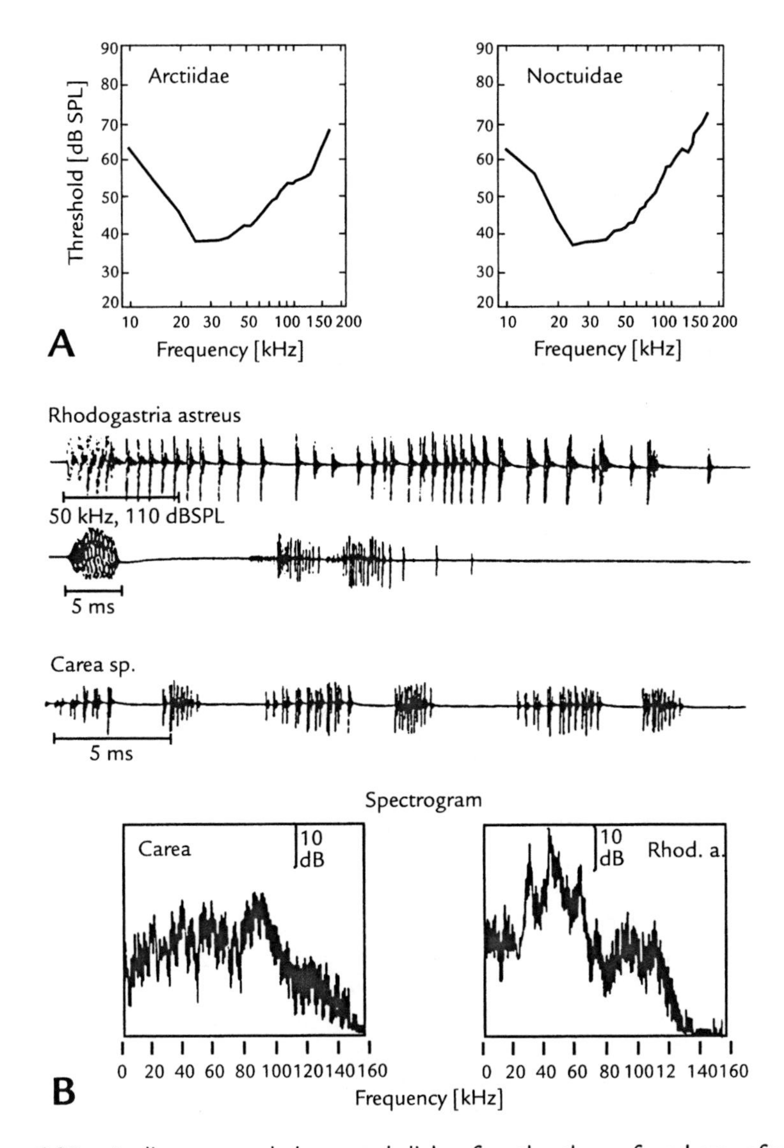

Figure 6.35 Audiograms and ultrasound clicks of moths, the preferred prey of many echolocating bat species. (A) Average audiograms of two families of moths. (B) *Top*: Oscillograms of moth ultrasound clicks. Upper trace: Clicks from *Rhodogastria astreus*. *Middle trace*: The moths respond to an ultrasound stimulus (50 kHz, 110 dB SPL) by series of clicks. *Lower trace*: Clicks from *Carea spp*. *Bottom*: Spectrograms of the clicks, showing energy distribution as a function of frequency.

studied, all but three had good hearing for loud ultrasound (fig. 6.35a). If the ultrasound calls of bats are played to these moths, they react with a series of erratic avoidance maneuvers. Some fly in a wild zigzag course, others fold their wings abruptly and fall straight down, out of the predicted flight path. Moths resting on leaves freeze and remain motionless as soon as they hear ultrasound signals. These maneuvers must significantly reduce the capture rate of these moths by bats.

Some species of moths, especially those in the Noctuid family, emit their own ultrasound pulses (fig. 6.35b). When bats hear the clicklike sounds emitted by these moths, they abruptly turn around and discontinue their pursuit. The clicks produced by the moths do not seem to affect the bats' echolocation systems as was once thought. Presumably the bats interpret the clicks as a sort of warning signal indicating that the moth is not tasty. Many species of moths protect themselves by means of bitter-tasting or poisonous plant-derived substances. Because the ultrasound clicks of the moths seem to have a semantic meaning, it is likely that some species practice mimicry, emitting ultrasound clicks even though they do not have a bad taste. Female moths of the genus *Euproctis* seem to use this trick. During flight they emit clicks that cause pursuing bats to turn away. If the clicking mechanism on the wings is destroyed, the moths are captured and eaten by bats. Species of moths that are flightless or fly only for brief periods usually cannot hear and have no special defense mechanisms.

References

Anderson ME, Racey PA (1993). Discrimination between fluttering and nonfluttering moths by brown long-eared bats, *Plecotus auritus*. Anim Behav 46:1151–1155.

Bishop AL, Henson OW Jr (1987). The efferent cochlear projections of the superior olivary complex in the mustached bat. Hear Res 31:175–182.

Bruns V, Burda H, Ryan MJ (1989). Ear morphology of the frog-eating bat, *Trachops cirrhosus*. J Morphol 199:103–118.

Casseday JH, Covey E (1996). A theory of the operation of the inferior colliculus. Brain Behav. Evol. 47:311–336.

Dannhof BJ, Bruns V (1991). The organ of Corti in the bat, *Hipposideros bicolor*. Hear Res 53:253–268.

Dear SP, Suga N (1995). Delay-tuned neurons in the midbrain of the big brown bat. J Neurophysiol 73:1084–1100.

Gelfand DL, McCracken GF (1986). Individual variation in the isolation calls of Mexican free-tailed bat pups (*Tadarida brasiliensis*). Anim Behav 34:1078–1086.

Grothe B, Vater M, Casseday JH, Covey E (1992). Monaural interaction of excitation and inhibition in the medial superior olive of the mustached bat—an adaptation for biosonar. Proc Natl Acad Sci USA 89:5108–5112.

Guppy A, Coles RB (1988). Acoustical and neural aspects of hearing in the Australian gleaning bats, *Macroderma gigas* and *Nyctophilus gouldi*. J Comp Physiol 162:653–668.

*Hartley DJ, Suthers RA (1987). The sound emission pattern and the acoustical role of the noseleaf in the echolocating bat, *Carollia perspicillata*. J Acoust Soc Am 82:1892–1900.

Hartley DJ, Suthers RA (1990). Sonar pulse radiation and filtering in the mustached bat, *Pteronotus parnellii rubiginosus*. J Acoust Soc Am 87:2756–2772.

*Henson OW, Bishop A, Keating A, Kobler J, Henson M, Wilson B, Hansen R (1987). Biosonar imaging of insects by *Pteronotus parnellii*, the mustached bat. Natl Geogr Res 3:82–101.

Henson OW, Pollak GD, Kobler JB, Henson MM, Goldman LJ (1982). Cochlear microphonic potentials elicited by biosonar signals in flying bats. Hear. Res. 7:127–147.

Henson OW, Xie DH, Keating AW, Henson MM (1995). The effect of contralateral stimulation on cochlear resonance and damping in the mustached bat: the role of the medial efferent system. Hear Res 86:111–124.

Jones G, Rayner JMV (1988). Flight performance, foraging tactics and echolocation in free-living Daubenton's bats, *Myotis daubentoni*. J Zool 215:113–132.

Kalko E, Schnitzler HU (1989). The echolocation and hunting behaviour of Daubenton's bat, *Myotis daubentoni*. Behav Ecol Sociobiol 24:225–238.

Kössl M, Russell IJ (1995). Basilar membrane resonance in the cochlea of the mustached bat. Proc Natl Acad Sci USA 92:276–279.

Kössl M, Vater M (1990). Tonotopic organization of the cochlear nucleus of the mustached bat, *Pteronotus parnellii*. J Comp Physiol A 166:695–709.

Kössl M, Vater M (1995). Cochlear structure and function in bats. In AN Popper and RR Fay, eds. Springer Handbook of Auditory Research, Vol. 511; pp. 191–234, 293–303. Springer-Verlag, Berlin.

Lancaster WC, Henson OW, Keating AW (1995). Respiratory muscle activity in relation to vocalization in flying bat. J Exp Biol 198:175–191.

Lawrence BD, Simmons JA (1980). Measurements of atmospheric attenuation at ultrasonic frequencies and the significance for echolocation by bats. J Acoust Soc Am 71: 585–590.

Long GL, Schnitzler HU (1975). Behavioural audiograms from the bat, *Rhinolophus ferrumequinum*. J Comp Physiol 100:211–219.

Manley GA, Irvine DRF, Johnstone BM (1972). Frequency response of bat tympanic membrane. Nature 237:112.

Mittmann DH, Wenstrup JJ (1995). Combination-sensitive neurons in the inferior colliculus. Hear Res 90:185–191.

*Nachtigall PE, Moore PWB, eds., (1988). Animal Sonar, Processes and Performance. Plenum Press, New York.

Neuweiler G (1989). Foraging ecology and audition in echolocating bats. Trends Ecol Evol 4:160–166.

Neuweiler G (1990). Auditory adaptations for prey capture in echolocating bats. Physiol Rev 70:615–641.

Neuweiler G, Metzner W, Heilmann U, Rübsamen R, Eckrich M, Costa HH (1987). Foraging behaviour and echolocation in the rufous horseshoe bat of Sri Lanka. Behav Ecol Sociobiol 20:53–67.

Neuweiler G and Vater M (1977). Reponse patterns to pure tones of cochlear nucleus neurons in the CF-FM bat *Rhinolophus ferrumequinum*. J. Comp. Physiol. A 115:119–133.

Novick A, Griffin DR (1961). Laryngeal mechanisms in bats for the production of orientation sounds. J Exp Zool 148:125–145.

Ostwald J (1984). Tonotopical organization and pure tone response characteristics of single units in the auditory cortex of the greater horseshoe bat. J Comp Physiol 155:821–834.

Poljak S (1926). The connections of the acoustic nerve. J Anat 60:465–469.

*Pollak GD, Casseday JH (1989). The Neural Basis of Echolocation in Bats. Springer-Verlag, Heidelberg.

Pye JD (1980). Echolocation signals and echoes in air. In RG Busnel and JF Fish, eds., Animal Sonar Systems, pp. 309–333. Plenum Press, New York.

Radtke-Schuller S, Schuller G (1995). Auditory cortex of the rufous horseshoe bat. 1. Physiological response properties to acoustic stimuli and vocalizations and the topographical distribution of neurons. Eur J Neurosci 7:570–591.

Rübsamen R (1987). Ontogenesis of the echolocation system in the rufous horseshow bat, *Rhinolophus rouxi*. J Comp Physiol A161:899–913.

Rübsamen R, Betz M (1986). Control of echolocation pulses by neurons of the nucleus ambiguus in the rufous horseshoe bat, *Rhinolophus rouxi*. I. Single unit recordings in the ventral motor nucleus of the laryngeal nerves. J Comp Physiol 159:675–687.

Rübsamen R, Neuweiler G, Sripathi K (1988). Comparative collicular tonotopy in two bat species adapted to movement detection, *Hipposideros speoris* and *Megaderma lyra*. J Comp Physiol 163:271–285.

Rübsamen R, Schuller G (1981). Laryngeal nerve activity during pulse emission in the CF-FM-bat, *Rhinolophus ferrumequinum*. II. The recurrent laryngeal nerve. J Comp Physiol 143:323–327.

Rübsamen R, Schweizer H (1986). Control of echolocation pulses by neurons of the nucleus ambiguus in the rufous horseshoe bat, *Rhinolophus rouxi*. II. Afferent and efferent connections of the motor nucleus of the laryngeal nerves. J Comp Physiol 159:689–699.

Saitoh I, Suga N (1995). Long delay lines for ranging are created by inhibition in the inferior colliculus of the mustached bat. J Neurophysiol 74:1–11.

Schmidt S (1988). Evidence for a spectral basis of texture perception in bat sonar. Nature 331:617–619.

Schmidt S (1992). Perception of structured phantom targets in the echolocating bat, *Megaderma lyra*. J Acoust Soc Am 91:2203–2223.

Schmidt U, Schlegel P, Schweizer H, Neuweiler G (1991). Audition in vampire bats, *Desmodus rotundus*. J Comp Physiol A 168:45–51.

Schnitzler HU (1972). Control of Doppler shift compensation in the greater horseshoe bat, *Rhinolophus ferrumequinum*. J Comp Physiol 82:79–92.

Schuller G (1985). Natural ultrasonic echoes from wing beating insects are encoded by collicular neurones in the CF-FM bat, *Rhinolophus ferrumequinum*. J Comp Physiol 155:121–128.

Schuller G, Pollak G (1979). Disproportionate frequency representation in the inferior colliculus of Doppler compensating greater horseshoe bats. Evidence for an acoustic fovea. J Comp Physiol 132:47–54.

Schuller G, Rübsamen R (1981). Laryngeal nerve activity during pulse emission in the CF-FM bat *Rhinolophus ferrumequinum*. I. Superior laryngeal nerve (external motorbranch). J Comp Physiol 143:317–321.

Schuller G, Suga N (1976). Laryngeal mechanisms for the emission of CF-FM sound in the Doppler shift compensating bat, *Rhinolophus ferrumequinum*. J Comp Physiol 107:253–262.

Schweizer H (1981). The connections of the inferior colliculus and the organization of the brainstem auditory system in the greater horseshoe bat, *Rhinolophus ferrumequinum*. J Comp Physiol 201:25–49.

Schweizer H, Rübsamen R, Rühle C (1981). Localization of brain stem motorneurons innervating the laryngeal muscle in the rufous horseshoe bat, *Rhinolophus rouxi*. Brain Res 230:41–50.

Shannon-Hartman S, Wong D, Maekawa M (1992). Processing of Pure-tone and FM stimuli in the auditory cortex of the FM bat, *Myotis lucifugus*. Hear Res 61:179–188.

*Simmons JA (1979). Echolocation and pursuit of prey by bats. Science 203:16–21.

Speakman JR, Racey PA (1991). No cost of echolocation for bats in flight. Nature 350:421–423.

Stanek VJ (1933). K Topograficke a Srovnavaci Anatomii Sluchoveho Organu Nasich Chiroptar. Nakladem Ceske Akad Ved A Umeni, Praha.

Suga N (1970). Echo-ranging neurons in the inferior colliculus of bats. Science 170:449–452.

Suga N (1976). Disproportionate tonotopic representation for processing CF-FM sonar signals in the mustache bat's auditory cortex. Science 194:542–544.

Suga N, Neuweiler G, Moller J (1976). Peripheral auditory tuning for fine frequency analysis by the CF-FM bat *Rhinolophus ferrumequinum*. IV. Properties of peripheral auditory neurons. J Comp Physiol 106:111–125.

Suga N (1990). Cortical computational maps for auditory imaging. Neural Networks 3:3–21.

Suga N (1994). Multi-function theory for cortical processing of auditory information: implications of single-unit and lesion data for future research. J Comp Physiol A 175: 135–144.

Suga N, O'Neill WE (1979). Neural axis representing target range in the auditory cortex of the mustached bat. Science 206:351–353.

Surlykke A, Miller LA (1985). The influence of arctiid moth clicks on bat echolocation: Jamming or warning? J Comp Physiol 156:831–843.

Suthers RA, Fattu JM (1982). Selective laryngeal neurotomy and the control of phonation by echolocating bat, *Eptesicus fuscus*. J Comp Physiol 145:529–537.

Suthers RA, Hartley DJ, Wenstrup JJ (1988). The acoustic role of tracheal chambers and nasal cavities in the production of sonar pulses by the horseshoe bat, *Rhinolophus hildebrandti*. J Comp Physiol 162:799–814.

Vater M, Feng AS, Betz M (1985). An HRP-study of the frequency-place map of the horseshow bat cochlea: Morphological correlates of the sharp tuning to a narrow frequency band. J Comp Physiol 157:671–686.

Waters DA, Jones G (1994). Wingbeat-generated ultrasound in noctuid moths increases the discharge rate of the bat-detecting A1 cell. Proc R Soc Lond 258:41–46.

Waters DA, Jones G (1995). Echolocation cell structure and intensity in five species of insectivorous bats. J Exp Biol 198:475–490.

Wenstrup JJ, Grose CD (1995). Inputs to combination-sensitive neurons in the medial geniculate body of the mustached bat: The missing fundamental. J Neurosci 15:4693–4711.

Wilson JP, Bruns V (1983). Basilar membrane tuning properties in the specialized cochlea of the CF-FM bat, *Rhinolophus ferrumequinum*. Hear Res 10:15–35.

Wilson JP, Bruns V (1983). Middle-ear mechanics in the CF-FM bat, *Rhinolophus ferrumequinum*. Hear Res 10:1–13.

7

VISION, OLFACTION, AND TASTE

7.1 VISION

ARE BATS BLIND? The answer to this common question is an emphatic "no." Even the tiny eyes of echolocating Microchiroptera are sensitive to light. Megachiropterans, with the exception of the genus *Rousettus*, do not possess echolocation and are highly visual animals that use their eyes to locate fruit-bearing trees during their nocturnal flights. The large, round eyes of flying foxes point nearly straight ahead. The angle of divergence between the axes of the two eyes is on average 20°. In lemurs the angle of divergence is 10°, and in primates, the axes of the two eyes are parallel. Animals in which the axes of the eyes are close to parallel have a large binocular visual field. This is important for animals that perform visually guided reaching with their front limbs.

FUNCTIONAL ANATOMY OF THE EYE

Both the Microchiroptera with their small eyes and the flying foxes with their large eyes fly at night to seek food. In contrast to the insectivorous Microchiroptera, which use acoustic orientation to guide their acrobatic pursuit of prey, flying foxes use their vision to fly straight to the trees where they feed on fruits and flowers. These very different behavioral patterns are closely related to the structure and function of the chiropteran eye.

Optics. The imaging ability of the eye is determined partly by the mosaic of visual cells in the retina and their associated neural networks and partly by the optical properties of the lens, iris, pupil, and cornea. Three optical parameters determine the sharpness with which an image is focused on the retina: the size of the eye, its refractive power, and the aperture of the pupil.

Eye size. Just as in photography a large camera is able to produce a higher resolution image, in vision a large eye is able to produce a high-resolution image on the retina. Flying foxes, which depend on vision for orientation, have eyes that are about 12 mm in diameter (fig. 7.1a). The visual image is projected onto a hemispheric retinal surface that is about 225 mm^2. The eyes of the echolocating horseshoe bat, in contrast, are only 1.8 mm in diameter and the retinal area approximately 5 mm^2.

Refractive power. According to the laws of optics, the refractive power of a lens depends on the curvature of its surface (*r*), the refractive index (*n*) of the material through which the light is transmitted, and the distance between the refractive surfaces of the lens (see box 7.1). In the eye, there are four refractive surfaces that include the inner and outer surfaces of the cornea and the inner and outer surfaces of the lens (fig. 7.1a). The refractive power and focal distance of the eye can be calculated on the basis of the curvatures of the four refractive surfaces, the distances of the refractive surfaces from one another, and the refractive indices of the different components of the eye (see box 7.1). The refractive power of the eye in the flying fox is 144 diopters; in humans it is 48.8 diopters.

Pupillary aperture. The diameter of the pupil not only determines how much light falls on the retina, it also influences acuity and depth of focus. When the pupil is constricted, acuity decreases because under this condition a light source is not projected as a point, but rather as an airy disk whose diameter is inversely proportional to the pupillary aperture. In contrast, the depth of focus is inversely correlated with the square of the pupillary aperture. Thus, constriction of the pupil improves depth of focus but at the same time decreases acuity.

Every photographer knows that a lens system with high refractive power and short focal distance produces small, bright pictures with a large depth of focus. On the other hand, the pictures made using telephoto lenses with long focal length and low refractive power have high resolution, but they are poorly illluminated and have a small depth of focus. Because it is impossible to maximize both spatial resolution and illumination with a single optical system, a compromise must be found that will allow both parameters to be optimized.

The optical systems of both flying foxes with their large eyes and Microchiroptera with their small eyes are characterized by a high refractive power and short focal distance. These properties are characteristic of *nocturnal eyes*, providing a high degree of illumination at the expense of spatial resolution (table 7.1). The eyes of flying foxes are large relative to their body size, as are the eyes of many nocturnal creatures including owls, goatsuckers, and lemurs. Large eyes maximize the amount of light collected and permit reasonably good acuity even in low illumination. Typical features characteristic of nocturnal eyes are:

- The corneal angle is large, correspondingly to a wide visual field.
- The cornea and lens are nearly spherical in shape, which minimizes the radius of the refractive surface regardless of the absolute size of the eye.
- The lens is large, often occupying half of the eye's volume.
- The spherical shape and large size of the lens result in a high refractive power and short focal distance.
- The retina is made up of densely packed rods.
- The high density of rods together with a low number of ganglion cells means that many rods converge on a single neural element. This convergence increases light collection but decreases acuity (table 7.1).

Accommodation. In mammals, the focal plane of the eye lies at the retina (fig. 7.1a) so that light rays reflected from distant objects enter the eye in parallel and

Box 7.1 Formulas and Definitions

Refractive power: $D = n_2 - n_2/n_2\, r$ [1/m = diopters]
where r = curvature
n = refractive index
n_1 (air) = 1.00
n_2 (medium) > 1.00
Focal length: $f = 1/D$ [m]

Total refractive power of the eye: $D_{eye} = D_{cornea} + D_{lens} - \delta\, D_{co}\, D_{le}\, n_2$
where δ = reduced distance between cornea and lens.

form a focused image. To focus light from nearby objects, the refractive power must be increased. This is accomplished through the action of the ring-shaped ciliary muscle attached to the lens by the suspensory ligaments. The ciliary muscle maintains the suspensory ligaments in a variable state of mechanical tension. When the ciliary muscle contracts, tension on the suspensory ligaments decreases, causing the curvature of lens to increase according to its elasticity. In this way the radius of curvature becomes smaller, the refractive power increases, and nearby objects can be brought into focus. This process is called accommodation. Nocturnal eyes like those of bats have poorly developed ciliary muscles, so it is thought that

Table 7.1 Morphometric measures for nocturnal eyes and a diurnal eye (human)

	Myotis sodalis	*Carollia perspicicillata*	*Phyllostomus hastatus*	*Pteropus giganteus*	*Rattus norvegicus*	Human
Diameter (mm)	1.68	2.62	3.94	12	6.4	48
Cornea/eye radius	0.78	0.92	0.89	0.93	0.96	0.32
Corneal angle (°)	146	176	164	130	148	68
Lens/eye diameter (%)	36	49	49	61	62	10
Refractory power (D)	965	634	410	144	324	60
Focal length (mm)	1.37	2.1	3.24	9.23	4.1	22.2
Refractive error (D)	10–12	5–7	3.5–4.5	2		0
Receptor cell density (thousand/mm^2)			416	672	51	147
Receptor cells / ganglion cell					275–520	1 foveal

From Neuweiler (1962) and Suthers and Wallis (1970).

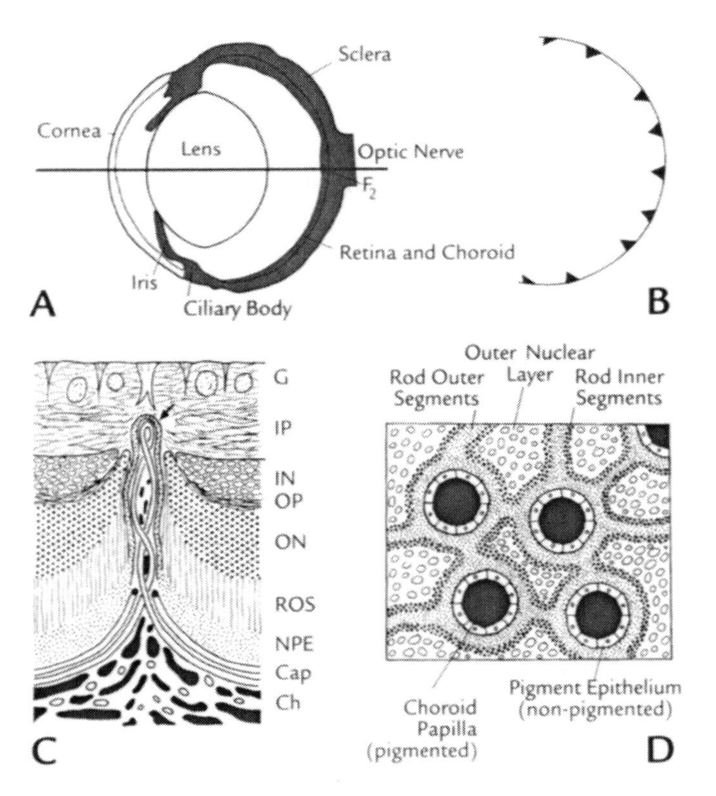

Figure 7.1 The eye of the flying fox (*Pteropus giganteus*). (A) Schematic drawing of the eye. Features in the structure of the eye that reflect its adaptation for night vision include a nearly spherical lens and a cornea with a high degree of curvature. The diameter of the eye is about 12 mm. F_2 = focal point. (B) The choroid papillae contain black pigment and are arranged so that when light shines on the retina they do not cast a shadow. The choroid cones in the middle of the retina are about 75 μm in height and about 85 μm wide at the base. (C) Schematic drawing of a choroid cone with capillary loop. ROS, rod outer segments; ON, outer nuclear layer (rod inner segments with nuclei); OP, outer plexiform layer (site of synapses between horizontal and bipolar cells); Ch, pigmented choroid, or vascular epithelium; Cap, capillary loop; G, ganglion cell layer; IN, inner nuclear layer, containing bipolar, horizontal and amacrine cells; IP, inner plexiform layer (site of synapses among amacrine, bipolar, and ganglion cells); NPE, nonpigmented epithelium. Arrow indicates turning point in capillary loop. (D) Horizontal section through the flying fox retina at the level of the rod outer segments. The usual uniform distribution of receptor cells is interrupted by the pigmented choroid papillae. From Neuweiler (1962) and Pedler and Tilley (1969).

they have little power of accommodation. In the flying fox *Pteropus*, physiological measurements demonstrate a change in refractive power of only 3 diopters, or 2% of the total refractive power of 144 diopters. In contrast, accommodation in humans is about 18% of the total refractive power of 48.8 diopters. However, because the lenses of flying foxes are already nearly round and thus provide excellent depth of focus, the lack of accommodation probably has little effect on the quality of the animals' vision.

Retina. The retina of bats is similar in structure to that of other nocturnal mammals. The receptor layer is composed of rods. In four genera of megachiropterans, the only visual pigment that has been found is rod rhodopsin. Thus, it seems improbable that bats have color vision.

Connections. The neuronal elements of the retina are organized into three synaptic layers: (1) a vertical "through pathway" from the photoreceptors, via the bipolar cells, to the ganglion cells. The axons of the ganglion cells pass across the retina to exit in the optic nerve. The optic nerve transmits to the brain information that has already been processed in the retina; (2) a tangential network of connections via horizontal cells, lying at the level of the synapses between the rods and bipolar cells; (3) a second network of lateral connections via amacrine cells, lying at the level of the synapses between the bipolar cells and the ganglion cells.

A large body of anatomical and physiological evidence shows that these lateral connections are crucial for the performance of the mammalian eye. They contribute to the ability to adapt the eye to different light levels (neuronal adaptation) and determine spatial resolution, contrast perception, and the perception of contours and movement. Each class of neuron accentuates and filters for specific parameters of the retinal image that are important for the animal's behavior.

Virtually nothing is known about the details of these interesting systems of retinal connections in bats. The retina of Microchiroptera contains a small number of neural elements and is only 100–150 μm thick. There are few bipolar and ganglion cells, with the result that many rods converge on a single ganglion cell. The retina of the echolocating horseshoe bat, *Rhinolophus rouxi*, contains only 4500 ganglion cells. The retina of a bat that gleans prey from the ground and has conspicuously large eyes, *Macroderma gigas*, contains 100,000–120,000 ganglion cells, approximately comparable to the eye of the rat.

The retinas of large flying foxes contain several hundred thousand rods, and those of humans more than 1 million. In all mammals the ganglion cells are most dense in the part of the retina that is most important for vision. This region of high visual acuity is called the *fovea*. Normally the horizon is represented on the foveal region, or area centralis, with high ganglion cell density. In Microchiroptera, the most important part of the retina is the lower temporal region. No retina of an echolocating bat contains a morphologically distinct fovea. Flying foxes fixate and follow a moving object with their eyes, and their retinas have an area centralis with a ganglion cell density of 10,000/mm^2, compared to only 1,200/mm^2 in the retinal periphery.

The lateral connections in the retina are dense and widespread. The dendrites of the horizontal cells overlap with one another in up to 80% of their total area. The fact that the eyes of bats contain few vertical connections but possess a dense system of horizontal connections suggests that they are specialized for detecting contours and motion in low-contrast nocturnal illumination, and that this is accomplished at the expense of acuity. In general, the pattern of neuronal connections in the chiropteran retina is what would be expected in an efficient nocturnal eye. The large, highly efficient eyes of flying foxes have a retina that is three to four times thicker and that contains many more neurons than the microchiropteran retina. Nevertheless, the connections are characterized by a high degree of convergence. In *Pteropus giganteus*, for example, the ratio of receptor cells to ganglion cells is 520:1 in the periphery and 271:1 in the center portion of the retina. In contrast, the human retina has a receptor to ganglion cell ratio of 125:1 in the periphery and 1:1 in the fovea.

Blood supply. Unlike most other mammals, bats do not possess the separate retinal circulation that normally enters the orbit with the optic nerve and then branches over the entire retina. The avascular retina of bats is supplied exclusively through the choroid, a highly vascular layer located behind the retina. The choroid is separated from the rod outer segments by a single-layered "pigment epithelium." In flying foxes, this epithelium does not contain any pigment, but the choroid is pigmented. The pigment epithelium plays an important role in the continual formation and breakdown of rod outer segment discs, which are made up of double-layered membranes and contain rhodopsin. The more active a visual receptor cell, the greater its membrane turnover and the more dependent this process is on the enzymes of the pigment epithelium. The pigment epithelium occupies a strategically important position between the choroid and the dynamic rhodopsin-containing membranes, transporting energy and materials from one side to the other.

The choroid of flying foxes. The structure of the choroid in flying foxes is unique among vertebrates (fig. 7.1b–d). Throughout the black-pigmented choroid are regularly spaced papillae that extend upward into the receptor cell layer to a height of about 130 μm and make the surface of the retina look like a volcanic landscape (fig. 7.1d). A blood vessel exits from the tip of each papilla. This vessel extends as far as the ganglion cells, turns around at that point, and returns to the tip of the papilla (fig. 7.1c). The rods that lie on the tips of the choroid papillae in flying foxes are up to 130 μm closer to the center point of the eye than are those that lie in the depressions between papillae. The choroid papillae and their looped capillaries are covered by the pigment epithelium which serves as a "metabolic interface." Because of the choroid papillae, the eye of the flying fox shows a marked departure from the usual retinal organization of meticulously maintained plane receptor sheets. A plane receptor surface is generally thought to be a requirement for a focused visual image.

The adaptive value of this remarkable structure remains a puzzle. The form of the papillae varies from the center of the retina to the periphery in such a way that no shadows are produced by light entering the eye (fig. 7.1b). This adaptation shows that the papillae are not optically insignificant. The distribution of the choroid papillae is such that continuous movements of retinal images are transformed into movements that rhythmically speed up whenever they pass over the slope of a papilla. It is possible that this transformation facilitates movement detection under low illumination. The capillary loops that run vertically through the retina provide a rich blood supply to the retina. This may be important because under the low illumination conditions of night vision, the rhodopsin content of visual receptor cells needs to be maximized. The presence of the choroid papillae results in a doubling of the contact surface between the pigment epithelium and the choroid capillaries. This increase in contact surface between the rhodopsin-containing elements and the pigment epithelium, which is metabolically active in rhodopsin turnover, could thus provide a great advantage for a nocturnally active, highly visual animal.

It is remarkable that neither the large and highly developed eye of flying foxes nor the small and primitive eye of Microchiroptera has any retinal blood vessels. If flying foxes really did evolve from primates as suggested by the organization of their visual pathways (reviewed below), it is hard to see how or why they would have lost the original characteristic retinal blood supply and replaced it with the complicated system of choroid papillae and looped capillaries. The avascular retina of flying foxes not only suggests that they are more closely related to the Microchiroptera than to primates, but it distinguishes them from most of the rest of the mammalian orders as well. The only other mammals that have no retinal blood supply are a few rodents, ungulates, and edentates.

PERFORMANCE OF THE VISUAL SYSTEM

There have been a few behavioral studies of visual capabilities in bats, but no physiological experiments. Flying foxes with their huge nocturnal eyes can see more sharply than humans in the dark, but are not any more sensitive (fig. 7.2). The absolute visual threshold of both flying foxes and humans is around 10^{-7} asb (Apostilb; measure of light density). At a low illumination level of 4.5×10^{-6} asb, flying foxes can discriminate 9-mm black and white stripes from a gray background, corresponding to a visual angle of 50'. Under the same condition, humans are able to discriminate only 15-mm wide stripes. *Myotis myotis* and *Megaderma* were able to distinguish vertical from horizontal black stripes on a white background, even when the illumination in the room was so low that the experimenter was unable to see the animal. *Eptesicus fuscus* maintained its ability to discriminate sharp contrasts until nearly absolute darkness.

Thus, flying foxes are better at resolving contours at near-threshold illumination levels than are humans. In diurnal animals, vision shifts from cones to rods at dusk (10 to 10^{-2} asb), and acuity decreases by orders of magnitude. In flying foxes, on the other hand, acuity remains at around 20' until the illumination level sinks below 10^{-4} asb, after which acuity rapidly falls off (fig. 7.2). Thus, the nocturnal eye is adaptive in that it provides sharpened acuity, and presumably also highly

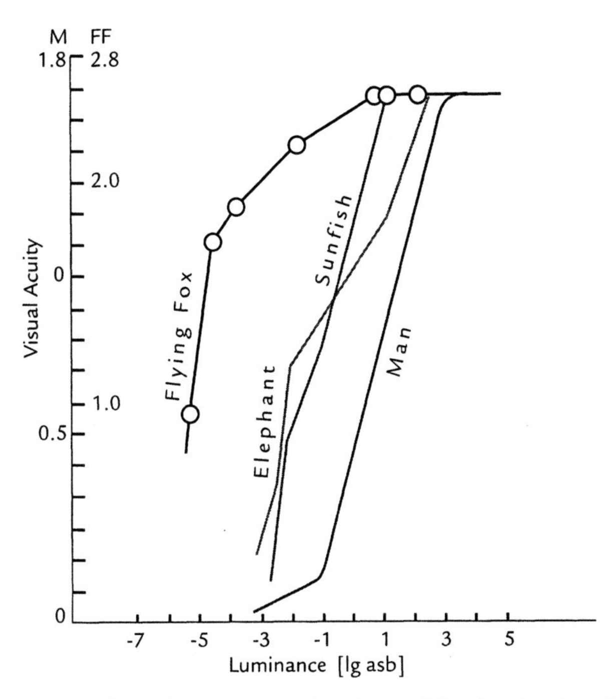

Figure 7.2 Relationship between visual acuity and illumination level in the flying fox *Pteropus giganteus* as well as humans , elephants, and sunfish. At low levels of illumination, the visual acuity of flying foxes is considerably better than that of diurnal animals. M, visual acuity scale for humans; FF, visual acuity scale for flying foxes. From Neuweiler (1962).

sensitive movement detection at low light levels, although there have been no experimental tests of this idea.

The spatial resolution for black and white stripes has been determined in a number of different bat species using behavioral experiments and the measurement of nystagmus in a rotating drum. These experiments demonstrated that resolution is better in bats with large eyes such as the flying fox where the limit of resolution is 18' than it is in bats with small eyes such as *Myotis lucifugus* where the limit of resolution is 360'. Among echolocating Chiroptera, the bats that prey on small animals and the frugivorous phyllostomids have large eyes. As a consequence, their visual acuity is better than that of vampires and insectivorous species (table 7.2).

There have been several behavioral studies of form vision in Microchiroptera that have never been systematically followed up. As a general rule, large-eyed phyllostomids perform better on tasks involving form discrimination than do other echolocating Microchiroptera. The few species that have been tested are all able to discriminate between vertical and horizontal stripes. Similarly, all are able to

Table 7.2 Visual acuity (in minutes of arc) in bats and other mammals

Species	Daylight	Dusk	Night
Flying fox			
Pteropus giganteus	18'	20–30'	47'
Frugivorous phyllostomids			
Carollia perspicillata		42' or less	
Saccopterix leptura		42' or less	
Anoura geoffroyi		42' or less	
Omnivores			
Phyllostomus hastatus		42–180'	
Artibeus jamaicensis		42–180'	
Vampires			
Diaemus youngi		180'	
Desmodus rotundus	48'	103'	158'
Insectivorous echolocating bats			
Myotis lucifugus		180–360'	
Myotis myotis		174–192'	
Chimpanzee	0.26"		
Cat	5.30"		
Rat	20'		

From Neuweiler (1962), Suthers (1966), and Manske and Schmidt (1979).

distinguish between erect and inverted triangles. Circles could be discriminated from squares regardless of their orientation, and all form discriminations could be made even when contrast was reversed. Vertical and horizontal contours and their position within the total visual field seem to play an important role in form discrimination. *Megaderma lyra* can distinguish a vertical bar from a tilted one, provided the angle of tilt is at least 15° from the vertical. *Myotis myotis* can also perform this task, but requires a tilt of at least 30° from the vertical. Both species could distinguish a horizontal bar from a vertical one provided the long side of the bar was at least 1.16 times as long as the short side. Neither *Megaderma lyra* nor *Myotis myotis* can discriminate between circles and squares of equal area; in contrast, the nectar-feeding bat *Anoura geoffroyi* makes this discrimination effortlessly even when the square is oriented in a diamond configuration.

In an experiment where a *Myotis myotis* was trained to make a visual discrimination between vertical and horizontal bars, it was able to generalize the learning of this discrimination to the modality of echolocation. In the visual experiment, the bat learned to discriminate between two metal bars measuring 3 × 6 cm, one of which was oriented vertically, the other of which was oriented horizontally. A glass plate was placed between the bat and the bars to prevent the use of echolocation. In absolute darkness with the glass removed, the bat was able to perform the discrimination on the basis of echolocation as well as it did using vision.

There are many observations to indicate that bats use vision for orientation. The African yellow-winged bat (*Lavia frons*) typically hangs in a bush and forages for large flying insects that would stand out against the night sky. Mice whose coats contrast with the background are caught significantly more often by *Megaderma* than are those whose coats match the background. Visual cues are used for general orientation by bats flying out of a cave, reentering a cave, and on long-distance flights. Echolocating bats that have been blinded fly closer to the ground than do bats with intact vision. This observation suggests that when foraging for insects, bats use their vision to control their flight altitude.

THE VISUAL PATHWAYS

In mammals, visual information is transmitted via the optic nerve to five different centers within the brain. In primitive mammals the majority of optic nerve fibers terminate in the *superior colliculus*, the midbrain visual center. In more advanced mammals this nucleus is less important, being concerned primarily with controlling movements of the head and eyes to track moving objects and with coordinating the direction of gaze with ear movements. In higher mammals, the majority of optic nerve fibers project to the *lateral geniculate body* in the thalamus. In the lateral geniculate, luminance contrast is enhanced, and in animals with color vision, color contrast is accentuated. The lateral geniculate projects in turn to the primary visual cortex in the forebrain (fig. 7.3). Visual information is also processed in (1) the *pretectal nuclei*. The pretectal nuclei are considered to be multimodal "data buses" for the association cortex. One of these structures, the nucleus of the optic tract, is part of a functional unit that also includes the nuclei of the accessory optic tract; (2) the three nuclei of the *accessory optic tract*. Neurons in the nucleus terminalis respond only to stimuli that move horizontally, in the temporal to nasal direction. Neurons in the lateral nucleus terminalis respond only to stimuli that move upward or downward. Neurons in the large medial nucleus terminalis respond only to vertical stimuli that move downward. These three nuclei are thought to play a role in coordination of head and eye movements to follow moving objects. In addition, the accessory nuclei transmit information to the cerebellum, a structure that is concerned with visuomotor coordination; (3) the *suprachiasmatic nucleus* (SCN in fig. 7.3). In mammals, this nucleus is a processing center for light signals that regulate circadian and seasonal rhythms. The pathway from the suprachiasmatic nucleus projects via the paraventricular nuclei to the pineal body. The pineal is entrained by light to produce melatonin on a circadian cycle. During the dark period, melatonin production is highest. It is thought that melatonin acts as a signaling molecule to regulate the circadian rhythms of many different bodily functions.

Lateral geniculate body. The visual pathways of bats are like those of other higher mammals (fig. 7.3). The majority of optic nerve fibers project to the lateral geniculate body. This thalamic nucleus is made up of two large subdivisions, the

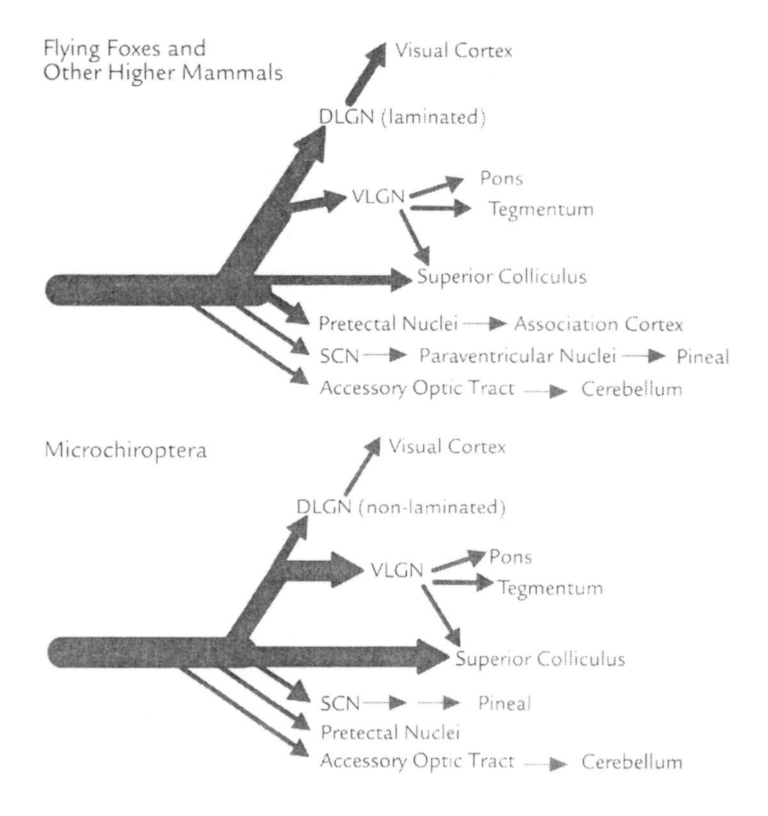

Figure 7.3 Diagrams comparing the visual pathways of microchiropteran bats with those of flying foxes (Megachiroptera) and higher mammals. The thickness of each arrow indicates the strength of the afferent projection. In Microchiroptera the majority of optic nerve fibers entering the thalamus project to the ventral lateral geniculate (VLGN); in Megachiroptera, to the laminated dorsal lateral geniculate (DLGN). SCN, suprachiasmatic nucleus.

dorsal lateral geniculate (DLGN) and the *ventral lateral geniculate* (VLGN). The dorsal lateral geniculate projects to visual cortex. The ventral lateral geniculate has connections with a number of different brain regions, including the tegmental thalamic nuclei, structures that are thought to play a role in the perception of illumination level, and to the visuomotor nuclei such as the pontine gray and superior colliculus (fig. 7.3). In echolocating Microchiroptera, the ventral lateral geniculate receives the bulk of the input from the optic nerve, suggesting that echolocating Microchiroptera use visual information mainly for orientation rather than for cognitive tasks. The pretectal nuclei of Microchiroptera are often so small that they are difficult to recognize.

In contrast to the Microchiroptera, most of the optic nerve fibers in flying foxes project to the dorsal division of the lateral geniculate (fig. 7.3). The highly structured organization of this input to the cortex reflects the importance of cognitive visual perception for flying foxes. The pretectal nuclei, especially the nucleus of the optic tract, are relatively large in both flying foxes and frugivorous phyllostomids such as *Artibeus jamaicensis*, a microchiropteran. In flying foxes, the accessory optic system is also prominent. These observations suggest that eye movements must be important for Megachiroptera as well as for frugivorous Microchiroptera.

In flying foxes the primary visual cortex (V1) extends over 120–140 mm^2 of the occipital forebrain. V1 contains a representation of the contralateral visual field and a narrow strip of the ipsilateral field along the vertical meridian (fig 7.4). The binocular field is organized along the vertical meridian and widens progressively from about 30° on the upper and lower zenith to 60° along the horizontal meridian.

The linear extent of a cortical region comprising all the neurons that respond to a given point in the visual field is called the "point image." The average point image decreases from 1.6 mm in the center to 1 mm at 20°–100° eccentricity. Large point images and smaller receptive fields in the central representation of the visual field indicate that this part of the visual field is magnified in the cortical representation and is larger than would be expected based on ganglion cell counts. Flying foxes track moving objects with eye and head movements to keep the image centered in the cortical area of high magnification. The smaller secondary visual cortex (V2) borders V1 and is a kind of mirror image of the map in the primary visual cortex. V2, however, emphasizes the lower visual field around the horizontal meridian where most of the flow of images occurs during flight. Thus, in flying foxes, the secondary visual field may be involved in monitoring the terrain below the flying animal.

Binocular connections. In primitive mammals the optic nerves of the left and right eyes remain separate. Each optic nerve crosses over completely to the contralateral side of the brain. Thus, subcortical nuclei receive no binocular information because they do not receive any significant input from the ipsilateral eye. In contrast, the dorsal lateral geniculate and superior colliculus of higher mammals receive binocular innervation.

In Microchiroptera, the organization of projections to the superior colliculus follows the pattern seen in primitive mammals (fig. 7.5). If an injection of a tracer substance such as horseradish peroxidase (HRP) is placed in the superior colliculus, it is taken up by axons that terminate at the injection site and is transported back to the neurons' cell bodies. In the case of microchiropteran bats, the labeled neurons are ganglion cells located in the contralateral eye. No labeled ganglion cells are found in the ipsilateral eye. Thus, each superior colliculus contains a representation of the complete contralateral visual field. If the same experiment is performed in a flying fox such as *Pteropus poliocephalus*, the results are quite different. HRP-labeled ganglion cells are found in the nasal half of the contralateral eye and in the temporal half of the ipsilateral eye (fig. 7.5). This means that the superior colliculus of Megachiroptera contains a binocular representation of the contralateral half of the visual field. This pattern is similar to that seen in primates.

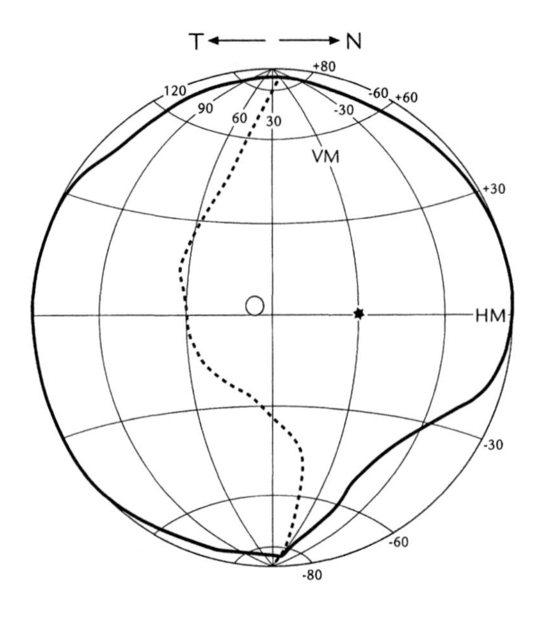

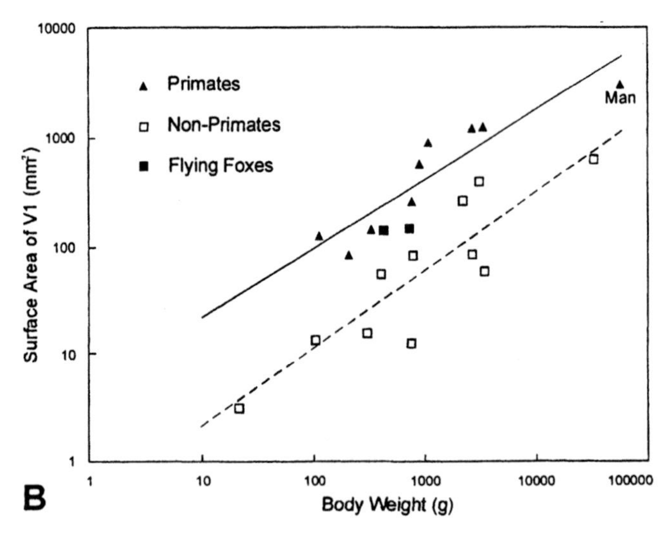

Figure 7.4 The binocular projections to the superior colliculus (midbrain) in Microchiroptera and flying foxes (*Pteropus*). In Microchiroptera, the optic nerves of both retinas cross completely. As a result, the object **G** is represented monocularly in both superior colliculi. In Megachiroptera, the optic nerves project as they do in primates, so that each superior colliculus receives input from either the temporal or the nasal half of the visual field, resulting in a binocular representation of half of the visual field in each superior colliculus. The object **G** is represented binocularly, but in only one superior colliculus, the left or the right, depending on its position in the visual field.

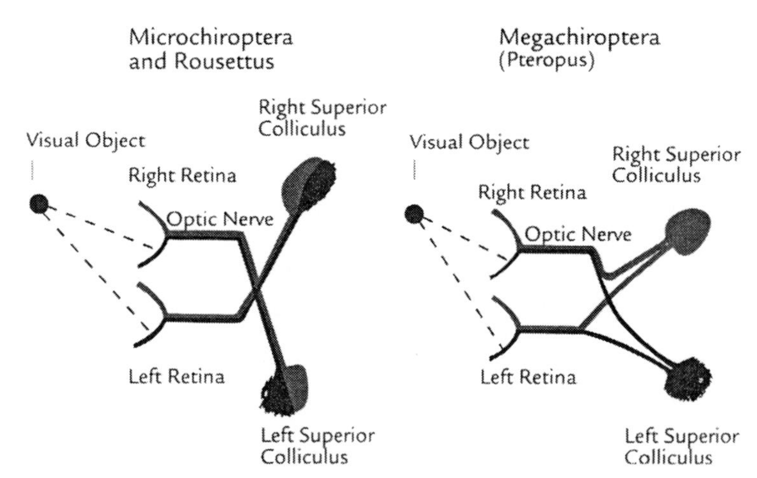

Figure 7.5 Schematic diagrams comparing binocular pathways to the superior colliculus (midbrain) in Microchiroptera and flying foxes (*Pteropus*). In Microchiroptera, the projections from the two retinas are completely crossed. Thus, a single object forms two monocular images, one in each superior colliculus. In Megachiroptera, the fibers of the optic nerve divide as they do in primates, so that each superior colliculus contains a binocular representation of the nasal or temporal half of the visual field. Thus, a single object forms a single binocular image in one of the two colliculi, depending on its position in the visual field.

The dorsal lateral geniculate of flying foxes also receives binocular inputs. This pattern of binocular connections, which is typical of tree-dwelling mammals, clearly distinguishes flying foxes from microchiropteran bats and reflects the fact that flying foxes are tree-dwelling animals whose feeding strategy requires precise visually guided reaching with the claw of the thumb.

Pettigrew published these neuroanatomical labeling experiments in 1986. At that time he suggested that on the basis of their primatelike visual pathways, flying foxes are derived from the primates and that the evolutionary origin of bats is diphyletic. Despite the fact that this conclusion seems so convincing given the very different organization of the visual pathways in Microchiroptera and Megachiroptera, it is necessary to keep in mind the large number of striking cases of evolutionary convergence in the animal kingdom. One example of convergence is the development of an acoustic fova in both Old World horseshoe bats and the New World mormoopid bat, *Pteronotus parnellii*, found in Central America. Thus, we cannot rule out the possibility that the binocular pattern of connectivity seen in the dorsal lateral geniculate and superior colliculus of flying foxes is not a pattern that has been handed down through phylogeny, but rather is a novel adaptation in response to the evolutionary pressure of visually guided reaching with the thumb. The idea that the similar patterns of visual projections in primates and flying foxes of the genus *Pteropus* represents evolutionary convergence is supported by visual projections in *Rousettus*, another genus of flying fox that lives in caves rather than trees. In *Rousettus*, the visual pathways resemble those of Microchiroptera.

7.2 OLFACTION

Mammals rely heavily on their sense of smell, as reflected in the progressive development of the nose during the course of evolution. The nose performs two important functions. First, as one of two respiratory passages, it regulates the temperature and humidity of the inspired air. Air entering the lungs is warmed and humidified; as air passes out of the body, heat and moisture are recovered and conserved. Second, the nasal olfactory epithelium signals the presence of odorant molecules in the inspired air.

Most mammals have a highly developed sense of smell. The earliest small mammals are thought to have possessed an olfactory system that made them superior to their late Mesozoic competitors in finding food sources. Thus, it is quite possible that their superior sense of smell was the deciding factor in the battle for survival. Scents that are specific to a given species or individual also play an important role in social communication, especially sexual behavior.

FUNCTIONAL ANATOMY OF THE NOSE

The development of the secondary palate in mammals resulted in a deepening of the nasal cavity so that it extends between the two orbits. This cavity, the "ethmoturbinal recess," is lined with the olfactory epithelium. The ethmoid bone at the rear of the nasal cavity is in direct contact with the braincase. The junction of these two bone layers forms the cribriform plate (fig. 7.6). This plate has perforations through which the fibers of the olfactory tract pass to reach the olfactory bulb, the first stage in the central olfactory pathway.

The two nostrils are separated from one another by a septum. The front, respiratory part of the nasal cavity turns downward to open into the throat; the more dorsal ethmoturbinal recess ends blindly as a cavity containing olfactory epithelium. During inhalation and exhalation, air passes over a labyrinth of bony lamellae, the turbinate bones (fig. 7.6b). These lamellae create eddies in the stream of air passing to the lungs and thus flush out the olfactory cavity. The turbinate ridges in the front part of the olfactory cavity are covered with mucosal epithelium, while those in the rear part of the cavity are covered with olfactory receptor epithelium. The respiratory portion of the turbinate region develops from the region of the septum and roof of the nasal cavity. In primates it is poorly developed, and in humans it is absent entirely. In Microchiroptera, the respiratory turbinate region is a smooth fold of skin with no underlying bony structure. The large maxilloturbinate bone (fig. 7.6a) extends from the lower rim of the lateral wall of the nasal cavity and is partly covered with olfactory epithelium.

The most complex lamellar organization is formed by the ethmoid bone at the base of the nasal cavity. The dorsal surfaces of these lamellae are completely covered by olfactory epithelium. The ethmoturbinate region occupies more than two-thirds of the nasal cavity (fig. 7.6a,b). Animals with a highly developed olfactory sense have up to nine lamellae in this region; bats have three to four main lamellae or endoturbinates, and one to three accessory lamellae or ectoturbinates. This pattern, also found in primates, is indicative of reduced olfactory capacity. It is

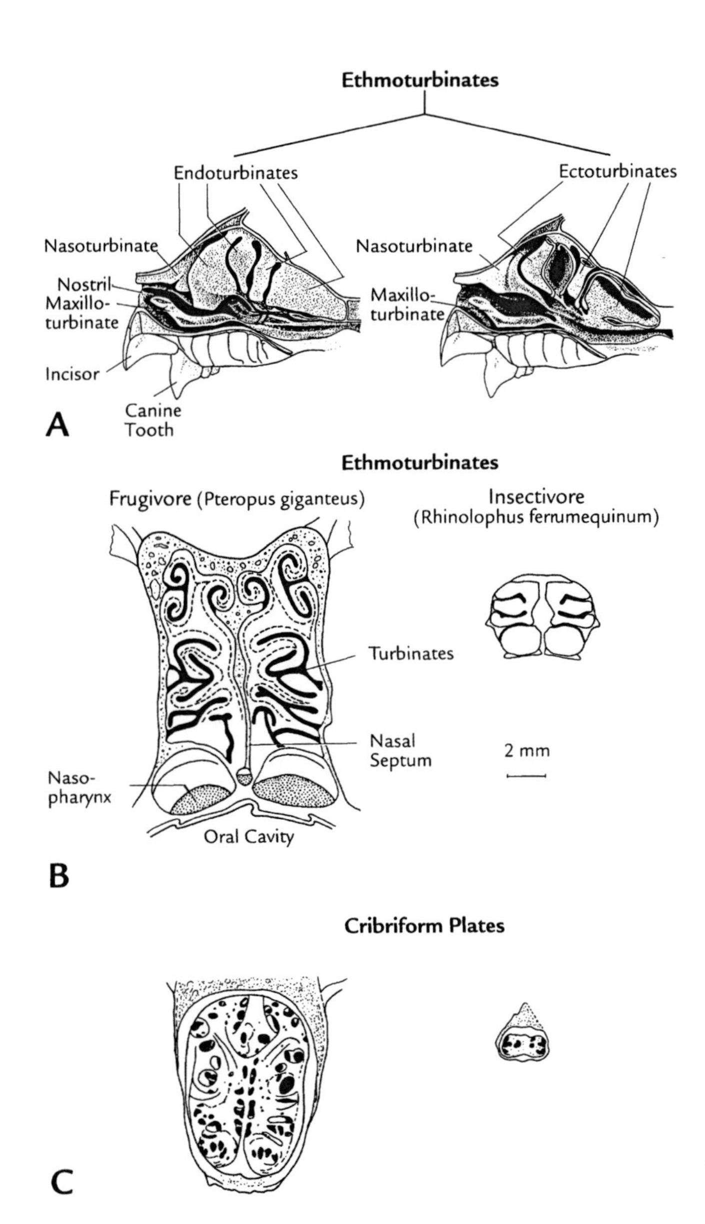

Figure 7.6 Nasal cavities in bats. (A) Nasal cavity of the vampire bat, *Desmodus rotundus*. Left: View of the endoturbinates as seen from the midline after removal of the outer wall of the nasal cavity. Right: View of the ectoturbinates after removal of the endoturbinates. (B) Transverse section through the nasal cavity (ethmoturbinates). *Left*: A frugivorous flying fox, *Pteropus giganteus*. *Right*: An insectivorous bat, *Rhinolophus ferrumequinum*. (C) View of the cribriform plate as seen from inside the skull. *Left*: *Pteropus giganteus*; *Right*: *Rhinolophus ferrumequinum*. Black areas indicate the perforations through which the olfactory nerve bundles pass. From Schmidt and Greenhall (1971), Bhatnagar and Kallen (1974), and Kamper and Schmidt (1977).

thought that the larger and more complex the ethmoturbinates, the better the sense of smell. In this regard, it is interesting to compare the ethmoturbinate structure of the fruit-eating bat *Pteropus* with those of the insectivorous horseshoe bat (fig. 7.6b).

The olfactory epithelium. The sensory epithelium also covers the septum and parts of the roof and sides of the olfactory cavity. Little is known about the structure and physiology of the olfactory epithelium in bats. The olfactory epithelium of mammals consists of a single layer of *supporting cells* and olfactory *receptor cells.* In bats, the receptor cells are 6–12 μm long and 4 μm thick. Beneath the epithelium lie many Bowman's glands, which empty onto the epithelial surface, coating it with a 50-μm thick layer of mucus. The mucus moves in the direction of the throat by the action of cilia. The supporting cells provide a carpet of microvilli out of which the tops of the receptor cells protrude. The peripheral process of each olfactory cell presents a tangential surface covered by a network of 5–20 nonmotile cilia about 200 μm in length, thus increasing the contact surface for odorant molecules by a factor of 100. Odorant molecules are adsorbed by the mucus coating and bind reversibly to the membranes of the receptor network, leading to a change in the receptor cell's membrane potential. This altered potential is called the receptor potential. The ciliary membrane contains specific proteins that act as binding sites for different classes of molecules. Each olfactory receptor cell has its own nonmyelinated axon that transmits excitation in the form of action potentials through the cribriform plate to the glomeruli of the olfactory bulb. The axons project in bundles through the perforations of the cribriform plate. Most olfactory receptors do not react to a single odorant, but to many different ones—sometimes even to many different classes of odorants.

Olfactory cells and taste receptor cells are the only sensory cells that continually die and are replaced. The life span of olfactory receptor cells is about 60 days.

The olfactory bulb. Millions of olfactory fibers converge on the mitral cells in the glomeruli of the olfactory bulb. The axons of the mitral cells are the only output of the olfactory bulb. They transmit information that has been processed in the olfactory bulb to the thalamus, the paleocortex, and the neocortical olfactory area. The olfactory bulb contains about 1000 glomeruli, each of which represents a unitary circuit made up of about 100 neurons. Each glomerulus is the target of about 10,000 to 100,000 afferent fibers, resulting in a convergence ratio of 100:1 to 1000:1. The net effect of the glomerular circuitry is to inhibit the activity of the mitral cells. Presumably it is the contrast effects produced by glomerular inhibition within the olfactory bulb that lead to specific olfactory perceptions.

Morphometric data. Although physiological data are lacking, there have been morphometric studies correlating the structure of the olfactory organs and the presumed olfactory capabilities of bats with their diet. It is thought that olfactory sensitivity is correlated with the thickness and area of the receptor epithelium and with the number and density of olfactory receptor cells. For example, the olfactory epithelium of "microsmic" insectivorous Microchiroptera is only 50-μm thick, while that of the frugivorous flying foxes is 250 μm. Although table 7.3 does not take into account differences in body size among the different species, it shows that the size of the olfactory epithelium and the number of receptor cells in bats is one to two orders of magnitude less than that of "macrosmic" animals such as dogs or rabbits. Among Chiroptera, the olfactory epithelium of frugivorous species is up to seven times the size of insectivorous species. For example, the olfactory epithelium in *Artibeus jamaicensis* measures 116 mm^2, whereas that of *Rhinolophus bocharicus* measures only 16 mm^2. Without knowing the sensitivity and specificity of olfactory receptor cells for behaviorally relevant odorants in the various species being compared, these morphometric correlations remain tentative.

If one makes the assumption that the size and number of perforations in the cribriform plate is correlated with the number of olfactory nerve fibers and that the number of olfactory fibers is correlated with olfactory performance, then a correlation can be drawn between morphologically determined olfactory capability and diet (figs. 7.6 and 7.7). In insectivorous bats, the number of perforations in the cribriform plate ranges from a few to about 75. The cribriform plate itself is small, about 15 mm^2. In carnivorous bats and those that feed on both insects and nectar, the cribriform plate is 12–23 mm^2 and has 30–100 perforations. In frugivorous bats, the cribriform plate may be as large as 54 mm^2 and have 75–145 perforations. In addition, the size of the olfactory bulb relative to the forebrain (in %) is correlated with diet. This relationship is greatest in frugivorous bats (41–60%) and smallest in the insectivorous species *Rhinolophus ferrumequinum* (25%).

Behavior. There is a wealth of behavioral observations indicating that even insectivorous bats use their sense of smell to locate and identify prey. For example, *Myotis myotis* can reliably locate and extract a dead insect or a paste made of macerated insects that has been hidden in moss or leaves. However, when the same insects are coated with plastic resin, the bats are unable to find them. Fecal analyses have shown that at certain times of year, *Myotis* feeds on crawling insects, spiders, and millipedes that it captures from the forest floor. Although there is no experimental evidence to prove it, it seems likely that the bats also use olfaction to detect their prey.

Potato beetles have a foul odor and are distasteful to *Myotis*. When normally palatable beetles are brushed with potato beetle paste, they are also avoided. Potato beetles are not the only insects that pump themselves full of bitter plant alkaloids; some moths do this as well, and are consequently unpalatable to bats. We have often observed a bat approach such prey, only to turn away at the last moment. It is likely that emitted odorants close to the prey signal its distastefulness.

Scent may be an important source of information used by fruit- and nectar-feeding bats and vampires to locate food sources. The Egyptian flying fox *Rousettus* is able to locate 0.1g of mashed banana paste while in flight and can distinguish it from banana oil. *Phyllostomus hastatus* is able to find pieces of banana hidden among the leaves of the jungle. Vampires exhibit distinct preferences for certain victims. The same animals within a large herd of cattle are visited by vampires night after night, and Swiss cattle are preferred over zebus. This selectivity is usually attributed to olfactory discrimination.

Table 7.3 Morphometric data on the olfactory systems of bats and other mammals

Species	Olfactory epithelium area (mm²)	No. of olfactory receptor cells (millions)	Diet	Source
Rhinolophus bocharicus	16		Flying insects	Bhatnagar and Kallen (1975)
Rhinolophus hipposideros	27.3	0.54	Flying insects	Kolb (1971)
Rhinolophus ferrumequinum	71.6	1.96	Flying insects	Kolb (1971)
Myotis lucifugus	18		Flying insects	Bhatnagar and Kallen (1975)
Myotis blythi	34		Flying insects	Bhatnagar and Kallen (1975)
Plecotus auritus	35	1.08	Insect gleaner	Kolb (1971)
Nyctalus noctula	176	5.37	Flying insects	Kolb (1971)
Myotis myotis	188	5.95	Insects, ground	Kolb (1971)
Artibeus jamaicensis	116		Fruit	Bhatnagar and Kallen (1975)
Human	1000	10		Shepherd (1983)
Cat	2080			Shepherd (1983)
Rabbit		100		Shepherd (1983)
Dog (German shepherd)	17,000	224		Shepherd (1983)
Deer	000	300		Starck (1982)

Olfactory thresholds. Olfactory thresholds have been measured in four species of bats that are thought to find prey on the basis of olfactory cues: vampire bats, the frugivorous species *Artibeus lituratus*, the omnivorous species *Phyllostomus hastatus*, and the insectivore *Myotis myotis*. Olfactory thresholds were measured based on changes in respiration rate, using an olfactometer and a negative conditioning behavioral paradigm. The bats' thresholds for short-chain alcohols, aldehydes, and acids was about the same as that of humans, but much less sensitive than those of dogs, which are macrosmic animals. For example, the thresholds for acids in bats are about six orders of magnitude higher than those of dogs. These data support the classification of bats as microsmic, based on morphometric data.

Among the four bat species tested, the olfactory thresholds of *Myotis*, an insectivore, were as much as eight orders of magnitude higher than those of omnivorous bats or those that feed on fruit or blood, especially for acids and alcohols. These threshold differences are also correlated with the number of olfactory receptor cells and the volume of the olfactory bulb, which in *Artibeus* is twice as large as in *Myotis*. Nevertheless, it is necessary to interpret these data cautiously

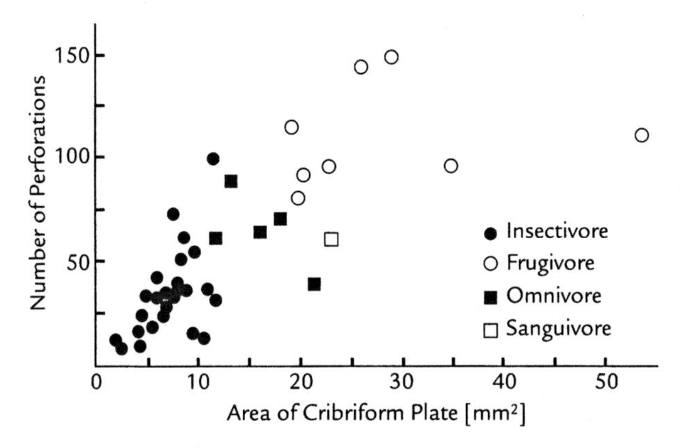

Figure 7.7 The size of the cribriform plate and the number of its perforations are correlated with the diet of different species of bats. It is assumed that the larger the cribriform plate and the greater the number of perforations, the better the sense of smell. From Bhatnagar and Kallen (1975).

because the differences observed might be due to differences in the spectrum of odorants detected by each species. This possibility was not considered in the above experiments, which are to date the only ones that have investigated olfactory sensitivity in bats.

CHEMICAL COMMUNICATION

In the social behavior of mammals, body odors and scent marks play an important role in identification of territory and sexual partners. When one visits a tropical cave, it is often possible to determine by smell, even before entering, which bat species will be found inside the cave. Hipposiderids, *Rhinopoma*, and *Taphozous* species all give off characteristic scents that are secreted by glands in the skin and in the urine.

Cutaneous glands that produce oily secretions are found in various places on the bodies of bats. Flying foxes have such glands on their shoulders, *Taphozous* under the chin, hipposiderids on the forehead behind the nose-leaf, horseshoe bats on the snout, *Saccopteryx* (Emballonuridae) on the lips, and *Noctilio* in the groin. The cutaneous glands are usually larger in males and are most active during mating season. In *Noctilio* the "scent pockets" in the groin region emerge at the same time of year that the testes descend into the scrotum. Because these scent pockets do not contain any obvious secretory glands, but do provide a substrate on which colonies of *Staphylococcus aureus* flourish, it is thought that the characteristic odor of *Noctilio* is produced by the bacteria.

Function of characteristic scents. No systematic experiments have been done on the function of characteristic scents, and there is only anecdotal evidence regarding their function. In large colonies and maternity roosts, it is likely that bats use scent marks to find their own roosting places and to identify other individuals. Some species of hipposiderids mark their special roosting place of the cave wall with their urine, and male noctules also mark the hollow trees where they roost with their own personal scent. Female flying foxes of the genus *Rousettus* probably use their sense of smell to identify their offspring, since in the dark caves where they live the only other means of mother–offspring identification and bonding would be a characteristic acoustic signal.

Large flying foxes of the genus *Pteropus* have patches of light yellow hair on their shoulders, beneath which lie glands that produce a brown, musky-scented secretion throughout the year. The animals rub this secretion over their fur and wings. Often so much secretion is produced that it drips from their fur. It is not known whether this secretion functions as an individual scent mark, whether it acts as a pheromone that synchronizes mating within the colony, or whether it acts as an antibacterial or antiparasitic agent.

Buchler observed that females and juveniles in a *Myotis* colony regularly visited an oak tree on the way to their foraging ground, rubbing their bellies against the bark of the tree trunk. Males always flew past the tree without paying any attention to it, whereas females and juveniles visited the tree for a short time every day. Within a short time, the bark of the tree had acquired the typical musky odor of the bats. Buchler suggests that this tree served as an olfactory landmark for young bats going out on their first flights between the colony and foraging ground. If this is indeed the case, it would be the first description in bats of the use of olfactory cues for orientation.

THE VOMERONASAL ORGAN

Many mammals possess a second, accessory olfactory organ that can be traced back in evolution through lizards and snakes to the urodeles. The components of this accessory olfactory system are (fig. 7.8):

- The vomeronasal organ, also called Jacobson's organ, located bilaterally on the lower part of the front wall of the nasal cavity,
- The vomeronasal nucleus, a specific cell group in the brain which is the source of unmyelinated afferents that innervate the vomeronasal organ,
- The accessory olfactory bulb.

Morphologically, the vomeronasal organ is a small, fluid-filled epithelial tube (fig. 7.8) containing sensory cells that resemble olfactory receptors. They differ from olfactory receptors in that they have long microvilli rather than cilia. It has been shown in hamsters that the vomeronasal epithelium is sensitive to odorants. The vomeronasal cavity opens via a small pore into the nasal cavity and is connected directly with the *nasopalatine duct* which opens into the mouth just behind the incisors. Through this oral pathway the vomeronasal organ receives nonvolatile protein-containing substances, requiring the ability to pump liquid into and out of the organ. The large veins (fig. 7.8) that extend the length of the organ on the side facing the nasal cavity act as hydraulic "vomeronasal pumps." Parasympathetic stimulation causes the veins to dilate, fill with blood, and press fluid

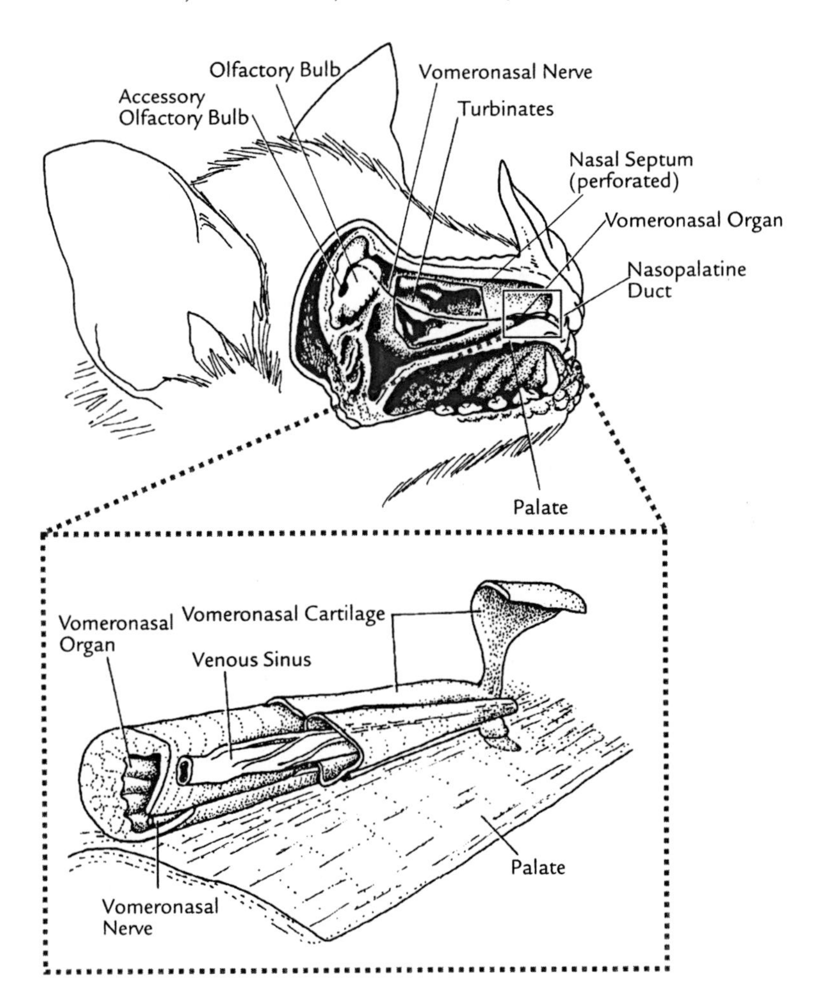

Figure 7.8 The vomeronasal organ (Jacobson's organ) in phyllostomids. From Cooper and Bhatnagar (1976) and Mann (1961).

out through the vomeronasal lumen. Sympathetic stimulation causes the veins to constrict, decreasing pressure within the organ, and drawing fluid in through the *vomeropalatine duct*. The entire tubular organ is surrounded by a cartilaginous cylinder, the vomeronasal cartilage, which opens onto the nasal cavity.

It is not known why some mammals have a functional accessory olfactory system and are thus *diosmic* while in other, monosmic mammals the vomeronasal organ is vestigial or absent. Insectivores, rodents, ungulates, lagomorphs, lemurs, and New World primates all possess a functional vomeronasal organ with an accessory olfactory bulb. Old World primates, anthropoid apes, and humans lack a vomeronasal organ. In bats, the situation is confusing. The vomeronasal organ and accessory olfactory bulb are completely absent in flying foxes, but the nasopalatine duct is conserved. In *Rhinopoma*, *Megaderma*, and the only two species of rhinolophids that have been examined, the accessory olfactory bulb is absent and the vomeronasal organ is more or less regressed depending on species. Among the 3 species of mormoopids and 13 species of vespertilionids that have been examined, only 1 species in each group has a functional vomeronasal organ: *Pteronotus parnellii* and *Miniopterus schreibersi*. In contrast, all 23 species of phyllostomids that have been examined to date have a fully functional vomeronasal system. The largest is found in the vampire.

Function. One can only speculate regarding the function of the accessory olfactory system in bats. In hamsters and rabbits it is known to mediate endocrine responses and behavioral reactions to sex *pheromones*. Pheromones are released from the vagina and through the urine of receptive females and are sensed via the vomeronasal organ of the male. During mating, males engage in a behavioral pattern referred to as "flehmen," in which they raise the upper lip, press the nostrils closed, and draw in air between the teeth of the exposed upper jaw. This reaction opens the nasopalatine duct and facilitates the entry of pheromones into the vomeronasal organ.

It is not understood how the vomeronasal organ regulates sexual behavior. Male mice respond to female pheromones with ultrasound vocalizations. Sexually experienced males vocalize even after removal of the vomeronasal organ, but if the vomeronasal organ is removed in an inexperienced male, it will never produce ultrasound vocalizations, even when caged continuously with females. In general, removal of the vomeronasal organ seems to depress certain forms of sexual behavior, but not to suppress them entirely. "Releasing hormones" that affect the secretion of hypophysial hormones have been found in the accessory olfactory bulb of rats and hamsters. Thus, it appears that afferent sensory input from the vomeronasal organ can modulate the endocrine control of certain behaviors. This interpretation is strengthened by the observation that the accessory olfactory bulb projects to the amygdala, a part of the limbic system known to play a role in sexual behavior.

Nothing is known about the role of the vomeronasal organ in the sexual behavior of bats, and "flehmen" has never been described in any species of bat. It is possible that pheromone sensitivity of the vomeronasal organ plays a role in synchronizing the reproductive cycles of individuals within a colony. Nevertheless, it is not clear whether the vomeronasal organ in *Miniopterus* performs such a function or how the other vespertilionid species that do not possess a functional vomeronasal organ manage to synchronize their reproductive cycles so precisely.

A more general function that has been suggested for the vomeronasal organ has to do with the examination of food in the mouth. This idea would be compatible with the highly developed vomeronasal organs found in vampires and frugivorous phyllostomids. The accessory olfactory bulb in these species are approximately twice as large as in the insectivorous species with functional organs. According to this hypothesis, however, one would expect the frugivorous flying foxes to have functional vomeronasal organs, which they do not.

7.3 TASTE

The nose leads the way to a source of food, but it is the papillae or taste buds of the tongue and soft palate that evaluate the quality of the food. Gustatory receptor cells do not have their own axons and, like the olfactory receptor cells, are constantly replaced. The life span of a gustatory receptor cell is about 10 days. Taste buds on the front of the tongue are innervated by the *chorda tympani* nerve, a branch of the facial nerve (cranial nerve VII). The cell bodies of these nerve fibers lie in the geniculate ganglion of the medulla. The caudal third of the tongue is innervated by the glossopharyngeal nerve (cranial nerve XI). The cell bodies of the nerve fibers are located in the petrosal ganglion. The taste buds of the soft palate and esophagus are innervated by a branch of the vagus nerve via the nodose ganglion. An important center for gustatory analysis is the *nucleus of the solitary tract*, a cell group that contains a topographic representation of the taste buds. The projections of this nucleus ultimately terminate in that part of the amygdala that does not receive olfactory input. In mammals, an additional taste pathway projects via the pons to the dorsal thalamus and cortex.

There are virtually no experiments on taste in bats. Recordings from the chorda tympani nerve in *Artibeus jamaicensis* have shown that the taste sensitivity of these frugivorous bats is similar to that of rodents and that they respond especially well to sodium chloride (NaCl). In contrast, the insectivorous bats, *Molossus* and *Myotis lucifugus* respond best to ammonium chloride (NH_4Cl) and potassium chloride (KCl), and less well to sodium chloride. Very little activity can be recorded in the chorda tympani in response to 0.5 M sodium nitrate or sodium sulfate; the response to sodium carbonate, however, elicits a response that is 10 times greater than that to sodium nitrate. The tip of the tongue is especially sensitive to sour stimuli. The high sensitivity to ammonium chloride is a property that the insectivorous bats share with carnivores.

It would be worthwhile to conduct further experiments on olfaction and taste in insectivorous and frugivorous bats, particularly because chemosensory experiences in mammals are stored for a long time in memory. Thus, it is possible that bats might learn very early in life, through one or a few contacts with odorants or taste stimuli, which insect species are distasteful or which fruits are especially nutritious. The study of chemosensory memory in bats would provide a useful general model for memory in mammals because bats live an unusually long time and are known to have excellent memories for cues used in orientation.

References

Vision

Chase J (1981). Visually guided escape responses on microchiropteran bats. Anim Behav 29:708–713.

Chase J (1983). Difference responses to visual and acoustic cues during escape in the bat *Anoura geoffroyi*: cue preferences and behaviour. Anim Behav 31:526–531.

Chase J, Suthers RA (1969). Visual obstacle avoidance by echolocating bats. Anim Behav 17:201–207.

Cotter JR (1985). Retinofugal projections of the big brown bat, *Eptesicus fuscus* and the neotropical fruit bat, *Artibeus jamaicensis*. Am J Anat 172:105–124.

Cotter JR, Pentney RJP (1979). Retinofugal projections of nonecholocating (*Pteropus giganteus*) and echolocating (*Myotis lucifugus*) bats. J Comp Neurol 184:381–400.

Manske U, Schmidt U (1979). Untersuchungen zur optischen Musterunterscheidung bei der Vampirfledermaus, *Desmodus rotundus*. Z Tierpsychol 49:120–131.

Neuweiler G (1962). Bau und Leistung des Flughundauges. Z Vergl Physiol 46:13–56.

Pedler C, Tilley R (1969). The retina of a fruit bat (*Pteropus giganteus*). Vis Res 9:909–922.

Pentney RJP, Cotter JR (1976). Retinofugal projections in an echolocating bat. Brain Res 115:479–484.

Pettigrew JD (1986). Flying primates? Megabats have the advanced pathway from eye to midbrain. Science 231:1304–1306.

Pettigrew JD, Dreher B, Hopkins CS, McCall MJ, Brown M (1988). The peak density and distribution of ganglion cells in the retinae of microchiopteran bats: Implications for visual acuity. Brain Behav Evol 32:39–56.

Rosa MGP, Schmid LM, Krubitzer LA, Pettigrew JD (1993). Retinotopic organization of the primary visual cortex of flying foxes (*Pteropus poliocephalus* and *Pteropus scapulatus*). J Comp Neurol 335:55–72.

Rosa MGP, Schmid LM, Pettigrew JD (1994). Organization of the second visual area in the megachiropteran bat *Pteropus*. Cerebral Cortex 4:52–68.

*Suthers RA (1966). Optomotor responses by echolocating bats. Science 152:1102–1104.

*Suthers RA (1970). Vision, olfaction, taste. In W.A. Wimsatt, ed. Biology of Bats, Vol. 2, pp. 265–309. Academic Press, New York.

Suthers RA, Chase J, Braford B (1969). Visual form discrimination by echolocating bats. Biol Bull 137:535–546.

*Suthers RA, Wallis NE (1970). Optics of the eyes of echolocating bats. Vision Res 10:1165–1173.

Thiele A, Vogelsang M, Hoffmann KP (1991). Pattern of retinotectal projection in the Megachiropteran bat *Rousettus aegyptiacus*. J Comp Neurol 314:671–683.

Olfaction and Taste

Bhatnagar KP, Kallen FC (1974). Cribriform plate of ethmoid, olfactory bulb and olfactory acuity in forty species of bats. J Morphol 142:71–90.

Bhatnagar KP, Kallen FC (1974). Morphology of the nasal cavities and associated structures in *Artibeus jamaicensis* and *Myotis lucifugus*. Am J Anat 139:167–190.

Bhatnagar KP, Kallen FC (1975). Quantitative observations on the nasal epithelia and olfactory innervation in bats. Acta Anat 91:272–282.

Bhatnagar KP, Matulionis DH, Breipohl W (1982). Fine structure of the vomeronasal neuroepithelium of bats: A comparative study. Acta Anat 112:158–177.

Buchler ER (1980). Evidence for the use of a scent post by *Myotis lucifugus*. J Mammal 61:525–528.

Cooper JG, Bhatnagar KP (1976). Comparative anatomy of the vomeronasal organ complex in bats. J Anat 122:571–601.

Defanis E, Jones G (1995). The role of odour in the discrimination of conspecifics by pipistrelle bats. Anim Behav 49:835–839.

Frahm HD (1981). Volumetric comparison of the accessory olfactory bulb in bats. Acta Anat 109:173–183.

Frahm HD, Bhatnagar KP (1980). Comparative morphology of the accessory olfactory bulb in bats. J Anat 130:349–367.

Gustin MK, McCracken GF (1987). Scent recognition between females and pups in the bat *Tadarida brasiliensis mexicana*. Anim Behav 35:13–19.

Kämper R, Schmidt U (1977). Die Morphologie der Nasenhöhle bei einigen neotropischen Chiropteren. Zoomorphologie 87:3–19.

Kolb A (1971). Licht- und elektronenmikroskopische Untersuchungen der Nasenhühle und des Riechepithels einiger Fledermausarten Z Säugetierk 36:202–213.

Laska M (1990). Olfactory sensitivity to food odor components in the short-tailed fruit bat, *Carollia perspicillata* (Phyllostomatidae, Chiroptera). J Comp Physiol A 166:395–399.

Loughry WJ, McCracken GF (1991). Factors influencing female-pup scent recognition in Mexican free-tailed bats. J Mammal 72:624–626.

Mann G (1961). Bulbus olfactorius accessorius in Chiroptera. J Comp Neurol 116:135–141.

Mendoza AS, Krishna A, Endler J, Kühnel W (1992). Die Regio olfactoria der Fledermaus Scotophilus heathi. Licht- und elektronenmikroskopische Studien. Ann Anat 174: 207–211.

Rieger JF, Jakob EM (1988). The use of olfaction in food location by frugivorous bats. Biotropica 20:161–164.

Schmidt U (1975). Vergleichende Riechschwellenbestimmungen bei neotropischen Chiropteren. Z Säugetierk 40:269–298.

Schmidt U, Greenhall AM (1971). Untersuchungen zur geruchlichen Orientierung der Vampirfledermäuse. Z. Vergl. Physiologie 74:217–226.

Shepherd GM (1983). Neurobiology. Oxford University Press, New York.

Starck D (1982). Vergleichende Anatomie der Wirbeltiere, Vol. 3. Springer-Verlag, Berlin.

Studier EH, Lavoie KH (1984). Microbial involvement in scent production in noctilionid bats. J Mammal 65:711–714.

Suthers RA (1970). Vision, olfaction, taste. In W.A. Wimsatt, ed., Biology of Bats, Vol. 2, pp. 265–309. Academic Press, New York.

Tuckerman F (1980). On the gustatory organs of some of the Mammalia. Z Morphol 4:151–193.

Tuckerman F (1988). Observations on the structure of the gustatory organs of the bats. J Morphol 2:1–5.

8

REPRODUCTION AND DEVELOPMENT

MAMMALS ARE SUPERIOR to all other animals in the extent to which they protect their embryos during development. Until birth, the mammalian embryo develops within the mother's body, shielded from outside influences, nourished by the mother, protected from infection by the mother's immune system, cushioned against disturbances in the environment, and maintained at an optimal temperature by the mother's own "thermostat." After birth, the young are fed by secretions of the mother's mammary glands until the time when their organ systems and behavioral programs are mature enough so that the young can obtain food on their own.

Lactation demands more energy from the mother than does pregnancy. For this reason the reproductive cycle is determined by the seasons so that the young are born and suckled during the time of greatest food availability. Although there is no way to extend or speed up the period of lactation, it is possible to alter the time course of pregnancy. Bats exhibit some especially interesting mechanisms for seasonal synchronization of their reproductive cycles, as well as prolongation or interruption of the sexual cycle at a variety of different functional stages.

8.1 MALE GENITAL ORGANS

About 90% of the volume of the testes consists of the seminiferous tubules, where sperm are produced. Between the tubules are the interstitial cells, also called *Leydig cells*, which secrete steroid hormones such as testosterone, androsterone, and estrogen. These hormones activate sperm production and the maturation of the secondary genital organs including the epididymis, the vas deferens, the seminal vesicles, and prostate gland, which secrete the seminal fluid, and the penis (fig. 8.1a). In all mammals the testes descend into the scrotum, outside of the body cavity, where the temperature is 1°–6°C lower than inside the body. This is important because sperm cannot form at temperatures as high as those in the body core. In bats, the testes descend only during the mating season.

The head of each sperm carries a haploid set of chromosomes. On the head of the sperm is a caplike structure, the acrosome, which plays an important role in penetration of the egg cell. The head of the sperm is connected by the neck to the middle piece and tail, which together form an organ of motility or flagellum (see fig. 8.6). The middle piece contains the mitochondria and thus acts as the energy source for the tail, which propels the sperm forward with beating movements. Mature sperm are stored in the epididymis, where they are maintained by the epithelial cells.

Mammals have many different ways of achieving the rigidity of the penis required for copulation. Some examples of these specializations are the fibroelastic penises of dogs, whales, and antelopes, and the baculum or penis bone of rodents. Bats are like primates and humans in that their penis contains erectile tissue, the corpora cavernosa and the corpus spongiosum. When the erectile tissue fills with blood, the penis stiffens. Many species of bats also have a small bone (i.e., a baculum) at the tip of the region of erectile tissue in the penis, and an additional area of erectile tissue at the tip of the penis, which, in some species, extends to the foreskin and causes it to stiffen as well. The result is that during copulation, the male and female are tightly locked together so that the male is unable to withdraw his penis until the erection subsides.

8.2 FEMALE GENITAL ORGANS

Female birds and mammals enter the world with a finite number of oocytes, which are stored in the paired ovaries (fig. 8.1b). A female bat has 3000–8000 oocytes. Oocytes are egg cells surrounded by a single layer of epithelium. As an egg matures, it becomes surrounded by a large, fluid-filled follicle with a multilayered wall that encloses the oocyte. The follicular epithelium produces estrogen, which promotes maturation of the egg. When ovulation occurs, the follicle bursts open and releases the egg into the Fallopian tube or oviduct, where fertilization takes place. After ovulation, the "egg-free" follicle temporarily becomes a hormone-producing endocrine gland, the corpus luteum. It continues to secrete some estrogen, but mainly secretes progesterone.

There is no direct contact between the ovary and oviduct. The paired oviducts are differentiated from proximal to distal into the Fallopian tube, uterus, and vagina (fig. 8.1b).

UTERUS

The uterus consists of a number of different, highly specialized layers. The outer layer or myometrium is made up of longitudinal and oblique layers of smooth muscle fibers that contract to expel the neonate. In *Myotis*, unlike other mammals, the myometrium receives dense adrenergic innervation. It is still an open question whether this is an isolated example, or whether it represents a general pattern of heightened neural control of the uterus in hibernating species. The inner uterine layer, or endometrium, consists of glandular tissue, blood vessels, and connective tissue. If pregnancy does not occur, the endometrium is either reabsorbed or shed.

Progesterone prepares the uterus for the implantation of the fertilized egg and induces the formation of the placenta, the organ that maintains the embryo. The

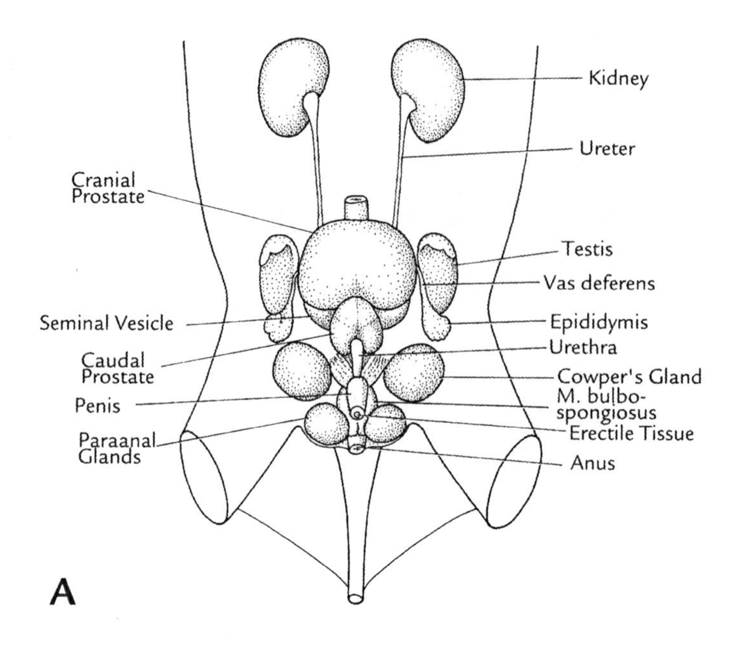

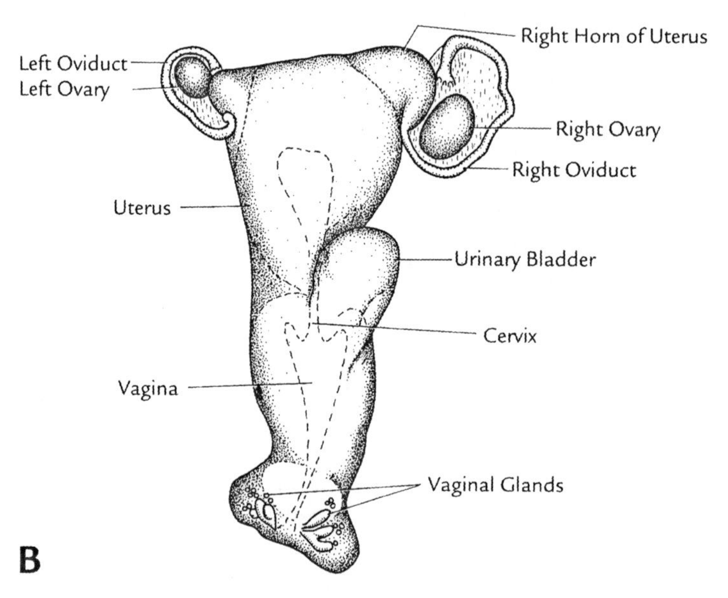

Figure 8.1 The reproductive organs of a bat (*Mormopterus planiceps*, Molossidae). (A) Male genitals; (B) female genitals. From Crichton and Krutsch (1987) and Krutsch and Crichton (1987).

mammalian egg contains little yolk and is surrounded by two membranes, the amnion and the chorion. The chorion and the allantois, which functions as the urinary bladder for the developing embryo, are connected to the placenta through dense interdigitations and vascularization of the endometrium. Although the embryo could potentially attach to any part of the endometrium, this does not happen. At the time when the fertilized egg is still making its way down the oviduct, a specific site is already being prepared for implantation of the blastocyst. The site and method of implantation varies from one bat species to another (fig. 8.2). The placenta produces its own hormone, chorionic gonadotropin, which maintains the corpus luteum and controls all functions having to do with maintenance of the embryo and preparation of the mammary glands for milk secretion. At birth, the placenta is expelled and eaten by the mother, and the corpus luteum disappears. It is interesting that menstruation (shedding of the previous endometrium shortly before ovulation) occurs in some species of bats. These include the vampire *Desmodus rotundus*, the two phyllostomids *Carollia perspicillata* and *Glossophaga soricina,* and the molossid, *Molossus ater*. Menstruation provides an efficient mechanism for bats to shed a well-developed endometrium in case of an embryonic loss or failure of fertilization. Menstruation is rare among mammals, except for primates and humans.

Only in primitive mammals such as monotremes and marsupials are there two separate vaginal canals; all other species have one. This fusion may also include to a greater or lesser extent the two uteri, resulting in the following morphological types (fig. 8.2):

- *Duplex uterus*. The two uteri are separate,
- *Bipartite uterus*. Only the part of the uterus that connects with the vagina is fused to form a single opening into the vagina,
- *Bicornuate uterus*. The caudal half of the uterus is fused and only the cranial parts of the two uteri remain separate, forming the two "horns" of the uterus,
- *Simplex uterus*. As in humans, both uteri are completely fused to form a single organ.

All of these forms are found in bats, from the duplex uterus in *Taphozous longimanus* (Emballonuridae) and the different species of *Pteropus*, to the simplex uterus in phyllostomids (fig. 8.2).

In mammals with bicornuate or double uteri and multiple embryos, the ovaries and uteri of both sides are active. In bats that give birth to a single offspring, often only the right side is active (fig. 8.2). Nevertheless, the inactive side remains potentially functional as shown by experiments in which the active side was removed.

Bats are generally monotocous, meaning that only one egg per cycle matures and is released for fertilization. However, three families of bats also include polytocous members. The flying foxes *Epomops dobsoni* and *Pteropus rufus* regularly give birth to twins, and twins have also been reported among *Rhinolophus* in India. The vespertilionids *Pipistrellus pipistrellus*, *Nyctalus noctula,* and *Vespertilio murinus* tend to give birth to twins in climates with harsh winters, whereas those that

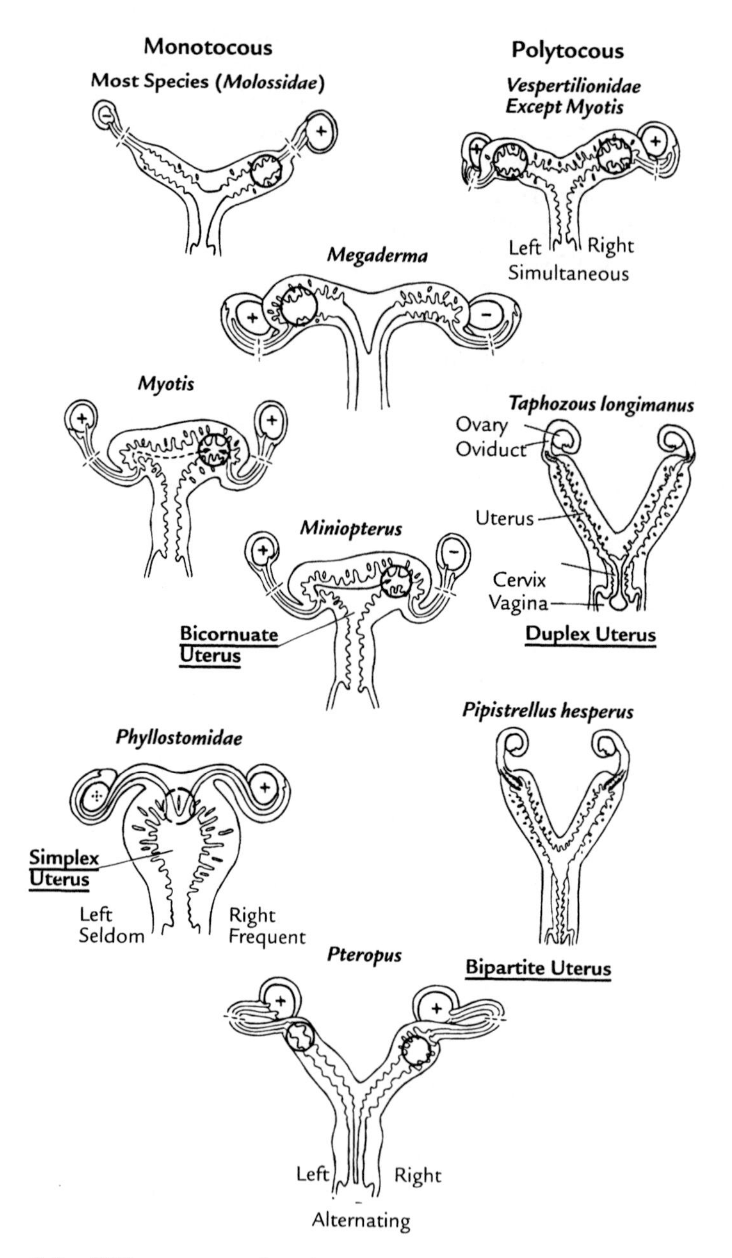

Figure 8.2 Different types of uteri, asymmetric ovulation, and implantation in bats. Ovulation and implantation of the egg usually occur on the right side in molossids, phyllostomids, and *Myotis*, on the left in *Megaderma* and *Miniopteris*. In *Taphozous longimanus* and *Pipistrellus hesperus*, both sides are active. Ovaries marked with a "+" produce mature eggs, ovaries marked with a "-" do not. Circled areas indicate sites of implantation of the fertilized egg. Names of the different types of uteri are underlined. From Wimsatt (1979) and Hill and Smith (1984).

live in mild climates usually give birth to only one offspring. *Lasiurus borealis*, a North American vespertilionid, gives birth to two or three young on average, but may bear as many as five. In biotopes where resources are unpredictable, it may be adaptive to fertilize multiple eggs (up to seven implanted embryos have been observed in *Eptesicus fuscus*), then, depending on food availability, to reabsorb the excess embryos. As a rule, *Eptesicus* give birth to only one or two young.

8.3 REPRODUCTIVE CYCLES

The precise timing of reproductive functions is controlled by sex hormones (figs. 8.3 and 8.4) that are produced by the testes, follicle, corpus luteum and placenta. The production of these hormones in both sexes is in turn regulated by two hormones produced in the adenohypophysis. Follicle-stimulating hormone (FSH) promotes the formation of sperm in the seminiferous tubules of the testes. In females, FSH maintains the follicle during its maturation, until the time of ovulation. It also promotes secretion of estrogen by the follicular epithelium and the corpus luteum. Luteinizing hormone (LH), together with FSH, stimulates the secretion of testosterone by the Leydig cells of the testes. In females, LH maintains the corpus luteum and stimulates progesterone production, thus allowing pregnancy to proceed. A third hypophyseal hormone, prolactin, acts synergistically with LH to stimulate the development of the mammary glands in preparation for lactation.

The hormones that control the reproductive cycle seldom act on a single organ or a single function. Usually their specificity and effectiveness depends on specific concentration ratios among the different hormones. During the period preceding ovulation, the main hormonal influence comes from FSH, during early pregnancy from LH, and during the second half of pregnancy and lactation, from prolactin. This schedule of events is valid for many nonhibernating bats.

The periodic nature of the reproductive cycle is especially obvious when one considers the ovulation cycle of female bats:

- *Proestrus*. An ovarian follicle begins to mature; the endometrium of the uterus begins to develop,
- *Estrus*. Ovulation occurs, either spontaneously or through an external stimulus such as copulation, e.g., in flying foxes and *Carollia* (Phyllostomidae),
- *Metestrus*. Fertilization occurs and the fertilized egg migrates into the uterus; the corpus luteum begins to develop,
- *Diestrus*. The fertilized egg implants; pregnancy occurs,
- *Anestrus*. Resting period during which the reproductive organs are inactive.

Animals that undergo only one such cycle per year are called monoestrous; animals that undergo multiple cycles are called polyestrous. In the tropics there are many polyestrous bat species (fig. 8.5). Some tropical bat species such as the molossid bat *Tadarida fulminans* ovulate while lactating. Such a "postpartum estrus" allows the bat to give birth to two or more litters during the hot, wet season

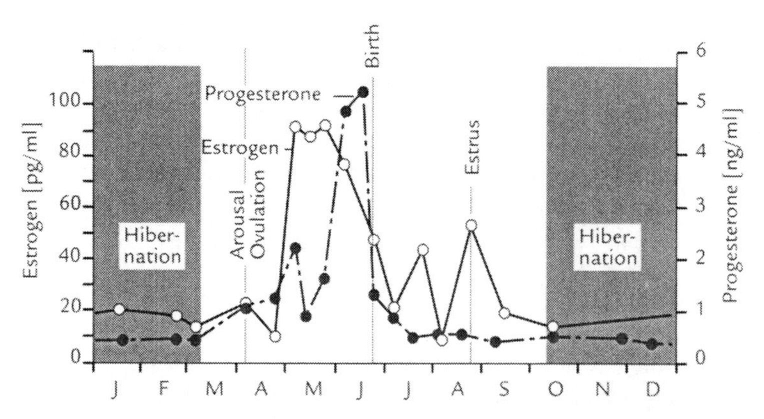

Figure 8.3 The female reproductive cycle and its hormonal control in a hibernating bat, *Antrozoous pallidus* (Vespertilionidae). Letters on the x-axis indicate months of the year. Bats mate during the autumn and winter. Solid line = blood levels of estrogen; dashed line = blood levels of progesterone. From Oxberry (1979).

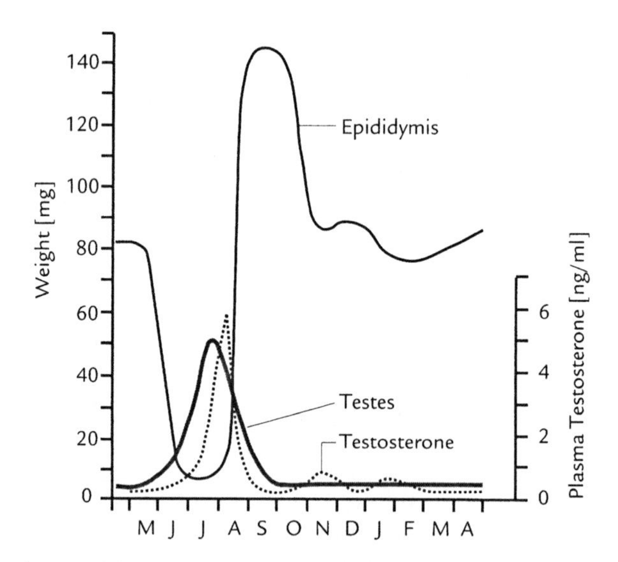

Figure 8.4 The asynchronous male reproductive cycle in *Myotis lucifugus* (Vespertilionidae). Letters on the x-axis indicate months of the year. Weight of the testes is correlated with spermatogenesis and testosterone production; weight of the epididymis is correlated with sperm storage. From Gustafson (1979).

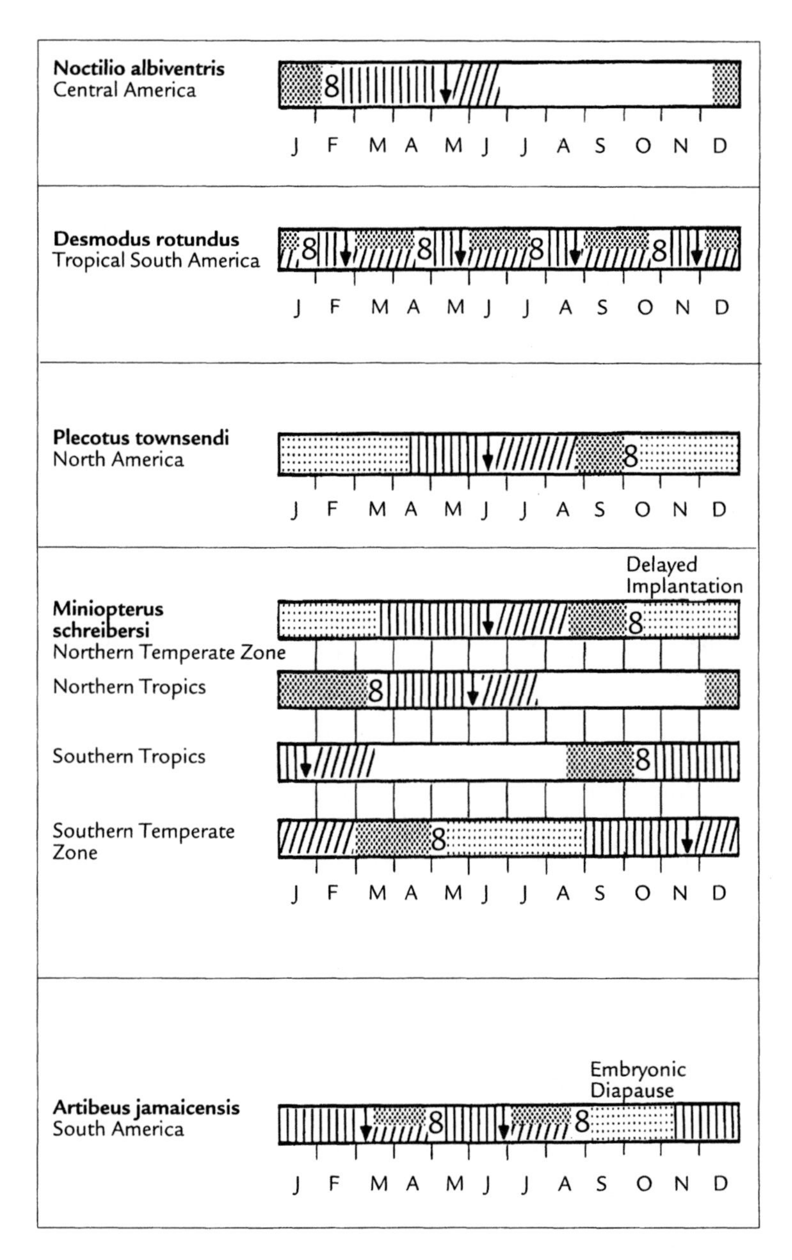

Figure 8.5 Adaptation of the reproductive cycle to seasonal changes in climate and food availability. Letters indicate months of the year. Dense stippling = estrus; 8 = mating; vertical stripes = pregnancy; diagonal stripes = lactation; white = anestrus; sparse stippling = hibernation with sperm storage, delayed implantation, or embryonic diapause; arrow = time of parturition. From Hill and Smith (1984).

when insects abound. Vampires, whose food source does not vary with the seasons, may reproduce continuously with an average of four estrous cycles per year. Phyllostomids have at most two estrus cycles per year, one before the start of the rainy season in March and April, and another at the end of the rainy season in July and August. During the dry season, the females are anestrous. The correlation of births and estrus with a particular season is especially obvious when one compares species that are distributed in both the Northern and Southern hemispheres. In Mexico, *Pteronotus parnellii* give birth from April through June, but in Peru, they give birth in October. This does not necessarily mean that biological cycles are determined by the geographical equator. *Taphozous melanopogon* in Bombay, India, give birth to their young in the spring; those in Sri Lanka give birth in the fall.

Most bats that live in the temperate zone are monoestrous (fig. 8.5). For temperate species, hibernation poses a serious problem; often the time of greatest insect availability occurs immediately following arousal from hibernation. This is the time when the young should be born because mothers need sufficient time to finish raising their young and then build up fat reserves to last through the next period of hibernation. Bats have solved this problem by timing the period of spermatogenesis in males and estrus in females so that it occurs in late summer or fall, and then interrupting the course of the reproductive cycle during the winter through various mechanisms (fig. 8.5).

DELAYED OVULATION AND FERTILIZATION

After copulation, sperm are stored in the uterus and oviduct. Ovulation and fertilization are delayed for weeks or months and do not occur until about 1–3 days after arousal from hibernation. Many hibernating species including vespertilionids and rhinolophids as well as tropical species such as *Pipistrellus ceylonicus chrysothrix* and *Noctilio albiventris* are capable of storing sperm.

The North American big-eared bat, *Plecotus townsendii* (Vespertilionidae) is a species in which delayed ovulation has been thoroughly studied. Copulation takes place in late fall and continues throughout the winter during temporary periods of arousal. Even though estrus begins in August and a fully developed follicle is present at the time of copulation, the sperm are unable to fertilize the egg. The egg remains enclosed in the follicle, surrounded by a "discus proligerus," a structure whose cells hypertrophy and store large quantities of glycogen. In the Japanese horseshoe bat, lipid is stored instead of glycogen. During hibernation, the egg uses the glycogen store for energy. During the entire hibernation period, the ovaries remain under the influence of FSH. In late February, LH secretion resumes and brings about ovulation. At this time, the sperm that have been stored in the uterus for many months can fertilize the egg.

On the basis of the enzymes present in the follicle during hibernation, it is possible to conclude that the reproductive cycle does not end, but is simply interrupted. Even though blood estrogen levels sink to a minimum during hibernation, the full complement of enzymes necessary for estrogen and progesterone synthesis are retained, disappearing only in the summer after the birth of the young.

THE PROBLEM OF SPERM STORAGE

Sperm are stored in either the caudal part of the epididymis or the vas deferens of the male (e.g., in *Tadarida brasiliensis*, *Rhinopoma microphyllum*, *Hipposideros speoris*, and *Rhinolophus capensis*) or after copulation in the uterus and oviduct of the female. Sperm storage in the female reproductive tract has been documented in 23 species of vespertilionids and rhinolophids as well as the nectar-feeding flying fox *Macroglossus minimus*. Whether in the epididymis or the uterus, sperm can remain alive and healthy for months. During the period of sperm storage, osmolality is extremely high in the caudal epididymis. How osmolality relates to longevity of sperm is not understood. Female noctules have been inseminated with sperm that was stored for 7 months in the epididymis of a male; these females subsequently gave birth to healthy pups. In *Pipistrellus pipistrellus,* sperm are known to survive in the uterus for up to 7 months. These observations raise two questions: Where do the sperm obtain the energy needed to remain intact and functional? When sperm are stored in the uterus, what mechanisms prevent an immune reaction of the host against "foreign protein" from occurring? After insemination of bats that do not have delayed implantation as well as in most other mammals, large numbers of white blood cells gather and phagocytize the sperm cells. It is not known what mechanisms prevent this immune reaction by the white blood cells in bats that store sperm in their uteri.

In hibernating female bats that have been inseminated, the uteri are often densely packed with sperm (6–10 million spermatozoa/μl in the noctule). The walls of the uterus or oviduct are lined with a special epithelium bearing many microvilli. This epithelium is covered with a thick carpet of sperm, each anchored head downward (fig. 8.6). The tip of the head of the sperm is surrounded by an outpouching of the uterine epithelial cells and is surrounded by many microvilli which wrap around it (fig. 8.6). It is thought that there are channels connecting the uterine epithelial membrane and the tip of the sperm cell. There is experimental evidence for glucose exchange between the uterine epithelial cell and the sperm cell. It seems unlikely, however, that any special nourishment of the sperm cell is required because the uterine fluids contain high concentrations of fructose and glucose, and the middle part of the sperm with its dense mitochondrial population probably has enough reserves of its own.

If the sperm must receive nutrients from the uterine epithelium in order to survive, a new problem arises due to the fact that most sperm cells float freely in the lumen, and only a small fraction are actually in contact with the epithelium. This means that the sperm would constantly have to take turns binding to their uterine "service stations." This is not likely to be the case because the high partial pressure of CO_2, together with special proteins in the uterine lumen, are thought to immobilize the spermatozoa, thus decreasing their energy demands.

Although evidence to support the idea that sperm are functionally anchored to the walls of the uterus and oviduct is increasing, the possibility that the close contact observed really represents a slow process of phagocytosis cannot be ruled out. It has been shown that in the Japanese horseshoe bat, as early as the beginning of

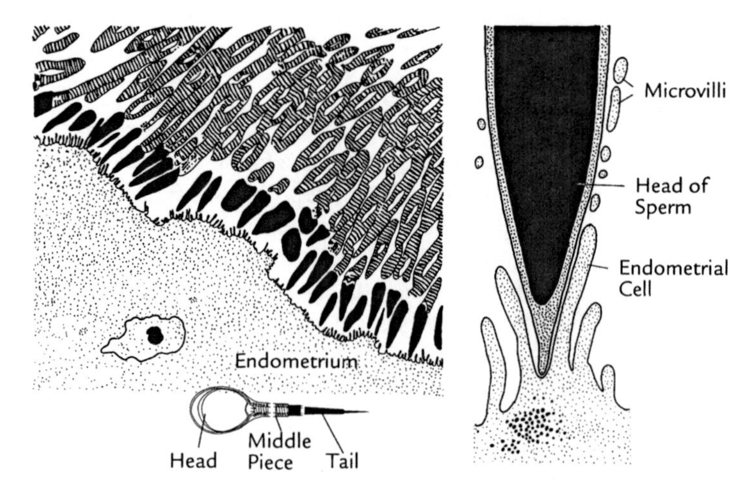

Figure 8.6 Sperm storage in the uterus of a hibernating bat (*Pipistrellus kuhli*, Vespertilionidae). Left: Sperm oriented with their heads (black) pointing toward the cilia-covered endometrium. The shaded structures are cross-sections through the mitochondria-rich middle pieces of the sperm cells. Right: At higher magnification (68,000×), it can be seen that an outpouching of the endometrium surrounds the head of the sperm and that microvilli wrap around it. Lower inset: Schematic view of a sperm cell. From Andreuccetti et al. (1984).

winter, sperm begin to penetrate the wall of the oviduct, where they are phagocytized by endometrial fibroblasts. Dead material is expelled into the lumen of the uterus, where it is taken up by leukocytes. In this way, phagocytosis is slowed down enough so that enough sperm survive in the uterus to fertilize the egg in the spring. The topic of sperm conservation is still controversial.

DELAYED IMPLANTATION

In species with delayed implantation, fertilization always occurs immediately after copulation, usually in the fall. The fertilized egg, or blastocyst, goes through several divisions and then enters an inactive state. Implantation is temporarily prevented, probably because LH and progesterone levels decrease rapidly following ovulation. After arousal from hibernation, LH levels once again increase, the blastocyst becomes implanted in the uterus, and the placenta forms. This method of interrupting the reproductive cycle has been described in the flying fox *Eidolon helvum*, in *Rhinolophus rouxi*, and in *Miniopterus schreibersi*. *Miniopterus*, which is distributed worldwide, provides a good example of the variability of the reproductive cycle, and of adaptation to the unique seasonal conditions of each biotope (fig. 8.5). Populations that live in the tropics have a monoestrous cycle that proceeds without interruption. In the tropical regions of the Northern Hemisphere, copulation occurs in February and March, and the young are born in June and

July. In the Southern Hemisphere, the cycle is reversed. Among hibernating populations in the Northern Hemisphere, copulation and fertilization occur in the fall, but implantation of the blastocyst is delayed until March, after arousal from hibernation. The same events take place, but in the opposite calendar sequence, in the Southern Hemisphere. Copulation and fertilization occur in March, implantation in September, and birth in December. Gestation time in this species is variable and depends on geographical latitude. The gestation period is 10 months in southern France (45°N), 8–9 months in Japan (32.5°N), 4.5–5 months in Malaysia (5°N), and 8 months in Australia (30°S).

EMBRYONIC DIAPAUSE

In the neotropical polyestrous species *Artibeus jamaicensis*, the egg fertilized on the first estrous cycle in the spring is implanted immediately, and embryonic development proceeds without interruption until birth of the young in July or August. A second estrus follows immediately. This time, however, the implanted blastocyst enters a resting phase with development resuming 2.5 months later so that the young are born in March, just before the large fruits ripen. In the nonhibernating bat *Macrotus californicus*, fertilization occurs immediately after copulation. However, embryonic development stops for 4.5 months during the cold season, so that the young are born in June after a gestation period of 8–9 months. The interruption is probably due to lowered levels of progesterone and estrogen, since the corpus luteum is virtually inactive during the cold season. In the flying fox *Haplonycteris fischeri*, development of the embryo can be interrupted for up to 8 months following implantation. Even though the actual time required for the embryo to develop is 3.5–4 months, the young are born only after a total gestation period of 11.5 months.

ASYNCHRONY OF THE MALE REPRODUCTIVE CYCLE

In flying foxes, as in most mammals, a large increase in plasma testosterone occurs during the mating season and is associated with increased testicular production. In contrast, in hibernating bats, plasma testosterone is at its lowest level during mating. Among hibernating bats, both males and females exhibit specializations in their reproductive cycles (fig. 8.4). In early summer, under the influence of increased LH production by the hypophysis, the Leydig cells increase their production of testosterone. This in turn stimulates sperm production. Testosterone production continues to increase rapidly so that by late summer the male bats' testosterone levels have reached the highest concentration found in any mammal. Then suddenly sperm production ceases. Perhaps this cessation is due to the extremely high testosterone levels because it has been shown in other mammals that high levels of testosterone inhibit sperm production.

By early fall, the secondary sex organs have reached their maximal size, and each epididymis is packed full of sperm. At this time testosterone concentration sinks to its lowest level and mating begins. Despite the animals' low testosterone levels, the secondary sex organs remain fully functional even during the many short periods of arousal from hibernation during the winter, when copulation continues to occur. In the spring, however, even though testosterone levels remain unchanged, the secondary sex organs undergo a regression. It is a mystery why the same hormone level that leads to a complete collapse of sexual function in the spring is sufficient to maintain sexual function during the winter.

In male bats that undergo the type of reproductive cycle just described, two processes that normally occur simultaneously are separated in time. In the summer, increased hormone production stimulates sperm production and sperm are stored in the epididymis. Only in late autumn, after sperm production has ceased and testosterone production is minimal, does copulation take place.

In the Indian bat *Rhinopoma microphyllum*, the asynchronous pattern is exactly the reverse. The testes of this species remain in the body cavity throughout the animal's whole life. Spermatogenesis takes place in the winter when the body temperature is low. Because sperm production does not occur at normal body temperature, it ends automatically when the male bats awake from hibernation, before the summer mating season.

As heterothermic mammals, bats are able to synchronize their reproductive cycles with hibernation or, in tropical regions, with periods of low food availability. They do this through a variety of mechanisms that include interruption of the reproductive cycle at different points and accelerating or slowing down embryonic development depending on energy supply. Interruptions in the cycle are controlled mainly through hibernation-induced changes in hormone production. This field of research, which is closely related to the study of other physiological rhythms, remains largely unexplored.

8.4 CONTROL OF RHYTHMS

The secretion of the hypophyseal hormones that control the production of sex hormones is in turn regulated by releasing hormones that originate in the neurosecretory centers of the hypothalamus. How do these hypothalamic neurosecretory areas obtain information about time of day or time of year? For all animals, sunrise and sunset are important timing cues or *zeitgeber* for daily (circadian) rhythms. The length of the daylight period and its continual change over the course of weeks are timing cues for yearly (circannual) rhythms.

CIRCADIAN RHYTHMS

If bats are kept in continual darkness, they sleep as usual during the day and attempt to fly out at night. This endogenous rhythm is independent of external cues and has a period of 23–26 h. Because the period is seldom exactly 24 h, animals kept in continual darkness either begin their nightly activity period slightly earlier every day, or slightly later every day. Under natural conditions, this internal clock is "reset" each day by the sun. In bats, sunset provides the cue that resets the clock. Bat colonies are very precise chronometers in that they leave their day roost every evening a set amount of time after sunset and are accurate to within a few minutes.

During the course of the year, the time bats fly out in the evening maintains a constant time relationship with the time the sun sets.

How does a bat hanging in a completely dark cave know when the sun sets outside? Currently there is no good answer to this question. Often part of the colony hangs near the cave entrance where they can perceive the amount of light over the course of the day. Presumably they are able to determine when sunset occurs by observing the gradient of daylight change, which is steepest around sunset. These animals could acoustically alert the rest of the colony hanging in the dark recesses of the cave through the increased activity that occurs just before they fly out in the evening. This type of synchronizing is called social synchronization. In other colonies, before the bats fly out in the evening, they all gather in indirectly lit areas in the front part of the cave that they use as a day roost. The parts of the cave that the bats use as "waiting rooms" have been called "light-sampling chambers," assuming that the animals use these areas to observe the light levels and thus set their inner clocks according to the sun. However, there is no experimental support for this hypothesis.

The epiphysis is an organ present in all vertebrates; its function is regulated by daylight. The neurosecretory cells of the epiphysis, the pinealocytes, produce the hormone melatonin, which has a synchronizing effect on many different vegetative functions. Melatonin secretion is suppressed by daylight and stimulated by darkness. Therefore, the shorter the day length, the greater the amount of melatonin secreted. The pinealocytes are derived from visual receptor cells. In nonmammalian vertebrates, a population of pinealocytes have retained their sensitivity to light. In mammals, the epiphysis receives information about light level from the visual system.

CIRCANNUAL RHYTHMS

Circannual cycles of body weight have been demonstrated in male *Antrozous pallidus* (Vespertilionidae). In September, about 1 month after the testes have reached their maximal volume, the males are twice as heavy as they are in February. This body-weight cycle is not correlated with testosterone secretion. Instead, like spermatogenesis, it can be induced as much as 1 month earlier by administration of melatonin. Thus, melatonin secretion by the epiphysis, together with endogenous rhythms, plays an important role in the control of circannual cycles. It has been shown in hibernating dwarf hamsters the epiphysis is involved in the regulation of many vegetative processes, especially in the control of reproductive cycles. Other evidence suggests that parameters such as seasonal changes in the quantity and quality of food that is available can control the synchronization of physiological processes.

In the equatorial phyllostomid *Anoura geoffroyi*, tests were conducted to determine whether the bats' annual reproductive cycle is synchronized by seasonal changes in the daily light regime. When the bats were kept under constant conditions (12 h 19 min light and 11 h 41 min dark), the testicular cycle free-ran with a period of 7.2–7.7 months. All light regimes that mimicked seasonal changes as they occur at the equator failed to entrain the seasonal reproductive cycle.

Apparently, in equatorial tropical mammals, the reproductive cycle may be governed by a truly endogenous rhythm that is not influenced by photoperiod. In bat species that live in temperate zones, the time of parturition may be determined by the climate. In horseshoe bats of Great Britain, birth timing varies considerably from year to year and is correlated with the mean temperature during spring. A rise of 2°C accelerated the date of birth by about 18 days (fig. 8.7). Bats are able to control gestation time quite precisely to synchronize the period of lactation with that of maximal food availability.

8.5 REPRODUCTIVE BEHAVIOR AND REARING OF THE YOUNG

"Genes are the replicators and individuals are their vehicles." This quote from Dawkins seems especially convincing when one considers the amount of time invested by animals in the selection of sexual partners and the rearing of their young, as well as the complexity and variety of the behaviors involved. In bats, the goal of producing offspring and rearing them to the age of sexual maturity is achieved through every imaginable form of sexual relationship, ranging from monogamy [e.g., *Saccopteryx leptura* (Emballonuridae), *Lavia frons* (Megadermatidae)] to harems or leks, to complete promiscuity. Promiscuous mating is often observed in bats of the genus *Myotis* (e.g., in Daubenton's bat, *Myotis daubentoni*). In this species, a male will mate with any available female. During periodic phases of arousal from hibernation, males will even mate with females that are still torpid, and pairs of copulating bats have been observed, both of which were hibernating. Promiscuity is most common among species that copulate during the winter, presumably because hibernating females are not able to reject males, and males are not active enough to seek out a particular female.

HAREMS

Reproductive activity can also take place in "harems," groups of females associated with a single male. Males of many species [e.g., *Nyctalus noctula* (Vespertilionidae), *Artibeus jamaicensis*, *Glossophaga soricina* (Phyllostomidae)] actively attempt to recruit females. They use display flights and special vocalizations to lure females into their day roosts. The males mark the day roosts as well as the females themselves with secretions from their scent glands.

Harems (fig. 8.8) of many tropical species (e.g., *Phyllosotmus hastatus*) gather passively. The *Phyllostomus* colony occupying a particular cave will divide into many different subgroups, each containing up to 25 females, a dominant male, and a large bachelor group made up of the nondominant males (about 80% of the males) and the young females. A dominant male will often preside over a harem for many years. In the frugivorous species *Carollia perspicillata*, however, the dominant male is replaced every 277 days on average. Females move freely from one harem to another, but mate only with dominant males. Thus, all of the offspring in a particular harem will inherit the genes of the dominant male. This

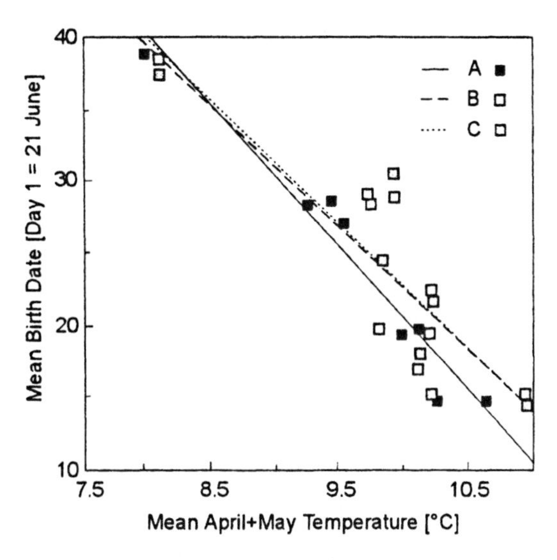

Figure 8.7 The effect of mean spring temperature on the mean birth date in *Rhinolophus ferrumequinum* of Great Britain. (A–C represent different colonies.) After Ransome and McOwat (1994).

Figure 8.8 A *Phyllostomus hastatus* harem. The large male hangs at the edge of a group of females. From McCracken and Bradbury (1981).

group organization obviously confers a significant reproductive advantage to the dominant male. Nevertheless, this system produces little genetic heterogeneity associated with the genome of the dominant male because the offspring leave the group and new females enter the harem randomly.

Even though the dominant males may defend their harems against other males, this is not because they have assembled the harem through their own efforts. In reality it is the groups of females that assemble first and are then joined by an older male. It is not clear what motivates the females to form groups. Harems are not groups of related individuals; females do not groom or feed one another or provide help in raising young. However, the females that belong to a harem all forage in neighboring areas, so it is thought that they interact according to the principle of reciprocity, alerting one another to the presence of food resources. When compared to a system of mutualism, in which conspecifics simultaneously help one another regardless of whether they are acquaintances or strangers, harem partners are better able to defend themselves against "cheaters" who take but do not give in return.

The term *harem* does not adequately describe the type of mutualistic reproductive community formed by *Carollia perspicillata.* In these groups, individuals reciprocate by helping one another, and the harems are typically actively formed by a group of usually unrelated females. The older males use these groups as a mating reservoir but do not appear to confer any additional benefit on the females.

Among vampires, reciprocity is clearly a basis for harem formation. In this species, too, the social unit is based on a group of females and their young. Such a group may use several tree cavities as their day roosts. The hollow trees are also occupied by a few males. The highest ranking male, who has risen to his position through fighting with other males, hangs in the most secure place, at the top of the tree cavity. He has the most access to the females, his copulation rate being about twice that of the other males. At night the hollow tree is often watched by males who defend it against intruders. Nevertheless, even though the female offspring are recruited by the males into their own harem, no heterogeneity is developed. The degree of relatedness among harem members remains less than 0.1 because both males and females frequently move from one hollow tree to another. Members of the group fly out alone at night to feed on blood from their victims, but they feed one another during the day when they have returned to their roost. Mothers regurgitate blood and feed it to their young. Any adult that has failed to obtain a blood meal during the night is also fed. Mutual grooming is thought to serve as a means by which a hungry bat can determine whether his neighbor has a full belly. Feeding lasts about 6–7 min and allows the recipient to survive for an additional 12 h. Thus the reproductive communities of vampires are based on mutual help among members of the colony and not on degree of relatedness as would be predicted by Dawkins's "selfish gene" hypothesis.

Saccopteryx bilineata (Emballonuridae), an insectivorous neotropical species, form true harems. Males defend the group of females that they have assembled in their day roosts by means of scent marking and visual displays, as well as vocal threats. The whole harem flies out as a group to forage for insects. The male defends the foraging territory against strangers, and the females thus receive some nutritional benefit. The young are all born synchronously, so that they become able to forage independently at the time when the supply of insects reaches a maximum. All of the young females leave the harem; about three-fourths of the young males remain in the harem, but in a state of sexual inactivity. These "cryptic males" may at some point have the opportunity to take over a group of females belonging to the harem of a male who disappears or dies. The degree of success depends strongly on the age of the males involved. This reproductive strategy is reminiscent of the strategies employed by many species of birds, in which the young males remain close to the nest, awaiting the chance to take over the nest and the female that occupies it.

SOCIAL FACILITATION

In the large, tree-dwelling colonies of the Indian flying fox *Pteropus giganteus*, there is a vertically organized social hierarchy among the males. The strongest males hang at the top of the tree, while males that are young or weak are allowed to hang only in the lowest branches. Often, sexually immature males are banished to a separate "bachelor tree." Within this male hierarchy, sexually mature females are accepted at any level. During mating season, a remarkable transfer of motivation to other males in the colony occurs. Following the first period of sleep in the morning, a few males start attempting to copulate with females. This activity is accompanied by long and loud intimidation cries. These cries stimulate nearby males to copulate as well, causing even more noise. In this way, copulation activity spreads like a wildfire through a colony of several hundred individuals, until the entire colony is a mass of screeching and copulating pairs. A half hour later, the colony is again peaceful and quiet. By means of such acoustic communication, all copulation within the colony is synchronized so that it occurs within a 30 to 40-min period each morning. This strategy decreases the chances of a female mating with multiple males and results in all the births in a colony of several hundred individuals occurring synchronously over a period of less than a week.

THE LEK SYSTEM

The African hammerhead bat, *Hypsignathus monstrosus*, has developed an unusual mating system. These bats mate twice a year, during the two dry seasons. During the mating season, all the males assemble in a particular place, known as a lek. In the lek, the males advertise for females and mating takes place there. In the evening the males fly to the river bank, where they light on trees. Within the lek, each male defends a territory approximately 10 m in diameter and returns each evening to his own place. Until about midnight the males tirelessly engage in courtship behavior meant to attract females. This behavior includes monotonous honking sounds and beating of the wings. It is the females who seek out their partners, based on some as yet unknown criteria. The most attractive males copulate with many different females, while about 70% of the males fail to find a partner. The males at the center of the lek are most attractive, while males at the periphery are less likely to be visited by females.

This unique mating system is related to a sex-specific diet and a sexual dimorphism that is rare among bats. Females are already sexually mature at 6 months of age, but the prominent secondary sexual characteristics of males do not appear until 1 year of age. The male secondary sexual characteristics include an enormous snout, an enlarged larynx for producing the honking sounds, and a weight gain of 60%. Mature sperm are not present in the males until 1.5 years of age.

In Gabon, where Bradbury discovered this lek system, *Hypsignathus* lives on two types of fruits: (1) fruits that are found on low trees that grow in groups. These trees bear fruits in various stages of ripeness over a period of many weeks and thus provide a source of food that is long term but relatively low quality. These trees are visited by females and sexually immature males; (2) fruits of different species of *Ficus* trees that grow in isolation and bear highly nutritious fruit for a short period of time. These trees are visited only by sexually mature males. Thus, the peculiar mating behavior is correlated with sex-specific feeding behavior.

MATING STRATEGIES AND FOOD AVAILABILITY

Bradbury, who has published most of the work on social behavior in bats, suggested that the different reproductive strategies employed by different species are specific adaptations to their different diets and foraging strategies. On the basis of a study of four sympatric insectivorous species of emballonurids, Bradbury concluded that in species that depend on periodically emerging insect populations, the time of lactation coincides with the time of maximal food availability. This situation is comparable to that in the frugivorous African flying foxes. Among species whose food supply is subject to negligible seasonal variations, it is pregnancy, not lactation, that occurs during the time of highest food availability.

The African yellow-winged bat, *Lavia frons* (Megadermatidae), spends its days beneath the leaves of *Acacia tortilis*. The reproductive cycle of *Lavia frons* is coupled to the vegetation cycle of this tree (fig. 8.9). During dry periods, the acacia loses its leaves, but after every rain it quickly produces new leaves and flowers. These sproutings attract many insects, which *Lavia frons* easily captures from its perch on the acacia branch. These bats form monogamous pairs, and the male defends the pair's roost and foraging territory. The birth of the young is synchronized with the beginning of a rainy period and thus with a rich supply of insects, especially swarms of termites.

Carioderma cor, another African species of megadermatid bat, also forms monogamous pairs. The male marks a territory by singing; only he and his mate are allowed to forage in this territory. Mating and marking of territories are only observed in coastal regions during the long dry periods when insects are scarce. In inland regions where the insect supply is more evenly distributed over the year and the population density of *Carioderma cor* is lower, each individual has its own territory, but does not mark or defend it.

The formation of monogamous pairs and defense of territory are presumably an adaptation to a food source that is stable in terms of its spatial distribution but highly variable in terms of its temporal distribution. In any case, the reproductive strategies of the species that have been studied to date are more closely correlated

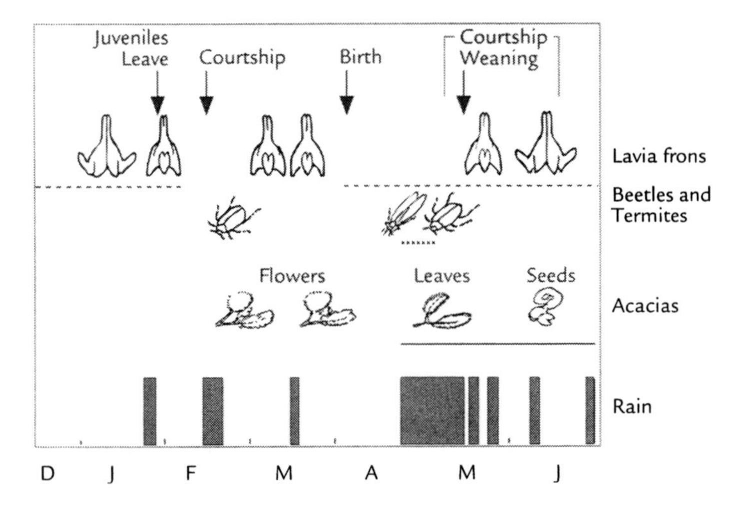

Figure 8.9 During the day the African bat *Lavia frons* (Megadermatidae) hangs in the branches of an acacia tree. In the evening it flys out to feed on flying insects. These bats coordinate their reproductive cycles with the seasons, so that all of the young are born during a rainy period when insect availability is high. At this time the trees are covered with dense foliage. Dashed line: Mothers and young are together. Dotted line: Time when termites swarm. Solid line: Flowering of acacia trees. From Vaughan and Vaughan (1986).

with diet and foraging strategies than with degree of relatedness. The comparative study of social behavior in bats would be a worthwhile field for social biologists.

PREGNANCY AND LACTATION

Pregnancy costs a female bat surprisingly little energy. Pregnant females of the species *Myotis lucifugus* (10 g bw) eat about 5.5 g of insects per night; lactating females eat about 6.7 g. The embryos of bats develop more slowly than those of any other mammal; at birth young bats are smaller than any other mammal relative to the length of the gestation period. However, if the weight of the young is compared to that of the mother, bats produce relatively larger offspring than do other mammals. The weight of the infant is typically 12–43% of the mother's weight. In comparison, the young of other small mammals are typically 7.8% of the mother's weight. Thus the relative biomass of a single bat pup exceeds that of an entire litter of young in other small mammals. Among bats, infant *Phyllostomus* are the most mature at birth. When they are born they already have fur, and their eyes are open. Newborn bats have good neuromuscular control, which they use immediately on the first day to cling to the ceiling of the nursery or to their mothers' fur.

The mother's energy costs for raising young are due to growth of the fetus during pregnancy and milk production during lactation. Heterothermic species such as *Plecotus auritus* (Vespertilionidae) compensate for the extra requirements of the fetus by going into torpor during the day, thus conserving energy. This strategy is abandoned during the later stages of pregnancy, at which time the mother's energy comsumption exceeds that of a homeothermic animal. During lactation, however, the mother is able once again to offset some of the ever-increasing energy demand due to milk production by entering a state of torpor during the day. Milk production and the associated energy turnover reaches a maximum just before the young are weaned. Based on the analysis of stomach contents, it was estimated that in female Mexican free-tailed bats (*Tadarida brasiliensis*) the average nightly energy intake ranges from 57 kJ/day in early lactation to 104 kJ/day in mid-lactation. This change in energy turnover corresponds to an increase in feeding rates of 39–73% of a female's body mass from parturition to mid-lactation.

Insectivorous species produce a highly nutritious milk and nurse their young for 4–8 weeks. Frugivorous species, on the other hand, nurse their young for up to 5 months. The mother's milk contains everything that the infant needs to maintain its metabolic rate and grow. The milk also contains symbiotic bacteria and immunoglobulins that confer the mother's immunity to disease on her offspring.

Because the composition of the mother's milk depends on her diet, it is not surprising that the milk of fruit- and nectar-eating bats contains more lactose and less protein than that of insectivorous species. In both cases the proportion of casein in the total protein content must be high because casein micelles bind large amounts of Ca^{2+} and phosphate. Table 8.1 compares the composition of bat milk with that of mice, cows, and humans. Bats generally secrete an energy-rich (i.e., fatty) milk. This type of milk is typical of mammals that nurse their young according to a set daily rhythm rather than at random times as needed. Because the mammary glands of bats can only synthesize short-chain fatty acids [≤ 14 carbon atoms (14C)], the palmitic acid (16C) and unsaturated 18C fatty acids that are the main lipid components of bat milk presumably come from the diet. During lactation, times of food scarcity such as would occur during cold weather are especially critical. On cold days *Myotis myotis* (Vespertilionidae) mothers do not return to the nursery colony. The young are left hanging alone, where they go into torpor until the return of the mother after the period of cold weather is over. This strategy allows both mother and offspring to subsist during the cold period with minimal energy expenditure. For a lactating female to maintain her body temperature when the ambient temperature is 14°C, she must increase her metabolic rate by 200%—a risky undertaking on cold days when few insects are available. Thus, heterothermy in bats is not only an important survival mechanism during the winter, but also during the rearing of young.

NURSERY COLONIES

Among many bat species, pregnant females assemble in a particular day roost, the nursery colony. An occasional lone male may be encountered at the periphery of the colony. Pups are raised in the nursery colony until they are old enough to fly, then the colony disbands. It has been shown that when lactating mothers and pups mass together, both expend less energy. A reduced energy requirement is a good

Table 8.1 Composition of milk in bats compared to that of mice, humans, and cows.

Species (diet)	Fat (g/100 g)	Lactose (g/100 g)	Casein (g/100 g)	Whey protein (g/100 g)	Citrate (g/100 g)	Energy (kcal/g)
Leptonycteris sanborni (nectar)	18.5	4.8	2.5	1.8	0.15	2.1
Vampyrodes caraccioloi (fruit)	29.0	4.1	0.83	2.3	0.09	3.0
Artibeus jamaicensis (fruit)	18.6	7.3	1.1	3.6	0.11	2.3
Artibeus cinereus (fruit)	23.0	3.8	0.57	3.4	0.06	2.5
Myotis lucifugus (insects)	6.0	3.1	3.8	3.5	0.19	1.1
Tadarida brasiliensis (insects)	16.3	2.8	3.0	3.2	0.21	2.0
Mouse (ground-dwelling omnivore)	13.1	3.0	7.0	2.0	0.005	1.9
Man (omnivore)	3.8	7.0	0.4	0.6	0.05	0.68
Cow (herbivore)	3.7	4.8	2.8	0.6	0.17	0.73

From Jenness and Studier (1976).

argument for why nursery colonies would form in temperate climates, but it does not explain why they would be adaptive in the tropics. Perhaps in the tropics pregnant bats assemble in a day roost that is close to an attractive source of food. Nursery colonies must have some special meaning for the bats because animals travel to the colony from a wide geographical area and stubbornly persist in returning to the same nursery roost for decades. A nursery colony in Poona, India, populated by an uncountable number of *Miniopterus schreibersi*, is reported to recruit animals from an area of 15,000 km^2.

MOTHER-INFANT AND ADULT SOCIAL INTERACTIONS

Although rearing young in large nursery colonies has many advantages, it also creates its own set of problems. When the mothers fly out to feed in the evening, they leave the young alone in the cave. In some species the young are left alone from the day of birth, in other species, beginning at a certain age. When the mothers return in the morning they must have some means of identifying their own offspring from a mass of pups, all of which are about the same age. The only species in which mothers have been observed to nurse any available pup is *Miniopterus schreibersi*. Mothers presumably recognize their own young on the basis of their characteristic isolation calls and their odor. Nursery colonies of Mexican free-tailed bats (*Tadarida brasiliensis*) in the southwestern United States often comprise millions of bats. The mothers deposit their young on the cave ceiling, where the pups aggregate in so-called creches with a density of about 4000 pups/m^2. Within a creche, the pups move freely. A mother usually feeds her own offspring with a milk meal two to several times during the course of a day; therefore, she has to be able to locate her own offspring from among thousands of moving pups. Individual recognition between mother and offspring is mutual and is based on olfactory and auditory cues.

It has been shown experimentally that lactating Mexican free-tailed bats approach the odor of their own pup significantly more often than the odor of other pups. It is thought that mothers rub their young with secretions of glands located on their snouts and chins, thus marking their offspring.

A multivariate statistical discrimination analysis has shown that Mexican free-tailed mothers can recognize the characteristic vocalizations of their own pups with a probability of 60%. In eight mother–infant pairs of *Phyllostomus discolor*, each mother's directive calls were distinctly different, and the isolation calls of the young gradually adapted to the time-frequency structure of the mother's call. In isolation calls of young *Nycticeius humoralis*, seven individually different types of sound patterns were detected, on the basis of which the authors of this study could differentiate among 1800 pups. In the rich repertoire of communication sounds in adult *Megaderma lyra*, a pure-tone sound is especially well suited to function as an identity marker because this call distinctly differs among individuals in three independent parameters: frequency, signal duration, and interval duration.

Because the young crawl about on the ceiling of the cave at night, accidents frequently occur in which pups fall to the ground. Counts made in a nursery colony in Sri Lanka revealed that every juvenile bat in the colony must have fallen to the ground at least once before learning to fly. The young that have fallen to the ground call to the group on the ceiling; every morning females fly over the areas where the young have fallen and retrieve them. Some of the young remain on the ground, where they starve after 3 days. Infant mortality in this colony was about 33%; among other species it is much less: 1.3% in *Tadarida brasiliensis*, 2% in *Myotis lucifugus*, and 7% in *Eptesicus fuscus*.

The first flight by a 4 to 6-week-old juvenile bat is quite risky given its inexperience with flight technique and orientation. Young bats first fly around in the vicinity of the nursery colony, where many of their attempts to capture insects are unsuccessful. Observations to date suggest that young bats must learn to capture insects without any help from their mothers. For this reason, many species feed their young during this critical period. *Molossus ater* mothers fill their cheek pouches with chewed-up insects, which they bring back to the roost and feed to their offspring.

Males usually do not participate in the care of the offspring. A few recent studies have disclosed that bats may have a rich repertoire of social interactions and communication calls which have rarely been noticed due to the fact that many of these signals are in the ultrasonic frequency range. Specific communication sounds are often associated with specific flight displays put on by males to attract females or to establish and strengthen pair-bonds.

A unique case of altruistic behavior has been described in a flying fox island species (*Pteropus rodricensis*) of the western Indian Ocean. A pregnant female who apparently had difficulty in giving birth was assisted and tutored by a neighboring female throughout the prolonged parturition process, which lasted 2.5 h. The helper licked the anovaginal area of the expectant mother and constantly groomed her. Repeatedly, the helper "showed" the mother the typical feet-down position for parturition. Except for one occasion, the expectant mother assumed this position only after the helper had demonstrated it to her. The helper cradled the mother during labor and even took care of the newborn pup. So far such obstetrical assistance has been reported only in dolphins. Studies of social behavior and communication patterns in bats should be most rewarding because bat species offer a wide variety of social organizations from single pairs to the largest mass aggregations found in any mammal.

The reproductive behavior of bats raises a number of unanswered questions that are of general interest.

1. What are the specific factors that determine the multitude of very different social structures?
2. What is the relationship between social structure, reproductive cycles, and food availability?
3. Through what mechanisms do hormones control the different types of interruptions in the reproductive cycle, and what are the external stimuli that trigger them?
4. How is the epiphysis able to regulate seasonal cycles in an animal that is seldom, or never, exposed to daylight?

Due to the recent precipitous decline in bat populations, studies designed to answer these questions have virtually come to a halt.

References

Andreuccetti P, Angelini F, Taddei C (1984). The interactions between spermatozoa and uterine epithelium in the hibernating bat, *Pipistrellus kuhli*. Gamete Res 10:67–76.

Beasley LJ, Zucker I (1986). Circannual cycles of body mass food intake and reproductive condition in male pallid bats. Physiol Behav 38:697–702.

Bradbury JW (1977). Lek mating behavior in the Hammer-headed bat. Z Tierpsychol 45:225–255.

Buchanan GD, Garfield RE (1984). Myometrical ultrastructure and innervation in *Myotis lucifugus*, the little brown bat. Anat Rec 210:463–475.

Cotterill FPD, Fergusson RA (1993). Seasonally polyestrous reproduction in a free-tailed bat *Tadarida fulminans* in Zimbabwe. Biotropica 25:487–492.

Crichton EG, Krutzsch PH (1987). Reproductive biology of the female little mastiff bat, *Mormopterus planiceps* in southeast Australia. Am J Anat 178:369–386.

Dawkins R (1989). The Selfish Gene. Oxford University Press, New York.

Defanis E, Jones G (1995). Post-natal growth, mother-infant interactions and development of vocalizations in the vespertilionid bat *Plecotus auritus*. J Zool 235:85–97.

Gustafson AW (1979). Male reproductive patterns in hibernating bats. J Reprod Fertil 56:317–331.

Gustin MK, McCracken GF (1987). Scent recognition between females and pups in the bat *Tadarida brasiliensis mexicana*. Anim Behav 35:13–19.

Heideman PD, Bronson FH (1994). An endogenous circannual rhythm of reproduction in a tropical bat, *Anoura geoffroyi*, is not entrained by photoperiod. Biol Reprod 50:607–614.

*Hill JE, Smith JD (1984). Bats, a Natural History. British Museum (Natural History), London.

Jenness R, Studier EH (1976). Lactation and milk. In R. J. Baker, J. K. Jones, D. C. Carter, eds., Biology of Bats of the New World Family Phyllostomatidae. No. 10, Part 1, pp. 201–218. Spec. Publ. Mus. Texas Tech University, Lubbock.

Jerret DP (1979). Female reproductive patterns in nonhibernating bats. J Reprod Fertil 56:369–378.

King JC, Anthony ELP, Gustafson AW, Damassa DA (1984). Luteinizing hormone-releasing hormone (LH-RH) cells and their projections in the forebrain of the bat *Myotis lucifugus*. Brain Res 298:289–301.

Krutzsch PH (1979). Male reproductive patterns in nonhibernating bats. J Reprod Fertil 56:333–344.

Krutzsch PH, Crichton EG (1987). Reproductive biology of the male little mastiff bat, *Mormopterus planiceps* in southeast Australia. Am J Anat 178:352–368.

Kunz TH, Allgaier AL, Seyiagat J, Caligiuri R (1994). Allomaternal care: helper-assisted birth in the Rodrigues fruit bat, *Pteropus rodricensis*. J Zool 232:691–700.

Kunz TH, Whitaker JO, Wadaloni MD (1995). Dietary energetics of the insectivorous Mexican free-tailed bat (*Tadarida brasiliensis*) during pregnancy and lactation. Oecologia 101:407–415.

Kurta A, Bell GP, Nagy KA, Kunz TH (1989). Energetics of pregnancy and lactation in freeranging little brown bats (*Myotis lucifugus*). Physiol Zool 62:804–818.

McCracken GF (1993). Locational memory and female pup reunions in Mexican free-tailed bat maternity colonies. Anim Behav 45:811–813.

McCracken GF, Bradbury JW (1981). Social organization and kinship in the polygynous bat *Phyllostomus hastatus*. Behav Ecol Sociobiol 8:11–34.

*McCracken GF, Gustin MK (1991). Nursing behavior in Mexican free-tailed bat maternity colonies. Ethology 89:305–321.

Mori T, Uchida TA (1980). Sperm storage in the reproductive tract of the female Japanese long-fingered bat, *Miniopterus schreibersii fuliginosus*. J Reprod Fertil 58:429–433.

Oxberry BA (1979). Female reproductive patterns in hibernating bats. J Reprod Fertil 56:359–367.

*Racey PA (1979). The prolonged storage and survival of spermatozoa in Chiroptera. J Reprod Fertil 56:391–402.

Ransome RD, McOwat TP (1994). Birth timing and population changes in greater horseshoe bat colonies (*Rhinolophus ferrumequinum*) are synchronized by climatic temperature. Zool J Linn Soc 112:337–351.

Rasweiler JJ (1974). Reproduction in the long-tongued bat, *Glossophaga soricina*. II. Implantation and early embryonic development. Am J Anat 139:1–36.

Rasweiler JJ (1979). Early embryonic development and implantation in bats. J Reprod Fertil 56:403–416.

Rasweiler JJ (1993). Pregnancy in Chiroptera. J Exp Zool 266:495–513.

Richardson BA (1979). The anterior pituitary and reproduction in bats. J Reprod Fertil 56:379–389.

Thomas DW, Marshall AG (1984). Reproduction and growth in three species of West African fruit bats. J Zool 202:265–281.

Uchida TA, Mori T, Oh YK (1984). Sperm invasion of the oviducal mucosa, fibroblastic phagocytosis and endometrial sloughing in the Japanese greater horseshoe bat, *Rhinolophus ferrumequinum nippon*. Cell Tissue Res 236:327–331.

*Vaughan TA, Vaughan RP (1986). Seasonality and the behavior of the African yellow-winged bat. J Mammal 67:91–102.

Wilkinson GS (1984). Reciprocal food sharing in the vampire bat. Nature 308:181–184.

*Wilkinson GS (1985). The social organization of the common vampire bat. I. Pattern and cause of association. Behav Ecol Sociobiol 17:111–121.

*Wilkinson GS (1985). The social organization of the common vampire bat. II. Mating system, genetic structure, and relatedness. Behav Ecol Sociobiol 17:123–134.

Wilson DE, Findley JS (1970). Reproductive cycle of a neotropical insectivorous bat, *Myotis nigricans*. Nature 225:1155.

Wimsatt WA (1979). Reproductive asymmetry and unilateral pregnancy in Chiroptera. J Reprod Fertil 56:345–357.

9

ECOLOGY

ECOLOGY IS CONCERNED with the relationships and interactions within or among animal species, as well as interactions between animals and the surrounding vegetation, geographical features, and climate. Although physiology and behavior have been dealt with separately throughout most of this volume, the study of ecology unites them into a complex, dynamic system. In some chapters, such as those on thermoregulation, water balance, and echo-imaging, some of the ecological implications of the systems under consideration were described. For this reason, the present chapter will be concerned with those topics that were not dealt with in previous chapters, specifically the geographical distribution, roosts, and migrations of bats as well as their relationships with plants.

9.1 GEOGRAPHICAL DISTRIBUTION

The world can be divided into six basic zoogeographic zones:

1. The Palearctic region includes the temperate regions of Eurasia and Africa north of the Sahara,
2. The Neoarctic region includes all of North America and the subtropical part of Mexico,
3. The Indomalaysian or oriental region includes tropical Asia and the Pacific islands,
4. The Ethiopian region includes Africa south of the Sahara, Madagascar, and the southern Arabian peninsula,
5. The Neotropical region includes South and Central America and the tropical part of Mexico,
6. The Australian region includes Australia, New Guinea, and nearby islands.

Because each of these regions has its own geological past, it also has its own unique fauna. The species in different regions overlap to a greater or lesser extent, depending on whether the regions were ever connected to one another through land bridges. Today, the characteristic fauna of each region has been "adulterated" due to the mobility of humans. In fact, humans have become one of the dominant factors in the reduction of species and in the dispersal of fauna.

The degree of relatedness among bat genera in the six geographic zones is not high. There are only five cosmopolitan genera, found in all six regions. These are the vespertilionid bats *Myotis*, *Pipistrellus*, *Eptesicus*, and *Nycticeius*, and the molossid bat *Tadarida*. All five genera forage for insects in flight. In contrast, the large phyllostomid family is restricted to the neotropical region, whereas the many species of Pacific island flying foxes of the genus *Pteropus* are restricted to Old World tropical regions and Australia. The generic similarity, defined as the percentage of genera common to two regions, is about 7.6% for the neotropical and Australian regions, which were separated by an ocean early in their geological history, and 47% for the Ethiopian and oriental regions which even today are connected through land bridges. The paleoarctic and neoarctic regions were connected via a land bridge over what is now the Bering Strait, and the bat species in the two regions have a generic similarity of 40%.

DIET

The similarity among bat faunae of the different regions appears considerably higher when the comparison is based on diet and foraging habits rather than on taxonomic criteria. This observation demonstrates the power of the ecological pressures that have led to convergent evolution, a factor that is often underestimated by taxonomists. In every geographic region, the majority of bat species forage for insects in flight. The percentage of species using this feeding strategy ranges from 44% in the neotropical region to 84% in the paleoarctic region. Presumably, the original foraging strategy of Microchiroptera was to hunt insects in flight. In any case, 80 of the 186 genera of bats feed mainly or exclusively on insects that they capture during flight.

Every geographic region also contains a minority of genera that mainly glean insects from the vegetation and from other surfaces. Because insects are frequently found resting on and around vegetation, it seems surprising that this feeding strategy has not become more common. Gleaning requires especially slow flight and a degree of maneuverability that sets strict limits on the adaptive usefulness of this strategy.

The piscivorous or fishing bats are considered to be a special group. The term *fishing bats* is somewhat misleading because these species actually forage over the surface of the water, using the claws of their feet, their tail membrane, or their jaws to capture organisms of an appropriate size. Their prey consist mainly of arthropods that live at the surface of the water, but also includes small fish and amphibians that are at the surface with some part of their body sticking out of the water. Every geographic region contains a small number of bat species that "fish" at the surface of the water. Fishing bats include the New World species *Noctilio leporinus* and the Old World bat *Myotis daubentoni*.

Strangely enough, carnivorous species —large bats that feed on vertebrates as well as insects—are only found in tropical regions. Despite the fact that the temperate zones offer an attractive selection of potential prey, no carnivorous bat species have evolved in these regions, probably because the energy demands of flight and prey capture in a carnivore would be incompatible with energy and thermoregulatory constraints during the winter. There are three species of vampires which feed on blood. Although this does represent a rather extravagantly specialized dietary strategy, it is not clear why sanguivorous species evolved only in South America and remain confined to the neotropical region.

It is easier to imagine why species that feed on nectar or fruit are restricted to the tropics. Phylogenetically, bats that feed on plants are divided into two groups. In the New World, they belong to the microchiropteran family *Phyllostomidae*; in the Old World, they belong to the suborder Megachiroptera. In the neotropical region and Australia, fruit-eating bats make up about one-third of the total bat fauna. In Australia this number is disproportionately high because so many different endemic species of *Pteropus* have evolved on the many isolated islands. Species that have become specialized for feeding on nectar and pollen are more common among the phyllostomids (13%) than among the Megachiroptera (5%).

9.2 BAT MIGRATIONS

Because their flight is so energy intensive and their body size so small, Microchiroptera are forced to ration their energy. The ability to go into daytime torpor or wintertime hibernation is one answer to this problem. Another answer to the problem is migration to a more favorable climate. A large number of bat species employ this latter strategy.

Just as in the case of birds, numerous banding studies of bats have shown that some populations make seasonal migrations over hundreds of kilometers. These banding studies were mostly abandoned in the early 1970s because it was feared that injuries due to banding could have contributed to the appalling decreases that were occurring in bat populations. Today there are better banding techniques that have allowed these studies to be resumed on a limited scale under the control of various conservation agencies.

There are two reasons a bat population will leave the territory in which it is living. The first is a seasonal lack of food; the second is the lack of appropriate winter living quarters. Bats avoid roosts where the temperature sinks below 0°C as well as roosts that are too warm. For these reasons bats from eastern and northern Europe migrate in the fall to middle Europe, Turkey, and western Europe. Bats even fly over mountain passes in the Alps to reach their destinations. In the spring the bats fly back to Russia or Scandinavia where the summer is short, but the insect supply is large. Depending on the species and climatic conditions, migration between summer and winter quarters may be undertaken by all individuals or by just a subpopulation. In middle Europe, for instance, certain populations of *Pipistrellus pipistrellus* are nonmigratory. The following are typical European migratory bat species: *Myotis dasycneme*, *Nyctalus noctula* (noctule populations in northern and eastern Europe), *Nyctalus leisleri*, *Pipistrellus nathusii*, and *Vespertilio murinus*. The following are species with migratory subpopulations: *Pipistrellus pipistrellus, Miniopterus schreibersi*, *Eptesicus nilssoni, Myotis myotis*, and *Myotis blythi.*

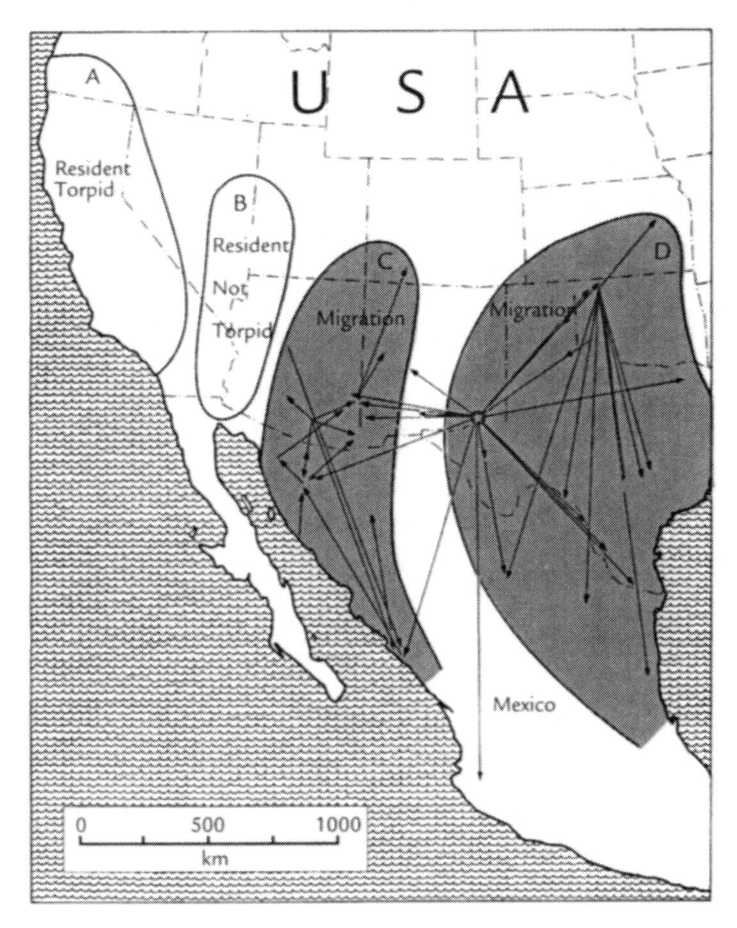

Figure 9.1 Resident populations (A and B) and seasonally migratory populations (C and D) of *Tadarida brasiliensis* (Molossidae). The migratory populations travel from eastern Arizona, New Mexico, Texas, and Oklahoma to Mexico. Arrows: direction of migration and destination. Colonies of *Tadarida* in California do not migrate, but become torpid in winter (A, torpid). Colonies in Utah and Nevada do not migrate either, but remain active throughout the winter (B, not torpid). From Hill and Smith (1984).

LONG-DISTANCE MIGRATIONS

The main factor that determines whether a species or subpopulation migrates is the climatic conditions. A good example to illustrate this point comes from populations of *Tadarida brasiliensis* that live in the southwestern United States (fig. 9.1). Colonies that live on the Pacific coast do not migrate, but remain there and forage during the winter, becoming torpid for days at a time. Colonies that live in eastern Nevada and western Arizona remain there during the winter, but never become torpid. Colonies that live in the eastern half of the range however, migrate to Mexico every year, flying distances up to 1000 km.

LOCAL MIGRATIONS

Not every movement of a population can be called a migration. For example, it was shown that a cave in New England, in which *Myotis lucifugus* hibernated, attracted animals from a surrounding area with a radius of approximately 70–80 km. However, a few individuals traveled to the cave from distances up to 260 km. Horseshoe bats in England provide yet another example of local migrations. These bats travel to various summer quarters lying within a 30-km radius of their winter quarters. Thus, in the winter the bats all congregate in the few suitable winter quarters, but in the summer they disperse to many different foraging biotopes.

The west European bat, *Myotis dasycneme*, leaves the chalk cliffs of south Limburg each spring and flies up to 300 km to reach its summer quarters in Champagne and in the Parisian basin. Observations over the last few years have revealed that the Rhine-Main area and more southern regions are a mating and hibernation center for *Pipistrellus nathusii*. In September 1986 a bat that had been banded 14 days earlier in southern Sweden was found in the Rhine-Main area. To cover this distance, the bat would have had to fly 774 km from north to south over a period of 14 days or less. This means that the bat must have flown an average of at least 55 km every night. As shown in figure 9.2, *Pipistrellus nathusii* migrates in late summer and early fall from eastern and northern Europe to Brittany, southern France, and Turkey. Noctules leave central Russia as early as August to spend the winter in hollow trees in southern Germany.

Even oceans do not present an insurmountable barrier to bats in their migrations between suitable summer and winter quarters. Bats have been known to land on oil drilling platforms in the North Sea, at least 300 km from the nearest coastline. *Lasiurus* and *Lasionycteris* have occasionally been sighted in the Bermudas, about 1100 km from the American east coast.

It is likely that bats are as stereotyped in their migratory routes as they are in their choice of living quarters. About 90% of the individuals belonging to the species *Myotis grisescens*, found mainly south of the Ohio River, hibernate in two caves in eastern Tennessee (375,000 animals) and in one cave in northern Alabama (1.5 million animals). The bats have used these caves as winter quarters for decades despite significant environmental changes. In England, the same individual *Myotis daubentoni* was found roosting at the same spot in the same cave for seven consecutive winters.

ORIENTATION DURING MIGRATION

To date there have been more studies in birds than in bats concerning the means by which animals find their way during migration and on their homing abilities. Experiments in which bats were transported to distant locations and released showed that *Eptesicus fuscus* were able to return to their home from a distance of 700 km, and *Myotis lucifugus* could return home from a distance of 430 km. A *Vespertilio murinus* was observed to return to its home within 48 h after being transported and released at a point 340 km away. Experience and practice seem to improve homing ability because in the release experiments young bats were found

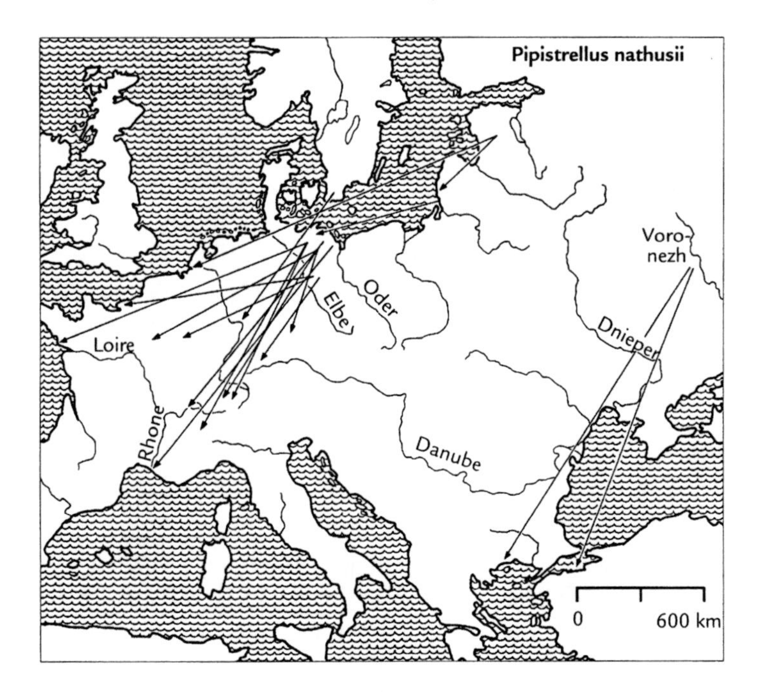

Figure 9.2 Late summer and fall migrations of *Pipistrellus nathusii* from Russia and northern Europe to southwestern Europe. From Roer (unpublished).

to return to their home roosts much less frequently than adults. However, this observation might also be due to the fact that young bats have a less well-developed perception of "home" than do adults. There is evidence to suggest that bats use visual landmarks to aid them in homing. The vespertilionid bats *Lasiurus borealis*, *Lasiurus seminolus*, and *Dasypterus intermedius* frequently join groups of migrating birds. In theory, bats could use the same orientation cues that have been described for birds that migrate at night. These include visual landmarks, the pattern of stars in the sky, olfactory gradients, infrasound from the surf on the coast or wind in the mountains, and the earth's magnetic field. There have been no studies to date on whether bats use any of these cues.

9.3 BATS AND PLANTS

Did bats and plants coevolve? In order for their seeds to be dispersed by frugivorous bats, no special adaptations on the part of plants would be required. Nevertheless, it is quite advantageous for the propagation of a plant species if its fruit is picked by bats and carried to a distant feeding place. Although many phyllostomids engage in this type of behavior, the numerous *Pteropus* species of Asia tend to consume the fruit in the same tree on which it has grown so that the seeds and pulp that are spit out fall close to the tree trunk.

The situation is different for plants that are pollinated by bats. Plants that depend on bats for pollination have special adaptations in the placement and form of their flowers that make them especially attractive and accessible to bats (fig. 9.3, see also fig. 9.6). The following features are characteristic of flowers that are pollinated by bats:

- The flowers of trees, woody shrubs, and vines are located at the edge of the foliage crown, facing outward (fig. 9.3).
- The flowers are large and the petals are sturdy. The stamens and pistil are often exposed (fig. 9.4).
- The flowers are often inconspicuously colored or green.
- The flowers have a very strong odor that is often unpleasant to humans.
- The flowers open in the evening and at night. Many are open for only a single night.
- The protein content of the pollen is higher than in flowers that are pollinated by insects, and the amount of nectar is greater.

Bats that visit flowers to feed on nectar and pollen characteristically have fewer and smaller teeth; elongated jaws and tongue (species with this feature include the phyllostomid *Glossophaginae* and the pteropid *Macroglossinae*; fig. 9.5); and a tongue covered with papillae and fine tufts of hair, which creates a spongelike mesh that sucks in nectar by capillary action. Many flower-visiting bats (e.g., glossophagine species) have the ability to hover in front of a flower just as hummingbirds do.

COEVOLUTION

It would be premature to conclude on the basis of mutual adaptations that bats and plants coevolved. Coevolution is a continual interactive process of adaptation by one species and counteradaptation by the other, leading to an ever-increasing dependence of each species on the other. It is unusual to find such a high degree of specialization in bat and plant species. Nearly 600 plant species in the neotropics and about 160 plants in the paleotropics, belonging to 270 genera, are known or suspected to be pollinated by bats. Plant families with a high percentage of chiropterophile species are *Cactaceae*, *Bignoniaceae*, *Lobeliaceae*, and *Caesalpinaceae*. Even though many of these plants are mainly pollinated by bats, only a few depend exclusively on bats. Among these are the wild banana and the calabash tree (*Crescentia alata* and *Parmentiera alata*). The columnar saguaro cactus, *Carnegiea gigantea*, is a typical "chiropterophile" plant. Although it is pollinated mainly by *Leptonycteris sanborni*, it is also pollinated by bees and doves. The small Malaysian flying fox *Eonycteris spelaea* and the plant *Oroxylum indicum* (Bignoniacea) are often cited as an example of coevolution because the flowers are adapted to the feeding behavior and head morphology of the flying fox (fig. 9.6). Nevertheless, *Eonycteris* is successful at obtaining nectar from other species of flowers as well.

At the time when the first bats appeared, probably sometime before or during the Eocene era, flowering terrestrial plants had already been in existence for 75

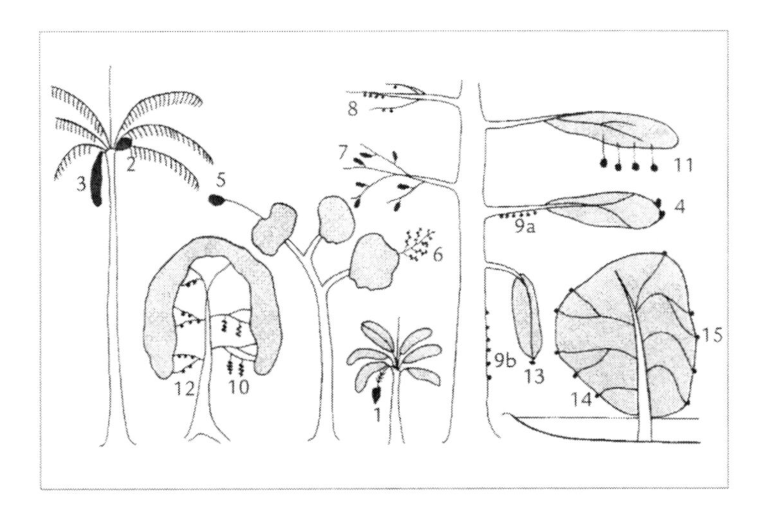

Figure 9.3 The exposure of flowers or flower stalks (black) in Malaysian plants that are visited by bats. Shaded area = foliage. 1, *Musa* species; 2, *Cocos nucifera;* 3, *Arenga* species; 4, *Mangifera* species; 5, *Oroxylum indicum*; 6, *Pajanelia multijuga*; 7, *Bombax valetonii;* 8, *Ceiba pentandra*; 9a,b, *Durio* species; 10, *Barringtonia* species; 11, *Parkia* species; 12, *Syzygium mallaccense;* 13, *Duabanga sonneratioides*; 14, *Sonneratia alba* (*ovata*); 15, *Sonneratia acida*. From Dobat and Peikert-Holle (1985).

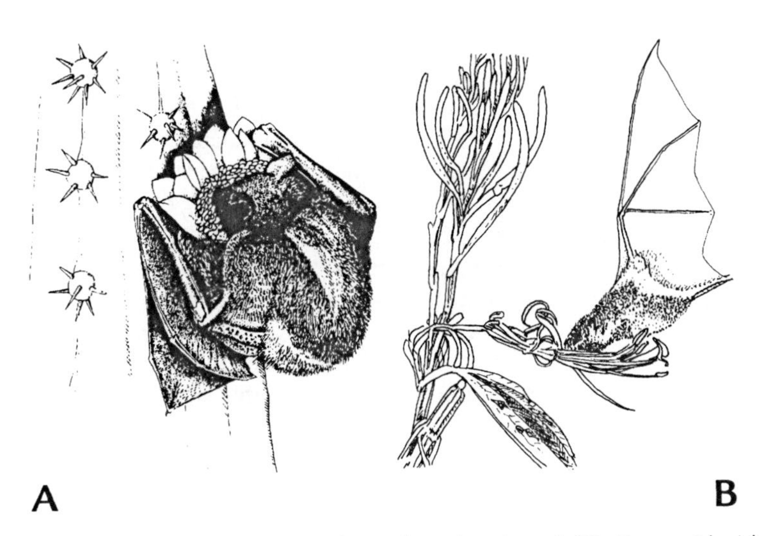

Figure 9.4 Bats visiting flowers. (A) *Glossophaga longirostris* (Phyllostomidae) landing on a cactus flower. (B) *Anoura geoffroyi* (Phyllostomidae) drinking nectar from a *Bauhinia rufa* (*Caeaslpiniaceae*) flower during hovering flight. During feeding the pollen rubs off on the hairs of the bat's chest. From Dobat and Peikert-Holle (1985).

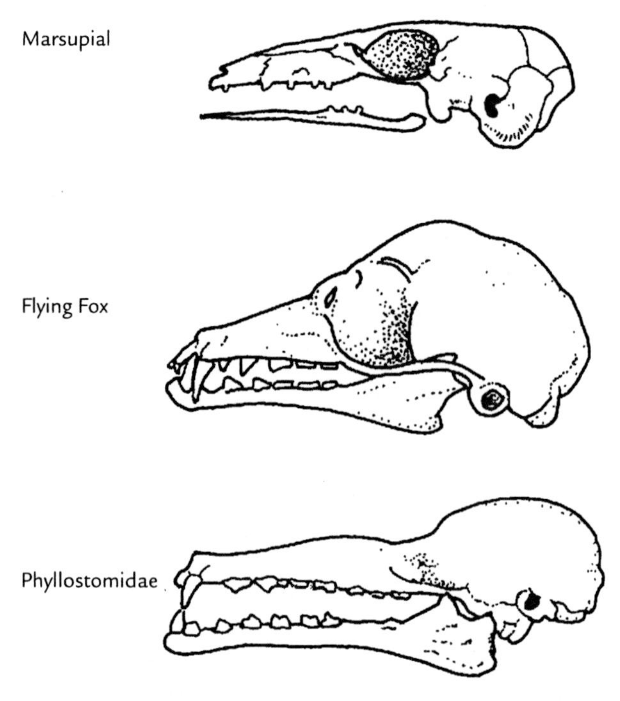

Figure 9.5 Convergent lengthening of the jaws in nectar-feeding mammals. Marsupial: *Tarsipes* species; flying fox: *Megaloglossus* species; Phyllostomidae: *Choeronycteris harrisoni.* From Kingdon (1984).

million years. It is thought that the fruit- and nectar-eating phyllostomids evolved from insectivorous phyllostomids in the late Myocene period, at a time when small terrestrial mammals were already well established. It seems probable, therefore, that bats discovered flowers that were already preadapted for visits from other vertebrates. Flight provided bats with a decided logistical advantage over tree-dwelling animals, resulting in a decrease in the importance of tree-dwelling animals as pollinators of flowers. The adaptation of flowers to visits by small mammals and bats occurred independently in at least 26 families of plants (table 9.1), including approximately 750 different species. Some of the most conspicuous adaptations occur in the *Sonneratiaceae*, seven species of which are pollinated mainly by bats, and many species of which disperse their seeds by means of frugivorous bats. Thus the coevolution of bats and plants is not narrow and specific, but wide and diffuse. It is therefore more accurate to refer to the relationship as mutualism than as coevolution. Pollination and seed dispersal in most chiropterophile plants do not depend exclusively on one species of bat, nor for that matter, on bats in general. Furthermore, bats do not depend on a particular species of plant for their food. Thus, the mutualistic relationship between bats and plants is not threatened by changes in the population sizes of different species. For example, among the phyllostomids there are only two species that feed exclusively on nectar and pollen: *Leptonycteris sanborni* and *Choeronycteris mexicana*. Both

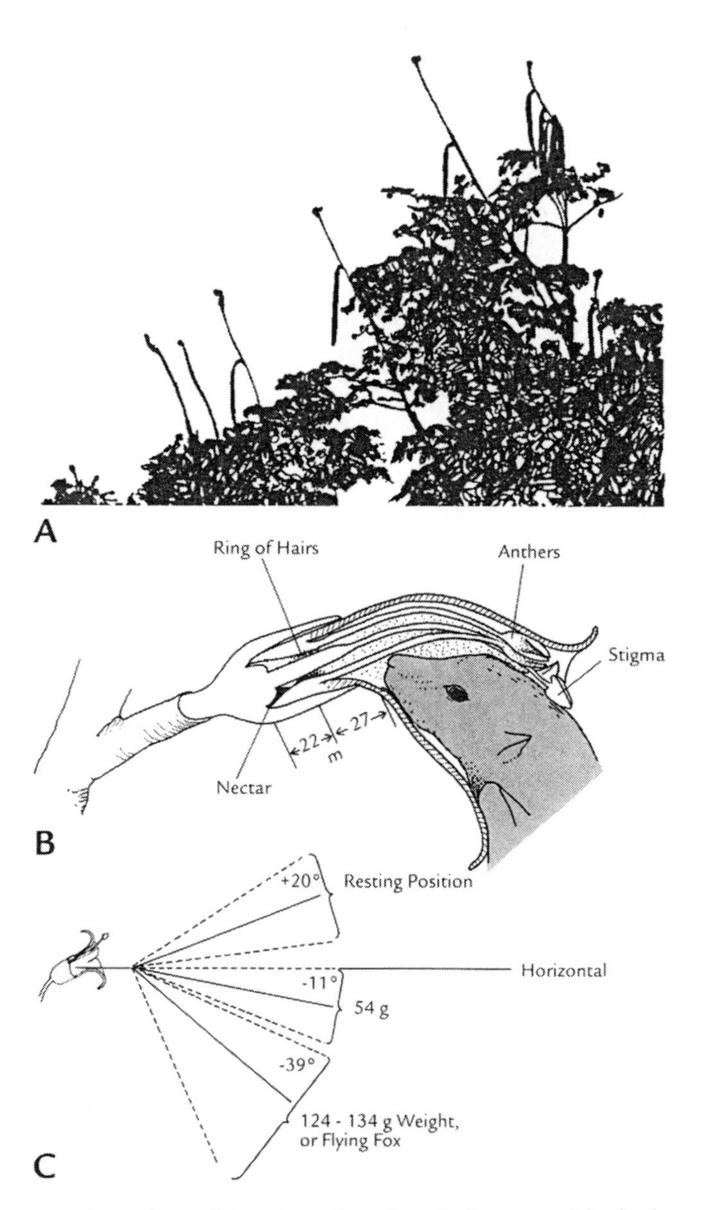

Figure 9.6 Adaptation of the plant *Oroxylum indicum* to visits by bats. (A) The flower stalks extend far beyond the tree's foliage. (B) Adaptation of the flower to the shape of the flying fox's head. The flower is pollinated by movements of the head. When the flower first opens at night, the flying fox's tongue can only reach a distance of 27 mm inside the flower; an hour later the tongue can reach an additional 22 mm, providing easy access to the nectar. (C) Position of the nectar receptacle in the resting position before a bat has lit on the flower, with a weight of 54 g placed on it, and during a visit by a flying fox weighing 124–134 g. When the flower is bent below the horizontal line, a row of hairs at the base of the flower becomes saturated with nectar, dispensing it in small portions. From Gould (1978).

Table 9.1 Plant families which include species visited by and adapted for pollination by bats

Family	Examples	Region
Acanthaceae	*Tricanthera gigantea*	NW
Agavaceae	Many species	NW, OW
Anacardiaceae	*Anacardium occidentale* (cashew)	
	Mangifera indica (mango)	NW, OW
Barringtoniaceae	*Barringtonia racemosa*	OW
Bignoniaceae	Many species, *Kigelia africana* (sausage tree)	
	Adansonia digitata (monkey-bread tree)	
	Bombax ceiba (Indian silk tree)	
	Ceiba pentandra (kapok tree)	
	Ochroma pyramidale (balsa tree)	OW, NW
Bromeliaceae	*Vriesea gladioliflora*	NW
Cactaceae	Many species, *Carnegiea gigantea* (saguaro)	NW
Caesalpiniaceae	*Bauhinia* species	NW
Capparaceae	*Cleome moritziana*	NW
Caricaceae	*Carica papaya* (papaya)	OW, NW
Caryocaraceae	*Caryocar villosum*	NW
Chrysobalanaceae	*Maranthes polyandra*	NW
Convolvulaceae	*Ipomoea arborescens*	NW
Curcurbitaceae	*Edmondia spectabilis*	NW
Euphorbiaceae	*Mabea occidentalis*	NW
Fabaceae	*Mucuna* species, many others	NW, OW
Gentianaceae	*Symbolanthus latifolius*	NW
Liliaceae	*Astelia fragrans*	New Zealand
Lythraceae	*Lafoensia* species	NW
Malvaceae	*Hibiscus elatus*	NW
Marcgraviaceae	*Marcgravia rectifolia*	NW
Mimosaceae	*Parkia* species	NW, OW
Moraceae	*Ficus benghalensis* (banyan tree)	OW
Musaceae	Many species, *Musa* species (banana)	NW, OW
Myrtaceae	Many species, *Eucalyptus* species	Australia, OW
Orchidaceae	*Vanilla chamissonis*	NW
Pandanaceae	*Freycinetia insignis*	OW
Passifloraceae	*Passiflora mucronata*	NW
Proteaceae	*Banksia* species	Australia, OW
Sapotaceae	*Bassia latifolia*	OW
Solanaceae	*Markea* species	NW
Sonneratiaceae	*Sonneratia* species	OW
Tiliaceae	*Luehea speciosa*	NW

OW, Old World; NW, New World.

There is evidence that plants that depend on bats for pollination and seed dispersal belong to 64 different families, about 270 different genera, and 750 species. Most of these (590 species) are found in the New World.

of these "beelike" bats feed on a wide variety of plants, even though they rely heavily on saguaro and agave for the sheer quantity of nectar provided. *Leptonycteris sanborni* eats pollen from at least 28 different species of plants. All of the other phyllostomid species that visit flowers supplement their diet to various extents with insects and/or fruits.

The phyllostomid subfamily *Glossophaginae*, which includes more than 30 species, feed mainly on nectar and pollen. Among the 175 different species of megachiropterans, only the 14 species of *Macroglossinae* feed mainly on flowers. None of these species are adapted for a specific plant. *Eonycteris spelaea*, often described as a specialist, visits at least 31 different species of plants. Nevertheless, each species has its favorite food source. *Artibeus* are especially fond of figs; *Carollia* prefer fruits of the genus *Piper*. The selection of flowers and fruits is determined more by seasonal availability than by nutritional content. In experiments in which *Carollia perspicillata* were offered a variety of different fruits, they did not choose those with the highest energy content, but rather those that were in season. There is hardly any competition for fruits between birds and bats, as exemplified in a study conducted in the Peruvian Amazon area. The few species of fruit eaten by both bats and birds composed less than 4% of most bats' diet.

Since the development of sociobiology as a science, the question of how animals obtain their food has become a favorite research topic. Optimal foraging theories are all based on the hypothesis that animals seek to maximize their energy intake with minimal energy expenditure. In all of the species that have been studied, ranging from ants to chimpanzees, this truism has been proven. This finding in itself is not very exciting; what is fascinating, however, is the wealth of different strategies used by animals to achieve the goal of optimal foraging. To optimize energy balance, animals have to make compromises. For example, the area surrounding optimal living quarters may not necessarily offer the best selection of food sources. In a colony of the hammer-headed fruit bat, *Hypsignathus monstrosus*, males commute about 7 km between their lek arena, where they make a communal effort to attract females and the fruit-bearing trees where they feed. The two flying foxes, *Eidolon helvum* which weighs about 300 g, and *Eonycteris spelaea* which weighs about 60 g, often fly 20–40 km from their day roosts to the trees where they feed. Smaller frugivorous bats often fly with a fruit to a protected location where they can devour it without being threatened by enemies. This strategy for reduction of raptor pressure costs *Carollia perspicillata* about 6% of its daily energy budget.

Related to the attractiveness of flowers for bats is the need to achieve a balance between the best interest of the plant and that of the bat. To maximize the effectiveness of pollination by an animal species, it is necessary to maximize the number of flowers of the plant species that are visited by the pollinator. The following conditions help bring about this ideal situation: only a few flowers per plant should be open at any given time, and the amount of nectar that can be obtained from one flower should be small. On the other hand, the amount of nectar must be large enough to attract bats and make visiting the flower worthwhile. The amount of nectar per flower varies between 0.1 ml and 9.0 ml, corresponding to a caloric value of up to 1.7 kcal for *Oroxylum indicum*.

A study by Gould on the flying fox *Eonycteris spelaea* (fig. 9.6) provides a good example of the relationship between bats and flowers. In Malaysia there are three species of flying foxes that feed mainly on nectar: *Macroglossus minimus*, *Macroglossus sobrinus*, and *Eonycteris spelaea*. The two *Macroglossus* species mainly visit *Musa* and *Sonneratia* flowers, whereas *Eonycteris* prefers the flowers of *Oroxylum indicum*, a tree that grows up to 26 m tall. Among the 31 different "chiropterophile" plants of Malaysia, *Durio*, *Ceiba*, and *Parkia* have a short period of massive flowering, *Sonneratia* blooms intensively twice a year, and *Musa* and *Oroxylum* bloom more or less continuously throughout the year. *Oroxylum* bears up to 40 long flower stalks that extend far out from the foliage; each stalk bears 100–200 flower buds (fig. 9.6). Every night at around 9:00 PM, one to four buds per stalk open up. They begin to close again within 4–5 h, so the time during which bats can visit them is short.

When an *Eonycteris*, which weighs about 50–80 g, lands on a bud that is ready to open, the five petals of the flower spring apart due to the weight of the flying fox, who is then able to insert his head deep into the cup of the flower. In doing so, he brushes his head against the anthers and the stigma (fig. 9.6). The nectar receptacle at the base of the flower is sealed by a row of hairs surrounding the ovary. Just before the flower opens, this receptacle contains up to 0.5 ml of nectar. The pressure exerted by the landing of the relatively heavy flying fox causes the flower to be momentarily pressed downward, so that the nectar flows downward and saturates the row of hairs. The flying fox is then able to lick the nectar from the hairs with his tongue. Before the flying fox takes off, however, the flower snaps back to an upright position, causing the nectar to flow back down into the receptacle. Thus, the hairs in the flower act as a saturated sponge, providing a simple method for dispensing the nectar in small portions. In this way the flower ensures that multiple visits occur and increases the chances of pollination.

Because of nectar rationing by the flower, it is not worthwhile for a flying fox to remain at the flower for more than about 1 s. During this time the flying fox obtains about 0.05 ml of nectar. Over the course of a night a flower secretes about 1.8 ml of nectar, which would be consumed after the visits of about 36 bats. In fact, it has been observed that each flower is visited approximately 33 times. Since the nectar of *Oroxylum* contains about 950 cal/ml, each animal obtains about 40–50 calories per visit. The result of this "nectar economy" is that flying foxes fly systematically, in a fixed order from one tree or flower to another, thus ensuring that the plants will be pollinated (fig. 9.7).

The strategies used by bats in visiting plants depends on the spatial distribution of the plants, the temporal distribution of their flowering or fruit-bearing periods, and their productivity. Trees that provide a large quantity of open flowers or ripe fruit for a short period of time are visited opportunistically by large swarms of bats. Examples of such plants are the kapok tree, *Ceiba pentandra*, *Lafoensia glyptocarpa*, various species of *Parkia*, *Durio zibethinus*, and various species of *Sonneratia*. Plants that provide a small amount of food over an extended period of time are visited by single animals or small groups in "traplines." This means that bats visit the same groupings of different species of plants every night in a fixed order. Examples of such plants are the calabash tree, *Crescentia cujete*, various

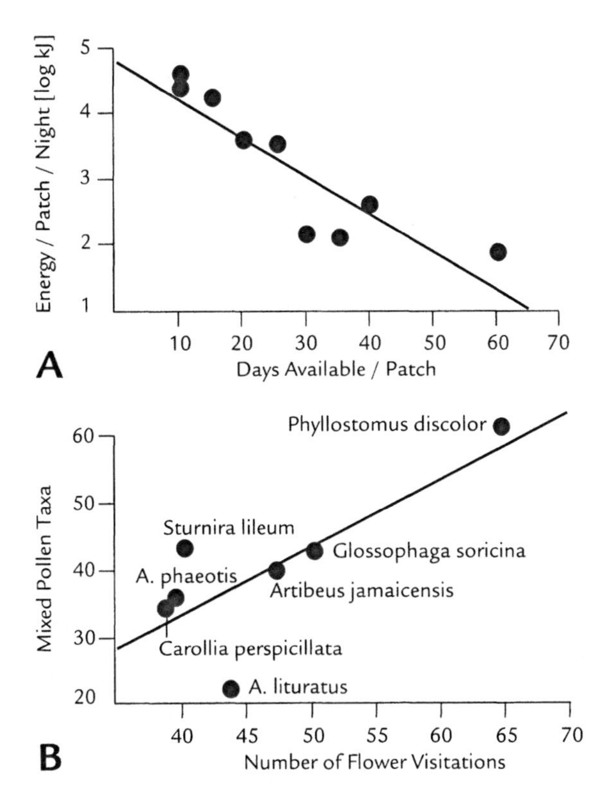

Figure 9.7 Food selection and plant visitation by bats. (A) Energy content of fruits on plants visited by *Carollia perspicillata* (Phyllostomidae) in Santa Rosa National Park, Costa Rica. For a given location, the energy per night available from fruits decreases as a function of the number of days fruits have been present on the plants. (B) The tendency of bats to visit different species of flowers per night (mixed pollen taxa) as a function of the bats' use of flowers (flower visitation). Flower visitation is measured as the proportion of bats captured that carried pollen. From Heithaus in Kunz (1982).

species of *Bauhinia*, and *Oroxylum indicum*. Some bats are sequential specialists, meaning that for a period of days or weeks they will fly mainly to one particular species of plant and then change to another species. In contrast, individual generalists visit multiple plant species every night. Generalists are mainly found among the frugivorous genera such as *Pteropodidae* and among the "trapline feeders" such as *Carollia perspicillata*.

Bats are flexible with regard to their feeding strategies and adapt their behavior to the selection available. For example, *Artibeus jamaicensis* flies every night to five fig trees on average, located within a radius of 25–400 m from its day roost. In regions where fig trees are not as common, the diameter of the bats' feeding territory expands to 8±2 km. *Phyllostomus discolor* visits flowers and fruit-bearing trees in groups of 2–15 individuals, the number per group varying according to food availability. Groups of 10–15 animals visit *Lafoensia glyptocarpa* when more than 60 flowers are open; when fewer than 10 flowers are open, the groups decrease to only 1–2 individuals. *Leptonycteris* visit agave flowers in large groups. Each bat visits the flowers in a set order. The swarm leaves an agave plant when each individual has visited about 35 flowers. Out of these 35 flowers, approximately 23 contain nectar and 12 do not. It is thought that a single bat will give the signal to leave the plant when it has encountered 8 flowers in a row that contain no nectar. By the time the group of bats leaves the agave plant, about 75% of the flowers have been emptied of nectar. After 20 min of feeding, the whole group rests for about half an hour in a night roost. Presumably the rest period spent within a tight group minimizes the "heating costs" associated with flying in the relatively cool night air of the forests in tropical mountain areas.

Bats seldom defend food resources against other individuals. Flight provides so much flexibility that it would hardly pay to defend any local food source. Only the relatively heavy species of flying foxes defend their fruit or flower-bearing branch by making buzzing noises, alternate waving of the wings, or aiming at the intruder with the sharp claw of their thumb. Frequently the social status of an individual within the group determines the energy costs associated with foraging. In colonies of *Phyllostomus hastatus*, the dominant male of a harem forages for fruit near the roost, while bachelor males may have to fly as far as 9 km to find food.

Many interesting problems related to bats that feed on plants remain to be studied. It is not clear how bats find food sources. It seems likely that they use olfactory cues for orientation, but this has never been convincingly demonstrated. The floral scent of eight chiropterophile plants from six different families has been shown to contain sulfuric compounds and fatty acid derivatives with a mushroom-like odor (1-octen-3-ol, 1-octen-3-one, 3-octanone). These odorants are not found in scents of related plants that are not pollinated by bats.

It is highly unlikely that echolocating species that feed on plants could use echolocation to find their way to distant sources of food, although it is possible that they could locate flowers on plants through acoustic imaging, since petals of chiropherophile flowers strongly reflect ultrasound. However, species that feed on flowers have relatively large eyes, so it is also possible that they use vision to detect flowers. Once food-bearing plants have been discovered, it seems likely that bats depend on their spatial memory to return to them. This is especially important for species that are trapline feeders. Another largely unexplored topic is the metabolism of nectar and pollen feeders. For example, it is not clear to what extent species that feed primarily on fruit and nectar depend on supplementary sources of protein. It is conceivable that these species also consume other plant material as a secondary food source.

Interaction between bats and plants is an important component of all tropical ecosystems. About 0.5–1% of neotropical angiosperms are pollinated by glossophagine bats. The ability of these species to hover and to be content with small quantities of nectar rewards has allowed many plant families to evolve bat-pollinated flowers in the Neotropics. Glossophagines could therefore be thought of as the "hummingbirds of the night." Because of the low degree of specificity in the mutualistic and complementary relationships between different species, bat–plant interactions remain relatively stable. Nevertheless, many species play key roles in the community. For example, in some biotopes and at certain times of year, *Ceiba acuminata* is virtually the only source of nectar. Without these plants, it would be necessary for the bats, which are in turn essential for other plants, to migrate to another source of nectar. These types of ecological relationships, with a few exceptions, have not yet been studied. Given the rapid disappearance of tropical rainforests worldwide, this gap in our knowledge of field ecology will probably never be filled.

The important role of frugivorous flying foxes in plant seed dispersal was especially obvious during the repopulation of the Krakatoa archipelago after a volcanic eruption in 1883 destroyed all life forms. Microchiropteran bats reappeared on the islands about 50–70 years after the eruption, but flying foxes appeared earlier, about 20–30 years after the eruption. In 1985 on the island of Anak Krakatoa, flying foxes were found whose feces contained fig seeds, even though no fig trees were as yet growing on the island. A year later, fig trees were found sprouting on bare lava. It is obvious that flying foxes always have been and still are of great importance in repopulation and dispersion of vegetation.

9.4 LIVING QUARTERS

DAY ROOSTS

As nocturnal animals, bats need a place where they can rest during the day, thus minimizing their energy expenditure and protecting themselves against predators. Although flight guarantees a quick and effective route of escape, a flying animal is highly visible and thus subject to predation. Because even the fastest flying bats are significantly slower than birds, they would have no chance whatsoever if matched against a bird of prey. Because of the pressure to escape predation and the necessity of conserving energy, a day roost must meet the following two criteria: (1) it must either provide a stable daytime climate with a near optimal temperature, or it must be cool enough to induce torpor; (2) it must be secure enough to exclude predators.

The best natural climate control is provided by caves, where extremes in the outside temperature are balanced out. When we measured the temperature of a tropical bat cave in India, we at first thought that our instruments had failed because for many months they registered a straight line at 27°C even though the temperatures on the stones outside varied from 20°C to 40°C (fig. 9.8). In the temperate zone, the temperatures in caves are lower, making them less well suited as day roosts. In the temperate zone, bats prefer roosts that heat up during the day to reach an optimal temperature. Thus many bats have become "domesticated" in that they prefer to roost in the attics of buildings, between outside walls and shutters, or behind plastic facing covering cement structures. The best bat roosts are structures that are dark, free of drafts, relatively constant in temperature, and easily warmed by the sun. Natural roosts in the temperate zone are provided by hollow trees, crevices in stones, and in the front parts of caves. In Europe, the most

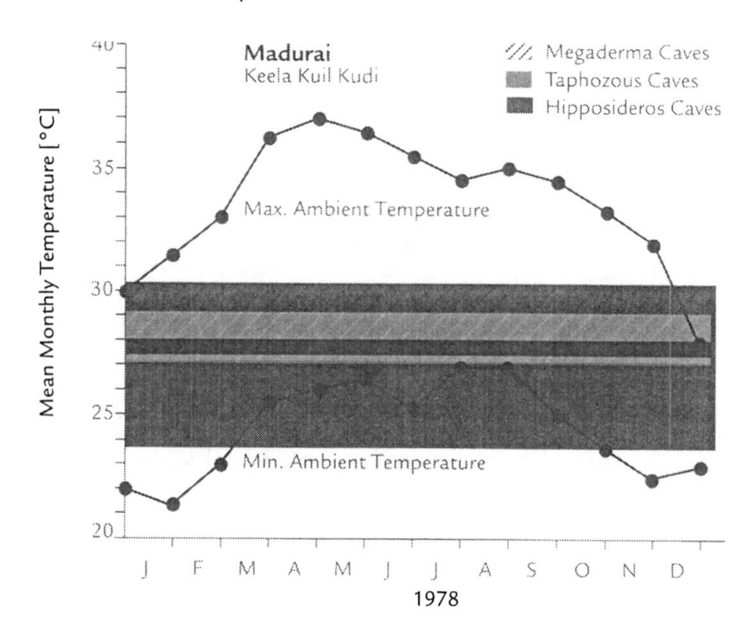

Figure 9.8 Constancy of temperature inside a cave, illustrated by an example from a cave in a rock formation in Madurai, India. The two curves show the monthly average maximal and minimal temperature outside the cave over the course of a year; the shaded areas show the temperature range in the part of the cave occupied by bats. In the part of the cave where *Hipposideros* were found (dark shading), the temperature remained virtually constant at 27°C, varying by less than 1°C over the course of a year.

familiar bat that still roosts in hollow trees is *Nyctalus noctula*, commonly found in forests and parks.

The period during which bats raise young is a critical phase in terms of energy balance. For this reason many species form nursery colonies in quarters that provide near-optimal temperature conditions. By crowding together so closely that their bodies touch, mothers and young can create extra warmth and insulation that will allow them to survive sudden short cold periods during the summer.

In the tropics, caves provide the ideal day roosts not only because the temperature in the caves (about 30°C) is optimal, but also because the relative humidity is high. In a warm climate there is always the danger that the fragile flight membranes will dry out, so for this reason many species seek out day roosts with high humidity. During dry periods in the desertlike region of Rajasthan in India, many mouse-tailed bats (*Rhinopoma spp.*) spend the day in cracks in the stone walls lining wells. Crevices in rocks are also favorite roosting places. Species that live in crevices in rocks such as the African species *Platymops setiger* (Molossidae), or that live under loose slabs of tree bark such as the two vespertilionid species *Mimetillus moloneyi* and *Laephotis wintoni*, are morphologically adapted to these quarters in that they have flattened skulls and bodies. The wings of *Mimetillus moloneyi* are also shortened so much that the bat must flap its wings vigorously and expend a great deal of energy to remain airborne. For this reason these bats return to a hiding place under tree bark to rest every 15 min or so during their nightly flights to forage for insects. Often groups of 9–12 animals rest together. These hiding places under tree bark are not especially safe, since hornbills have been observed to look for bats under loose bark and capture and eat them. Many bats that live in crevices have a highly sensitive tail. If *Rhinopoma* hanging free on cave walls are disturbed, they will quickly and skillfully withdraw into crevices in the rock. They crawl with their heads pointed downward and their tails pointed upward, feeling back and forth along the rock wall with their tails until they discover a crevice into which they can crawl.

In areas with lush tropical vegetation, there are many additional possibilities for day roosts, especially hollow trees and crevices in tangles of aerial roots. Many species that are solitary or that live in small groups do not put much effort into finding day roosts, simply disappearing into the thick foliage of the trees and lianas. One such species is *Epomophorus franqueti*, the singing fruit bat. These bats crawl under the loose bark of trees or simply hang on tree branches. Bats that roost in relatively conspicuous places are often camouflaged by cryptic coloration and patterning of their fur. For example, the fur of *Rynchonycteris naso* (Emballonuridae) is patterned to resemble the bark of a tree trunk. Bats of the North American genus *Lasiurus* hide among the thick needles of evergreen trees. Their fur is wooly, grayish in color, and has a speckled pattern. This coloration makes them almost invisible from the ground and also protects them from the cold. The African bat *Hipposideros cyclops* has fur that is patterned to resemble bark and is practically invisible when resting on a tree trunk. The African butterfly bat, *Glauconycteris variegata* (Vespertilionidae) and the flying fox *Micropteropus pusillus* hide among leaves during the day. Patterns on their wings mimic the patterns of veins on leaves so that from the ground the animals appear as dead leaves. *Kerivoula argentata* (Vespertilionidae) has an even more sophisticated method of camouflage. These bats hang together in clumps under the eaves of buildings, and the shape, color, and structure of the clump of bats resembles the nests of stinging wasps.

In the tropics some small species of bats use large-leaved plants such as palms, bananas, and various species of *Heliconia* as day roosts (fig. 9.9). For example, a favorite hiding place of the African banana bat, *Pipistrellus nanus*, is a young banana or heliconia leaf which is still rolled up to form a conical tube. Some species such as the neotropical bat *Thyroptera tricolor* have suction discs or pads on their wrists and ankles that enable them to cling to the wet, slippery inner side of the rolled-up leaf. The bats rest inside the leaf with their heads pointing upward toward the opening at the top of the leaf.

Bats also take advantage of nests and holes that have been made by other animals. For example, the African bat *Nycteris hispida* can be found not only in the crowns of papyrus plants, but also in termite nests or anteater burrows. *Kerivoula harrisoni* (Vespertilionidae) commonly lives in abandoned birds' nests. This bat's fur and the pattern on its wings are so well adapted to mimic the appearance of the inside of a bird's nest that the bat can only be seen if it moves. The two Indomalaysian species of *Tylonycteris* have the flattest skulls of any mammal and thus

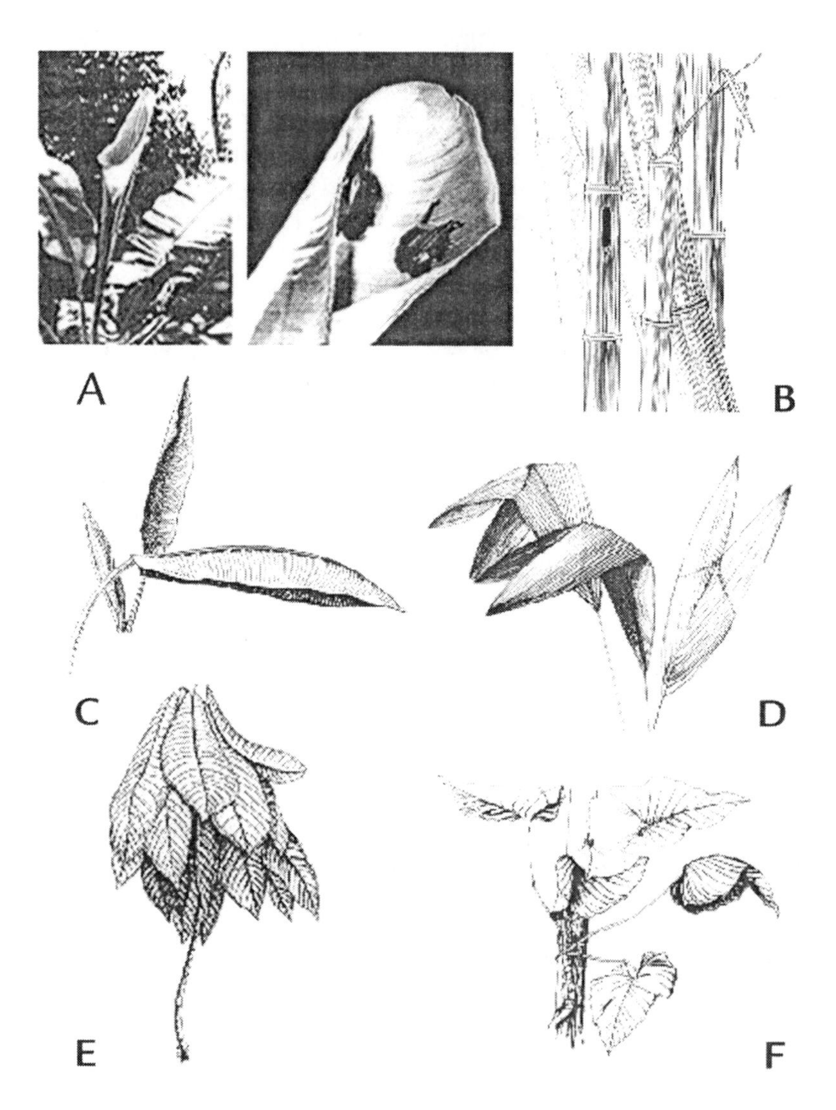

Figure 9.9 Leaf dwellings and tent-making. (A) Left: rolled-up banana leaf; Right: two bats (*Thiroptera tricolor*) clinging to the inside surface of the banana leaf. (B) Slits that beetles have bored in bamboo provide a protected hiding place inside the bamboo cane for *Tylonycteris pachypus* and *Tylonycteris robustula* (Vespertilionidae). (C) Tent fashioned by *Ectophylla alba* (Phyllostomidae) by crushing the horizontal ribs of a *Heliconia* leaf. (D) Bifid tents built by making oblique cuts through ribs are common types of shelters. (E) Conical tents are built by chewing through the midribs of several leaves along a tree trunk. (F) Apical tents are built by crushing the midrib of a long leaf in a more apical position. From Kunz (1982, 1994).

provide an extreme example of adaptation to specific living quarters. Every morning these small vespertilionids disappear with incredible accuracy into small slits in the sides of bamboo canes (fig. 9.9b). They cling to the inside of the hollow bamboo cane using suction pads on their thumbs and feet. Small groups of bats roost together in the internodal space of the bamboo cane. The slits in the bamboo canes are not natural openings, but are made by beetles (*Chrysomelidae*) or by woodpeckers. Other species of bats, the vespertilionids *Glischropus tylopus* and *Pipistrellus minimus*, are also reported to disappear without a trace among stands of bamboo.

Several species of fruit- and nectar-eating phyllostomids are not content simply to make use of what nature has to offer or what other animals have created. Instead, they build their own day roosts, in the form of shelters fashioned from large leaves. The favorite building materials for these shelters are the leaves of various *Heliconia* species and palms. For example, the phyllostomid *Ectophylla alba* chews through the horizontal ribs of *Heliconia* leaves from the underside so that the portions of the leaf on either side of the central rib collapse to form a simple tent (fig. 9.9c). *Artibeus watsoni* makes oblique cuts in the sides of palm leaves so that the front half of the leaf droops to form a tent (fig. 9.9d). Peter's tent-making bat, *Uroderma bilobatum,* chews multilobed palm leaves so that they form a triangular tent with sides like venetian blinds.This bat species also cuts the midribs of 6–14 leaves along a tree trunk (e.g., juvenile *Coccoloba manzanillensis*), thus creating a conical tent. The bat cuts the midrib of the lowest leaf about 12 cm away from the trunk in order to secure sufficient roost space. As the bat cuts leaves progressively higher on the tree, it chews the midrib closer to the trunk. In this way it seals up its roost, thus protecting it from rain and predators (fig. 9.9e). There are indications that bats might even repair such multileaf shelters.

At least 16 species belonging to 6 neotropical genera (*Artibeus, Uroderma, Ectophylla, Mesophylla, Vampyressa, Rhynophylla*) are known to make tents, and all are fruit-eating bats. In the Old World only two fruit-eating species, *Cynopterus sphynx* and *C. brachyotis*, and one insectivorous vespertilionid bat are known to make and use tents as day roosts.

Shelters made from leaves naturally last only a short time. The demand placed on the energy budget of these small bats by the continual task of rebuilding leaf shelters should not be underestimated. To make a single shelter, a bat must gnaw through as many as 80 ribs of a leaf. To date, no one has ever observed bats performing this task, but it is thought to be a community activity in which many animals engage simultaneously.

The only bat that constructs its own tree-hollows to use as day roosts is *Mystacina tuberculata*, native to New Zealand. These bats use their teeth and claws to excavate tunnels and chambers in dead, fallen trees (*Agthis australis*), which are then occupied by large colonies of more than 100 individuals. This example shows that even though bats' ability to perform "construction work" is limited by the adaptation of their hands for flight, they still have the potential for creative manipulation of their environment. This potential has been exploited very little during the course of evolution, even though the availability of suitable shelter is a factor that limits the distribution of bats. It seems likely that in most biotopes, vulnerability to enemies during tree excavation, together with the high energy demands of this activity, offset any adaptive advantage it might offer. Tent builders, on the other hand, engage in their activities from the protected undersides of leaves.

There is still no good understanding of why some bat species congregate in huge colonies while other species in the same biotope are solitary or live in small groups. Cave-dwelling groups of up to 20 million *Tadarida brasiliensis* represent the largest known colonies of any mammal. Energy conservation or lack of living space do not seem to be plausible explanations for the formation of extremely large bat colonies. The larger the number of bats in a colony, the smaller the risk of predator attack on any individual. However, this factor would not require aggregations of millions of animals.

NIGHT ROOSTS

Until recently little was known about night roosts, or resting places visited by bats during their nocturnal period of activity. Only since field observations have been carried out using ultrasound detectors and miniature radio transmitters has it become clear that many species of bats interrupt their nightly foraging for insects with rest periods. Carnivorous and frugivorous bats, especially phyllostomids, often transport their food to a night roost where they can consume it in relative safety. Lactating females of the species *Megaderma lyra*, the false vampire bat, use night roosts as a safe place to deposit their young while they go out to forage for food. They return to visit their young at intervals throughout the night.

With the exception of moths, flying insects are generally most plentiful at the beginning and end of the night. Many bats adapt their foraging activities to this pattern of activity on the part of their prey. During the middle part of the night, these bats rest in a night roost located in or near their foraging territory. This strategy saves the bats from having to commute back and forth to the colony, often many kilometers away from the foraging area. Ideal night roosts and day roosts share the same properties: they are secure from predators, draft-free, and stable in temperature. However, bats frequently alight on rock faces, walls of buildings, tree trunks, branches, etc., because the pressure from predators is rather small, especially on moonless nights.

See the section on temperature regulation for a discussion of winter quarters.

9.5 BATS AND HUMAN ACTIVITIES

The impact of humans on bat populations is mixed. The increase in agriculture and the clearing of wide forest areas in Europe during the Middle Ages might have increased the species diversity and population sizes of bats in the same way it did for songbirds. The meadows, orchards, and assorted fruit fields provided new biotopes that attracted a rich variety of bird species. The same process might have taken place for bat populations; however, no reliable records comparable to those for birds were kept for bats.

Human settlements and mining activities have supplemented natural roost facilities for bats and some man-made structures such as church attics and barns have become favorite roosts for maternity colonies due to the fact that they are heated quickly by the sun. Quite a number of bat species all over the world have become "synanthropic animals" and now preferably live in human settlements. Well-known examples are the big brown bat, *Eptesicus fuscus*, in North America and the mouse-eared bat, *Myotis myotis*, in Europe.

With the advent of industrialized agriculture and insecticides, the fate of the beneficial cohabitation of bats and humans has been reversed. Since the middle of this century, bat populations have declined dramatically throughout nearly the whole world for reasons not yet fully clarified and understood.

It is obvious that the "silent spring" resulting from indiscriminate and large-scale application of insecticides must have decimated those animal populations that depend on insects, as do the majority of bat species. The stabilization and slight recovery of bat populations in middle Europe after banning of the most harmful agents and the more controlled application of those still in use suggests that insect scarcity might have been the dominant factor responsible for the decline of bat populations.

Another killer of bat colonies roosting in buildings has been timber preservatives containing lindane or related compounds. It has been experimentally shown that such chemical preservatives are very effective contact poisons capable of exterminating bat colonies even weeks after application. Most of these aggressive chemicals are no longer on the market, and many building owners and preservation companies have been persuaded to use nonchemical methods of timber preservation.

Organochlorine compounds such as DDT and its derivatives are universal and persistent poisons that accumulate in fatty tissues of long-lived animals such as bats. DDT has been banned in many countries, but is still in use in others such as Mexico and those of Central America. These regions are wintering areas for the huge colonies of Mexican free-tailed bats, *Tadarida brasiliensis*, from Carlsbad Caverns in New Mexico. There, the population declined from more than 8 million in 1936 to about 200,000 in 1973; it has now recovered to about 700,000. In Europe, wintering bat populations declined to about 2–5% of their original numbers from the early 1950s to the late 1980s. Organochlorines are not only stored in the fatty tissue of adult animals but are also transferred to juveniles through the placenta and through the milk. Mexican free-tailed bats are born with pesticide residues that reach their lifetime maximum at the end of the lactation period. In young volant bats as well as in adults, organochlorines are mobilized when the bats use up their fat stores during winter migration. It is suspected that this flush of organochlorines during migration may be one of the major causes of the population decline in Carlsbad Caverns.

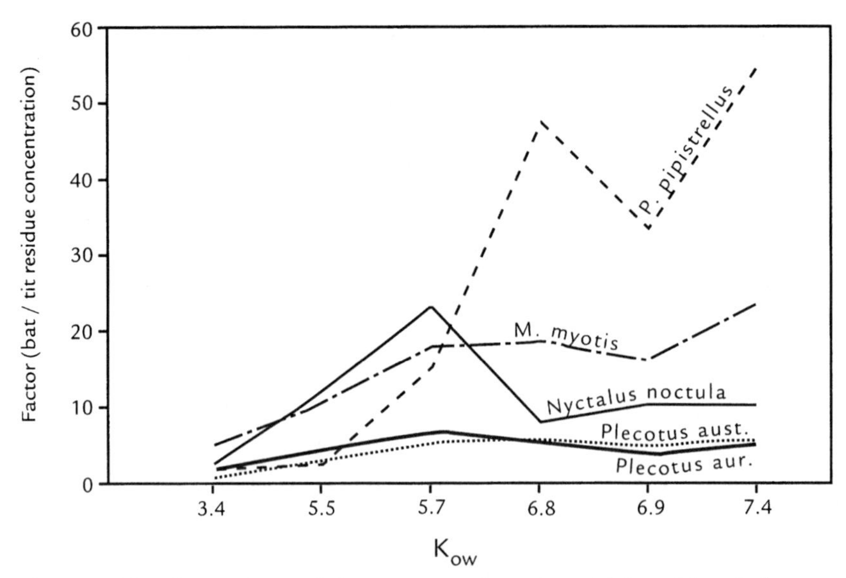

Figure 9.10 Comparison of the concentration of pesticide residues in bat species and great tits. The ratio "bat/tit" is especially high in the common pipistrelle (*Pipistrellus spp.*), which preferably roost and forage within and close to settlements. The relative contamination was lowest in *Plecotus* species, which prefer forests. Abcissa: K_{ow} is the *n*-octanol/water partition coefficient, an index for the lipophilic properties of a compound. After Streit et al. (1995).

Because males cannot transfer pesticide compounds to the juveniles, their organochlorine contamination is generally higher than that of females. A comparative study in populations of bats and great tits in southwest Germany demonstrated that the pesticide load in bats is at least five times larger than in birds (fig. 9.10; 38–260 ppm on a lipid mass basis for bats as compared to 8 ppm for great tits). However, at least in Europe, the decline in population numbers is very different from species to species. In middle Europe some species (e.g., rhinolophids) are nearly extinct, whereas others, including the common bat *Pipistrellus pipistrellus*, have fared rather well despite high levels of pesticides.

The disturbing disappearance of bats from barns, church attics, and caves, combined with reports in the popular science media on the many spectacular and fascinating achievements of bats has brought about a change in the attitude of society toward bats from one of repulsion or indifference to one of concern and care. In many countries bats are now a favorite cause of nature conservation groups. Specially designed bat houses have been placed in forests from which all of the dead and hollow trees had previously been removed; the renovation of church attics and buildings have been scheduled so as not to interfere with the reproductive cycles of bats living there. Rural development offices now make sure that hedges and river banks are maintained or replanted as attractive flyways for bats commuting to foraging areas at forest edges or within woods.

Perhaps one of the most heartwarming stories in conservation biology is that of BCI—Bat Conservation International—in Austin, Texas. This "global player" was founded single-handedly by a young scientist, Merlin Tuttle, who not only invented the "Tuttle trap" for catching bats but also started to educate the public through his photographs of bats in action. He raised private funds and within a short period built a powerful organization that is active all over the world. BCI has saved many flower-visiting bat populations and protected many wintering quarters from tourists and intruders. It created and encouraged local bat conservation groups on all continents, and now even owns bat caves. BCI demonstrates how a scientist-turned-entrepreneur can help change the attitude of people toward living nature through education, motivation, persuasion, and courageous action.

References

Audet D (1990). Foraging behavior and habitat use by a gleaning bat, *Myotis myotis*. J Mammal 71:420–427.

Avery MI (1985). Winter activity of pipistrelle bats. J Anim Ecol 54:721–738.

Bell GP, Bartholomew GA, Nagy KA (1986). The roles of energetics, water economy, foraging behavior and geothermal refugia in the distribution of the bat, *Macrotus californicus*. J Comp Physiol B 156:441–450.

Bonaccorso FJ (1979). Foraging and reproductive ecology in a Panamanian bat community. Bull Fla State Mus Biol Sci 24:359–408.

Brosset A (1962). The bats of central and western India. Parts I, II, III, IV. J Bombay Nat Hist Soc 59:1–746.

Choe JC (1994). Ingenious design of tent roosts by Peter's tent-making bat, *Uroderma bilobatum*. J Nat Hist 28:731–737.

Cox PA (1982). Vertebrate pollination and the maintenance of dioecism in Freycinetia. Am Nat 120:65–80.

Dinerstein E (1986). Reproductive ecology of fruit bats and the seasonality of fruit production in a Costa Rica cloud forest. Biotropica 18:307–318.

*Dobat K, Peikert-Holle TH (1985). Blüten und Fledermäuse. Waldemar Kramer Verlag, Frankfurt.

Estrada A, Coates-Estrada R, Meritt D (1993). Bat species richness and abundance in tropical rain forest fragments and in agricultural habitats at Los Tuxtlas, Mexico. Ecography 16:309–318.

Fenton MB (1990). The foraging behaviour and ecology of animal-eating bats. Can J Zool 68:411–422.

Fenton MB, Boyle NGH, Harrison TM, Oxley DJ (1977). Activity patterns, habitat use, and prey selection by some African insectivorous bats. Biotropica 9:73–85.

*Fleming TH (1993). Plant-visiting bats. Am Sci 81:460–467.

Freeman PW (1995). Nectarivorous feeding mechanisms in bats. Biol J Linn Soc 56:439–463.

Gorchov DL, Cornejo F, Ascorra CF, Jaramillo M (1995). Dietary overlap between frugivorous birds and bats in the Peruvian Amazon. Oikos 74:235–250.

*Gould E (1978). Foraging behavior of some Malaysian nectar-feeding bats. Biotropica 10:184–193.

*Heithaus ER (1982). Coevolution between bats and plants in T.H. Kunz, ed., Ecology of Bats, pp. 327–367. Plenum Press, New York.

*Helversen von O (1993). Adaptations of flowers to the pollination by glossophagine bats. In W. Barthlott, et al., eds., Animal–Plant Interactions in Tropical Environments, pp. 41–59. Museum Koenig Bonn, Germany.

Hill JE, Smith JD (1984). Bats, a Natural History. British Museum (Natural History) London.

*Howell DJ (1974). Bats and pollen: Physiological aspects of the syndrome of chiropterophily. Comp Biochem Physiol 46A:263–279.

*Howell DJ (1977). Time sharing and body partitioning in bat-plant pollination systems. Nature 270:509–510.

*Howell DJ (1979). Flock foraging in nectar-feeding bats: advantages to the bats and to the host plants. Am Nat 114:23–49.

Kingdon J (1984). East African Mammals, Vol. IIA. Insectivores and Bats. University of Chicago Press, Chicago.

Knudsen JT, Tollsten L (1995). Floral scent in bat-pollinated plants: A case of convergent evolution. Bot J Linn Soc 119:45–57.

Krull D, Schumm A, Metzner W, Neuweiler G (1991). Foraging areas and foraging behavior in the notch-eared bat, *Myotis emarginatus* (Vespertilionidae). Behav Ecol Sociobiol 28:247–253.

*Kunz TH, ed. (1982). Ecology of Bats. Plenum Press, New York.

Kunz TH (1994). The world of tent-making bats. Bats (Spring):6–12.

Morrison DW (1980). Foraging and day-roosting dynamics of canopy fruit bats in Panama. J Mammal 61:20–29.

Neuweiler G (1989). Foraging ecology and audition in echolocating bats. Trends Ecol Evol 4:160–166.

O'Shea JT (1980). Roosting, social organization and the annual cycle in a Kenya population of the bat Pipistrellus nanus. Z Tierpsychol 53:171–180.

Sazima I, Sazima M (1978). Bat pollination in the passion flower *Passiflora mucronata*, in southeastern Brazil. Biotropica 10:100–109.

Shore RF, Myhill DG, French MC, Leach DV, Stebbings RE (1991). Toxicity and tissue distribution of pentachlorophenol and permethrin in pipistrelle bats experimentally exposed to treated timber. Environ Pollut 73:101–118.

Streit B, Winter ST, Nagel A (1995). Bioaccumulation of selected organochlorines in bats and tits: Influence of chemistry and biology. Environ Sci Pollut Res 2:194–199.

Tamsitt JR (1967). Niche and species diversity in neotropical bats. Nature 213:784–786.

Thies ML, Thies K, McBee K (1996). Organochlorine pesticide accumulation and genotoxicity in Mexican free-tailed bats from Oklahoma and New Mexico. Arch Environ Contam Toxicol 30:178–187.

Timm RM (1984). Tent construction by Vampyressa in Costa Rica. J Mammal 65:166–167.

Timm RM (1987). Tent construction by bats of the genera *Artibeus* and *Uroderma*. Fieldiana: Zoology n.s. 39:187–212.

Willig MR, Moulton MP (1989). The role of stochastic and deterministic processes in structuring Neotropical bat communities. J Mammal 70:323–329.

Wilson DE (1973). Bat faunas: A trophic comparison. Syst Zool 22:14–29.

10

PHYLOGENY AND SYSTEMATICS

10.1 PHYLOGENY

Where and when does the evolutionary history of bats begin? Although textbooks state that bats are descended from insectivores, no one really knows for sure when or how they originated. To date no fossil dig has yielded an *Archichiropteryx*, and even the rich paleontological material from the Eocene period 40–55 million years ago, when bats first appear in the fossil record, contains only fully differentiated bats complete with wings and large cochleas. It seems that by the Eocene period bats had already mastered the task of echolocation.

Nevertheless, fossils from the early and middle Eocene period contain skeletal characteristics that are not consistent with the anatomy of more recent, currently living, species. Species that presumably became extinct during the Myocene period, 10–25 million years ago, have been grouped together in a single family, Paleochiropterygoidea, that does not belong to any living taxonomic group. These fossil bats include *Icaronycteris*, found in Wyoming, *Palaeochiropteryx*, found in France, *Archeonycteris* and *Hassionycteris*, found in France and in the Messel quarry in southern Germany, *Ageina*, found in France, *Cecilionycteris*, found in middle Eocene deposits in Germany, and *Matthesia*, found in the Geiseltal (Germany). The frugivorous "flying fox" *Archeopteropus*, which lived in Italy during the Oligocene period 35 million years ago, also belonged to this extinct microchiropteran-like family. The first true megachiropteran is thought to be the extinct African species *Prototto leakeyi* from the early or middle Myocene period.

The earliest insectivorous bat is thought to have been *Icaronycteris index*, the only fossil bat that has a claw on the second finger as do the present-day flying foxes. Nevertheless, even this well-preserved specimen from Wyoming does not provide any clues as to which tree-dwelling or ground-dwelling mammal might have been the first to take to the air. As will be discussed later, bats possess some characteristics that resemble those of primates and gliders (Dermoptera). The teeth of the Paleochiroptera appear to have been similar to those of the tree shrew (*Tupaia*). For this and other reasons, the present-day Chiroptera, Tupaidae, Dermoptera, and primates are all grouped together under the taxon Archonta. This is the group from which bats are most likely to be descended—possibly from a *Tupaia*-like ancestor—and not from the insectivores. Whatever their ancestor was, bats apparently evolved rapidly and changed little thereafter. This is attested to by the molossids, a family that is considered to be highly derived and specialized for flight. Molossids are known to have already been differentiated in the middle Oligocene era, about 30 million years ago. This rapid evolutionary development probably took place very early, in the Paleocene era about 70–100 million years ago, at a time when flowering plants were becoming widespread and flower-visiting insects such as beetles, butterflies, and flies were developing. However, once bats reached a high degree of differentiation, their evolution appears to have come virtually to a standstill.

10.2 SYSTEMATICS

In 1758, Carl Linnaeus, the founder of modern taxonomy, knew of only seven species of bats. He grouped these bats together under the genus *Vespertilio*, into the order of primates. In 1780, Blumenbach classified bats in a separate order, Chiroptera, the name that is used today. Nevertheless, as recently as the last century, there was still some uncertainty as to how to classify bats, so they were placed in various different mammalian orders (for a summary of the different families of bats, see pp. 5–8). For a long time the flying lemur *Cynocephalus* was classified under the order Chiroptera. Since the Australian neurobiologist Pettigrew described the Megachiroptera as "flying primates" based on their midbrain visual pathways, some bats appear to have wandered back into the primate group. The traditional system of taxonomy which seems so well established is in upheaval. The idea that the Chiroptera have a monophyletic origin rests on the fact that they are the only mammals with wings. Nevertheless, it is possible that the wings of Megachiroptera and Microchiroptera developed through convergent evolution. The similarity of the wing structure in the two groups could simply be due to the constraints imposed by the mammalian body structure, which would allow few other options.

The goal of systematics or taxonomy is to characterize the various taxa and organize them on the basis of their evolutionary relationships. Classical taxonomy depended heavily on museum material, especially bones. In classifying the Chiroptera, the premaxilla and teeth were used as indicators of the diet; the bones of the limbs, especially the joint connecting the shoulder and upper arm, were used as indicators of the ability to fly. Joints that resemble those of flightless mammals are considered to be ancestral. This type of joint is found in flying foxes of the genus *Pteropus*. Joints that show clear adaptations for flight are considered to be derived. This type of joint is found in molossid bats, among others. Patterns of dentition that resemble those of insectivores are considered to be primitive. It is on the basis of collections of characters such as these that family trees are deduced. However, these family trees are by no means definitive, especially since there is a tendency to underestimate the possibilities for convergent evolution as well as specific functional adaptations in taxa of equal rank.

The system used today for classification of the Chiroptera is based on the organization proposed by Miller in 1907. Miller classified 16 different families of Microchiroptera on the basis of their bone structure (fig. 10.1). Although it is pos-

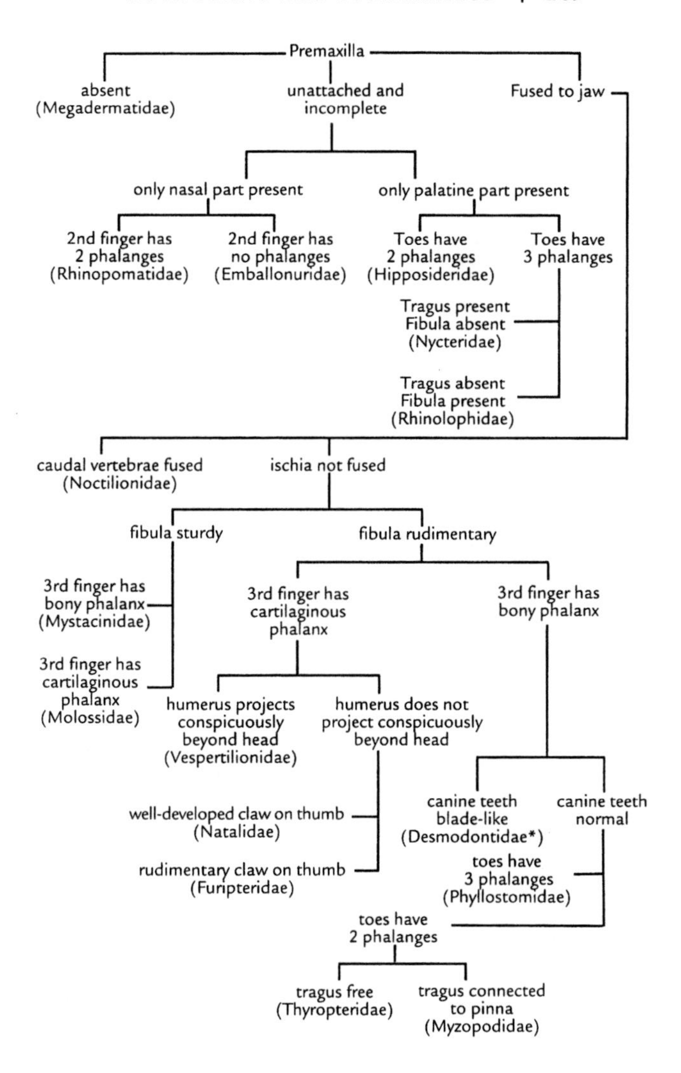

Figure 10.1 Characters used by Miller (1907) to classify the families of Microchiroptera. The system used by Miller forms the basis of the present taxonomy.

sible to divide bats into families based on these criteria, they do not provide any insight into how the different families are related to one another. Today, the issue of bat phylogeny is more controversial than ever, even though a number of new characters and new methods for studying this problem were introduced in the 1970s and 1980s.

CLADISTICS

Cladistic analysis, introduced by Henning in 1950, has been widely accepted as a method of evolutionary analysis, especially in the United States. Cladistic analysis attempts to identify a group of derived (apomorphic) characters and a group of primitive (plesiomorphic) characters for taxa that appear to be related. A hypothesis about phylogenetic relationships (cladogram) can then be formulated on the basis of derived characters that are shared (synapomorphic). The larger the number of independent characters that are included in the cladogram, the higher the probability that it will reveal true phylogenetic relationships.

An example of cladistic taxonomic analysis is the classification of phyllostomids based on the morphology of the uterus and oviducts. Six derived characters (fig. 10.2) were analyzed to determine which were shared (synapomorphic) among the genera studied. The result was that the two species of thyropterids, along with the phyllostomids, differ from all other Chiroptera in that they have an externally fused uterus and a simple connection between the oviduct and uterus. On the basis of this synapomorphy, the Thyropteridae should be classified in the superfamily Phyllostomoidea, even though they are placed in the superfamily Vespertilionoidea on the basis of their bone and skull morphology. The cladogram based on other characters related to the female sex organs (fig. 10.2) is not completely consistent with cladograms based on other features. This example demonstrates that various schemes of relatedness can be constructed depending on what features are selected for analysis. It is difficult to identify unequivocally synapomorphic traits because they may be mixed up with convergences or synplesiomorphic (shared, primitive) characters. In fact, one of the most difficult problems of evolutionary biology is that of arriving at congruent cladograms based on independent characters. Perhaps species would not need to be reclassified from one family to another so often if, in addition to morphological and molecular characters, functional and behavioral characters were also considered.

KARYOTYPING

The goal of karyotyping is to look for similarities in the morphology and number of chromosomes and to establish their banding patterns. The large amount of karyotypic variation that has been found among present-day species provides clues about the evolutionary distance of each species descended from a common ancestor with a single karyotype. All bats possess a set of diploid ($2n$) chromosomes which vary in number from 16–62. On average, bats have 36.8 chromosomes (fig. 10.3). This is fewer than other mammals. The mass of DNA in a haploid genome is called the genome size (C value). The C value of bats varies from 50–87% of that characteristic of other mammals and humans. However, different numbers of chromosomes do not necessarily mean that there are differences in the genome because chromosome segments can split off from one another or fuse together, resulting in what is called Robertsonian variation. For this reason, a further measure, the fundamental number (FN), has been introduced. The fundamental number is the total number of arms of the autosomal chromosomes. In bats, the FN is on average 51.6 and varies less than the number of chromosomes. In taxa with identical FN values but different numbers of chromosomes, it is possible that Robertsonian variation due to dissociation and fusion of chromosomal segments is responsible for the differences in chromosome number.

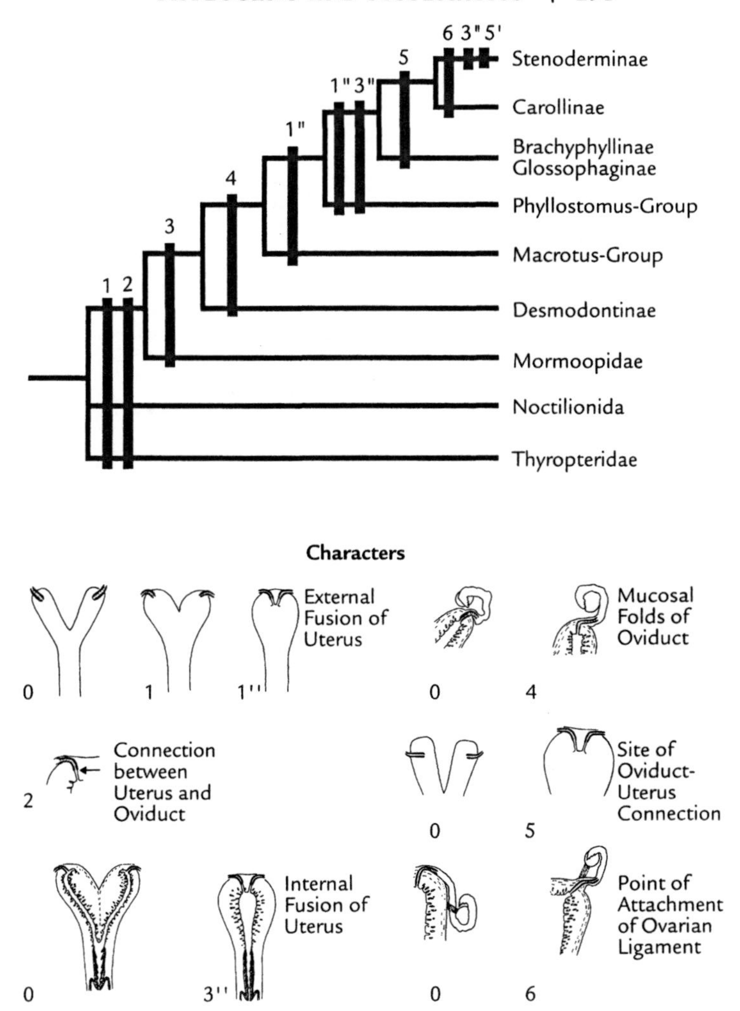

Figure 10.2 Cladogram of the Phyllostomoidea based on morphological characteristics of the uterus and oviducts (bottom). Characters 1–6, illustrated here, are designated according to their degree of differentiation, with 0 being undifferentiated or plesiomorphic, and 1" being apomorphic or highly differentiated. The black bars indicate synapomorphies, or shared derived characters. The arrow at the junction of the uterus and oviduct (2) indicates a junction that is smooth, without folds. From Hood and Smith (1982).

The chromosome counts of bats are not as variable as those of other mammals, and within a genus they are quite constant. For example, all but one of the 44 different species of *Myotis* that have been examined possess 44 chromosomes. The exception is *Myotis daubentoni*, with 42 chromosomes. All 8 species of *Eumops* (Molossidae) have 48 chromosomes. Comparison of the banding patterns obtained

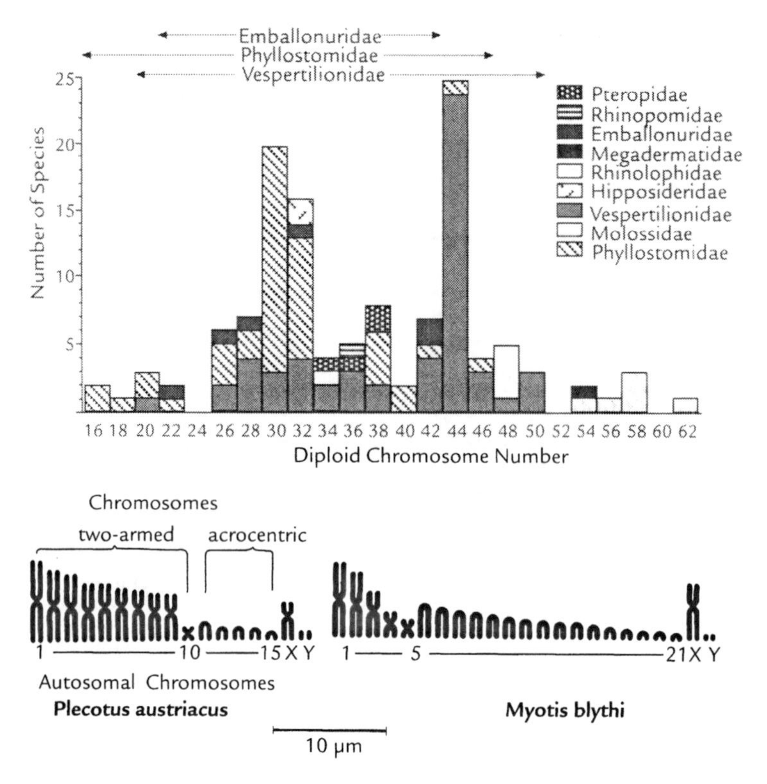

Figure 10.3 Chromosome numbers and karyograms. *Top*: Diploid chromosome numbers for the most important families of Chiroptera. In some families the range of chromosome numbers is wide (arrows). *Bottom*: Karyograms of *Plecotus austriacus* (Vespertilionidae) and *Myotis blythi* (Vespertilionidae) from Czechoslovakia showing chromosome morphology. From Wimsatt (1970).

using different staining methods, for example, G-bands revealed by Giemsa staining, provide a way to identify homologous chromosomes. These methods provide more reliable information about the similarities between two species.

The Central American species *Rhogeessa* (Vespertilionidae) provides a good example of the power of karyotyping (fig. 10.4). This genus has an FN of 50, corresponding to 25 pairs of chromosomes. In *Rhogeessa parvula*, the arms of the chromosomes are organized into 44 diploid chromosomes. *Rhogeessa tumida*, on the other hand, includes 5 different karyotypes with diploid ($2n$) numbers of 30, 32, 34, 42, and 44. The type $2n = 42$ has given rise to a new species, *Rhogeessa genowaysi*. Comparison of banding patterns can identify the homologous chromosome arms 1–50 for every karyotype. This analysis revealed that the karyotype $2n = 32$ arose from two different sets of chromosomes. In *Rhogeessa* from Nicaragua (32N), chromosomal arms had fused to form biarmed chromosomes different from those in *Rhogeessa* from Belize (32B).

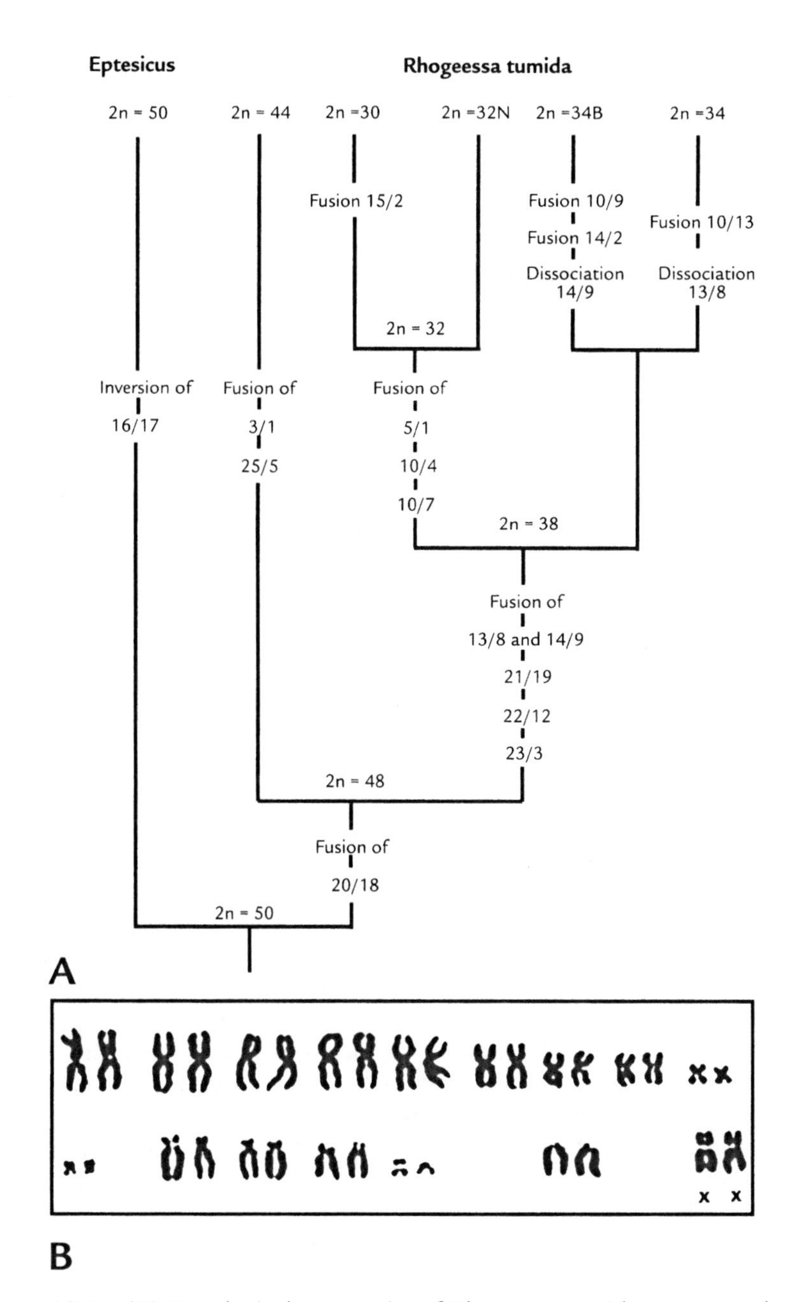

Figure 10.4 (A) Karyological systematics of *Rhogeessa tumida*. A presumed ancestor with 50 chromosomal arms (2*n*) gave rise to the genera *Eptesicus* and *Rhogeessa*. Numbers indicate identified chromosomal arms. 32N = *Rhogeessa tumida* from Nicaragua with 32 chromosomes. 32B = *Rhogeessa tumida* from Belize with 32 chromosomes. From Baker et al. (1985). (B) Karyotype (2N = 32, FN = 58) of a female *Rhogeessa* of the karyotype 32B, which is hypothesized to be separate from *R. tumida* as a new species, *Rhogeessa aeneus*. From Audet et al. (1993).

To obtain clues about phylogenetic relationships, the five different karyotypes of *Rhogeessa tumida* (30, 32N, 32B, 34, and 44) were organized into a cladogram (fig. 10.4), using Henning's method of cladistic analysis. A common ancestor with 50 chromosomes is assumed to have given rise to both *Eptesicus* and *Rhogeessa.* In *Eptesicus*, the fused chromosomal arms 16 and 17 are inverted. This inversion is considered to be a derived (apomorphic) character which only occurs in the genus *Eptesicus*. In all species of *Rhogeessa*, chromosomal arms 18 and 20 are fused, reducing the number of chromosomes to 48. This fusion is considered a synapomorphic character. Additional fusions and dissociations gave rise to the 5 different karyotypes of *Rhogeessa tumida* as shown in the tree of phylogenetic relationships based on derived and primitive characters (fig. 10.4). The advantage of this cladogram is that it requires relatively few assumptions. These assumptions are that fusion occurred between chromosomal arms 13 and 8 and between arms 14 and 9 in an ancestor that gave rise to the 5 different karyotypes; later, these chromosomal arms dissociated again to give rise to karyotypes 34 and 32B.

If we assume that offspring of individuals with different karyotypes are not capable of reproducing, the results of chromosomal analysis would imply that *Rhogeessa tumida* actually consists of five different species that are morphologically indistinguishable. This example illustrates how new species might arise through fusion and dissociation of chromosomes.

Phylogenetic trees based on one character or one method alone are generally useless. As an example, the two genera *Glossophaga* and *Erophylla* are clearly distinguishable based on morphological criteria; nevertheless, their karyotypes are identical. *Uroderma bilobatum* (Phyllostomidae), on the other hand, exhibits striking differences in karyotype even within the same species. Because there is no way to experimentally repeat evolution, subjective assessments are bound to influence the phylogenetic scenarios proposed.

Karyotyping has shown that marked morphological differences need not be accompanied by corresponding karyotypic differences. The wide range of morphological features found among mammals seems to be more due to point mutations than to chromosomal changes.

Chromosomal analysis is often unable to distinguish a derived pattern from an ancestral pattern and thus can provide no information about the direction taken by evolution. This remains its basic weakness. Nevertheless, taxonomists continue to be seduced into characterizing species whose chromosome number is close to the average of 36.8 as "primitive." *Rhinopoma* and *Pteropus* are two families that are considered primitive based on characters other than karyotype. The fact that *Rhinopoma* has 36 chromosomes and *Pteropus* has 38 seems to support this interpretation.

MOLECULAR BIOLOGICAL METHODS

Since the development of molecular biology it has become common to look for phylogenetic relationships in primary molecular structure using methods such as chromatography, immunology, protein sequencing, and comparison of DNA sequences in selected genes and mitochondrial RNA genes to identify similarities in enzymes and other proteins, as well as amino acid sequencing. It is assumed that the more similar the primary structure of a molecule in two species, the more recently they have diverged.

An example of how molecular biology can be used to construct phylogenetic trees comes from the species *Mystacina tuberculata*, native to New Zealand. Immunological comparison of the plasma albumins and transferrins of this species revealed a high degree of similarity between the albumin molecule in *Mystacina* and the superfamily Phyllostomoidea, but little similarity with other species. Transferrin molecules in *Mystacina* were strikingly similar to those of *Noctilio*. Up to the time of this analysis, *Mystacina* had been placed in six different families. However, on the basis of these molecular homologies, it was concluded that *Mystacina* is a descendent of the Phyllostomoidea. If one accepts the time scale attributed to the molecular clock, and if one assumes that random amino acid substitutions that do not affect function occur at a moderate rate, the measured immunological distance between transferrins indicates that *Mystacina* in New Zealand must have diverged from its South American ancestor about 35 million years ago.

The same immunological methods, together with an electrophoretic similarity analysis of structural genes for proteins, were used to study intrafamily relationships in three species of vampires and their classification within the phylogenetic system. The albumin molecules of the rare species *Diaemus*, which feeds preferentially on the blood of birds, and the common species *Desmodus*, which feeds preferentially on the blood of mammals, show a degree of similarity that is otherwise found only in subspecies that can interbreed. *Diphylla*, a species that is specialized to feed on the blood of birds, is immunologically more distant from the other two species. Although some morphological characteristics are consistent with the immunological cladogram, chromosomal analysis yields a different system of relationships. If the morphological analysis of the uterus and oviducts described earlier is performed on vampires, they appear to be most closely related to *Macrotus*, a group that is often considered to be close to the original ancestor of the phyllostomids. This means that the vampires probably diverged early from the other phyllostomids, but that the divergence of *Diphylla* and *Diaemus/Desmodus* occurred as late as 8 million years ago. *Diaemus* and *Desmodus* diverged even later, about 5 million years ago.

In spite of a growing number of studies, the phylogenetic relations among bat species are far from settled. Conflicts among immunologic, karyotypic, and morphological traits create ambiguities about phylogenetic relationships below the family level for many proposed classifications.

CLASSIFICATION OF THE MEGACHIROPTERA

The biggest problem in bat systematics remains the classification of the Megachiroptera. Fruit- and nectar-eating flying foxes, which are found only in the Old World, do not echolocate, have large eyes specialized for night vision, and spend their days hanging in well-lit trees. Thus, they have a very different lifestyle from the Microchiroptera, with one notable exception. The genus *Rousettus* lives in caves and uses echoes to orient in complete darkness. *Rousettus* does not use its larynx to produce echolocation signals; instead it produces sounds by clicking its tongue. Most flying foxes have a nonfunctional claw on the second finger. Classical systematics has regarded this claw as a vestige left over from the early evolution of the wing and on this basis classified flying foxes as the oldest group of bats. The fact that the joint between the shoulder and upper arm is not highly differentiated for flight strengthens this classification scheme.

An analysis of the amino acid sequence of hemoglobin in four flying foxes and six Microchiroptera suggests that all the Chiroptera evolved from a common primatelike ancestor. This hypothesis gains support from a morphological character that is totally independent of molecular specializations—the penis.

Neuroanatomy suggests a very different classification scheme. In 1986, Pettigrew made a surprising discovery. The retinotectal visual pathways of the Megachiroptera are like those of primates in that they have binocular connections, and the midbrain visual representation is of the contralateral field as transmitted by both eyes. However, in Microchiroptera, as in most other mammals, the visual pathways to the midbrain are monocular, and the midbrain visual representation includes the entire field of the contralateral eye. Pettigrew concluded from this that the Megachiroptera are flying primates and that their origin is therefore separate from that of the Microchiroptera. If this were the case, the wing would not be synapomorphic. Instead, it would be convergent, having developed independently in a primate and in the insectivore that became the ancestor of present-day microchiropterans.

Although this argument sounds convincing, it underestimates the possibilities for convergent evolution. A binocular visual projection to the midbrain is not the only feature that flying foxes and primates have in common; both engage in similar visually guided behavior. Flying foxes and many primates are tree-dwelling animals that grasp target objects with their fingers or the claw of their thumb. It is possible that selective pressure for stereoscopic vision would be just as effective as a common ancestor in bringing about binocular connections in both groups. Neuroanatomical studies have shown that the visual pathways in *Rousettus* are like those of Microchiroptera. This finding could be the downfall of the hypothesis that Megachiroptera are flying primates. In recent years, Pettigrew's thought-provoking hypothesis has initiated a number of genetic studies and reexamination of morphological characters. Most of these investigations favor a monophyletic origin of Microchiroptera and Megachiroptera (fig. 10.5). However, the files on this exemplary case are not yet closed.

The penis, premaxilla, and sperm. The mammalian penis consists of the shaft, glans, and foreskin or prepuce. During the course of evolution, the penis of the Archonta (Tupaidae, Dermoptera, Chiroptera, and primates) lost its attachment to the belly to become a freely hanging, movable organ. A freely movable penis is considered a synapomorphic characteristic of Archonta. When erectile tissue (corpus cavernosum and spongiosum) in the shaft and glans of the penis becomes engorged with blood, the normally flexible penis stiffens. Many primates and bats have a small rodlike bone, the baculum, that serves as an additional stiffening ele-

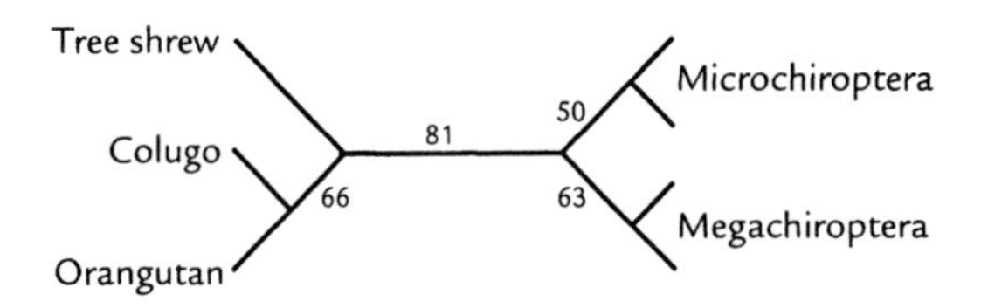

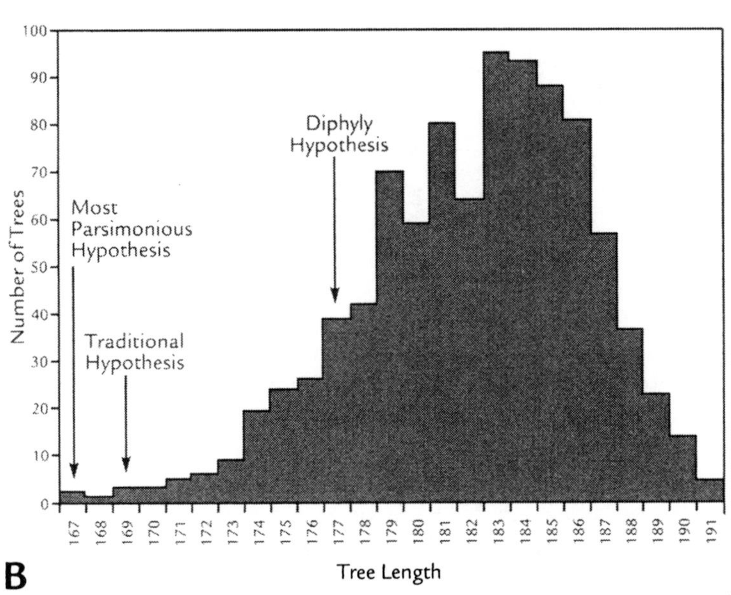

Figure 10.5 Phylogenetic classification of Megachiroptera. (A) The most parsimonious unrooted phylogenetic tree for bats and primates including tree shrews (*Tupaia*), based on an analysis of the mitochondrial 12S ribosomal RNA gene in 11 species. Numbers give "bootstrap values," an indicator of phylogenetic distance. (B) The distribution of all possible phylogenetic trees based on the same analysis as in panel A, but restricted to seven taxa. The most parsimonious tree, i.e., that which requires the smallest number of steps (tree length 167), is shown in panel A. From Ammerman and Hillis (1992).

ment. Although there is no good basis for it, many phylogeneticists consider the baculum a primitive character that is absent in highly developed primates and in humans and in highly derived bats such as the molossids. However, the baculum is also lacking in vespertilionids, phyllostomids, and noctilionids.

Under the assumption that the baculum and accessory erectile tissue are plesiomorphic, the following distribution of characters can be assigned:

- *Microchiroptera*: Baculum reduced or absent (apomorphic), unspecialized corpus spongiosum (plesiomorphic), and accessory erectile tissue (apomorphic).

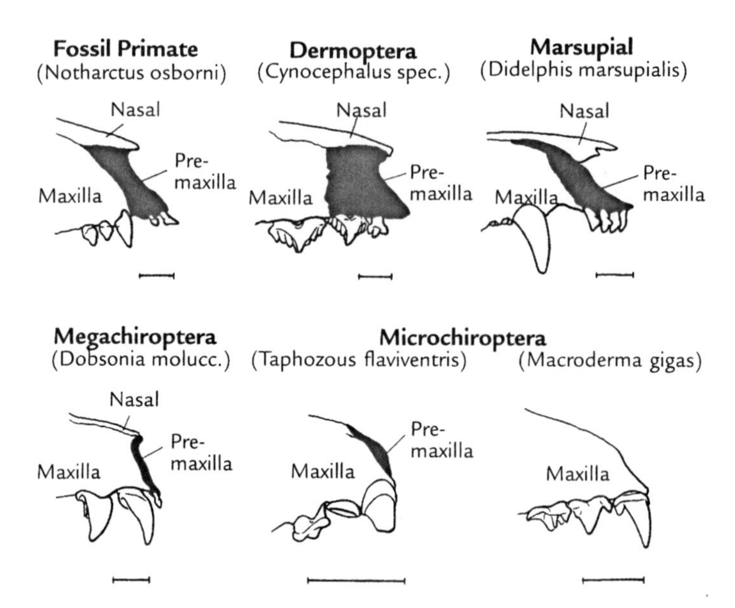

Figure 10.6 The premaxilla as a systematic character. The premaxilla (black) in Chiroptera (bottom row) is reduced compared to other mammals (upper row). Each calibration bar represents 5 mm. From Wible and Novacek (1988).

- *Megachiroptera*: Baculum reduced, corpus spongiosum expanded distally (apomorphic, a character shared by Dermoptera and primates), no accessory erectile tissue (apomorphic, a character shared by Tupaidae, Dermoptera, and primates).

The morphology of the penis in Chiroptera places them close to primates; the Megachiroptera somewhat closer than the Microchiroptera. Insectivores, generally regarded as the ancestors of the Chiroptera, do not even enter into this argument.

In the beginning it was shown that the premaxilla was the first feature used to differentiate bats from other mammals. On this basis the diphyletic origin inferred from the organization of the visual pathways reverts again to being monophyletic (fig. 10.6). Other features of the skull also support a monophyletic origin of the Mega- and Microchiroptera. A thorough analysis of many independent characters will be necessary to provide a more definitive answer to this controversial question.

The ultrastructure of sperm is a character that has been used successfully in primate classification. However, when this character is examined in the Chiroptera, the many possible phylogenetic relationships become even more confusing. The spermatozoa of the Megachiroptera resemble those of some insectivores (e.g., moles) and those of lemurs. The spermatozoa of Microchiroptera resemble those of haplorhine apes, tupaias, and hedgehog-like insectivores.

Independent of the relationship between Microchiroptera and Megachiroptera, it is clear that all Chiroptera possess a set of characters that indicate a close relationship with primates. These include the primary structure of hemoglobin, the frontal position of the eyes, and the morphology of the penis. The two cladograms for the Archonta summarize both of the hypotheses regarding the phylogeny of the Megachiroptera. The tree that is based on many different independent characters in which the Megachiroptera and Microchiroptera are sister groups (fig. 10.5) seems to be the more convincing one based on currently available evidence.

References

Ammerman LK, Hillis DM (1992). A molecular test of bat relationships: monophyly or diphyly? Syst Biol 41:222–232.

Audet D, Engstrom MD, Fenton MB (1993). Morphology, karyology, and echolocation calls of Rhogeessa from the Yucatan peninsula. J Mammal 74:498–502.

Bailey WJ, Slightom JL, Goodman M (1992). Rejection of the "flying primate" hypothesis by phylogenetic evidence from the γ-globulin gene. Science 256:86–89.

Baker RJ, Bickham JW, Arnold ML (1985). Chromosomal evolution in *Rhogeessa*: Possible selection by centric fusion. Evolution 39:233–243.

Friant M (1963). Les chiroptera (chauves-souis), revision des rhinolophidae de l'epoque tertiaire. Acta Zool 64:161–178.

Hill JE (1977). Areview of the rhinopomatidae (Mammalia, Chiroptera). Bull Br Mus Nat Hist (Zool) 32:29–43.

Honeycutt RL, Greenbaum IF, Baker RJ, Sarich VM (1981). Molecular evolution of vampire bats. J Mammal 62:805–811.

Hood CS, Smith JD (1982). Cladistical analysis of female reproductive histomorphology in phyllostomatoid bats. Syst Zool 31:241–251.

Jepsen GL (1966). Early eocene bat from Wyoming. Science 154:1333–1339.

Kleinschmidt T, Braunitzer G (1982). Die Primärstruktur des Hämoglobins vom Ägyptischen Flughund (*Rousettus aegyptiacus*). Hoppe-Seyler Z Physiol Chem 363:1209–1215.

Kleinschmidt T, Koop B, Braunitzer G (1986). The primary structure of a mouse-eared bat (*Myotis velifer*) hemoglobin. Biol Chem Hoppe-Seyler 367:1243–1249.

Kleinschmidt T, Sgouros JG (1987). Hemoglobin sequences. Biol Chem Hoppe-Seyler 368:579–615.

Mann G (1963). Phylogeny and cortical evolution in Chiroptera. Evolution 17:589–591.

Miller GS (1907). The families and genera of bats. Smithsonian Inst US Nat Mus Bull 57, Washington, DC.

Morales JC, Bickham JW (1995). Molecular systematics of the genus *Lasiurus* based on restriction-site maps of the mitochondrial ribosomal genes. J Mammal 76:730–749.

Padian K (1985). The origins and aerodynamics of flight in extinct vertebrates. Palaeontology 28:413–434.

*Pettigrew J (1986). Flying primates? Megabats have the advanced pathway from eye to midbrain. Science 231:1304–1306.

Pierson ED, Sarich VM, Lowenstein JM, Daniel MJ, Rainey WE (1986). A molecular link between the bats of New Zealand and South America. Nature 323:60–63.

Rouse GW, Robson SK (1986). An ultrastructural study of megachiropteran spermatozoa: Implications for chiropteran phylogeny. J Submicroscr Cytol 18:137–152.

*Smith JD (1980). Chiropteran phylogenetics: Introduction. In Proc. of the Fifth International Bat Research Conference, pp. 233–244. Texas Tech Press, Lubbock.

Smith JD, Madkour G (1980). Penial morphology and the question of chiropteran phylogeny. In Proc. of the Fifth International Bat Research Conference, pp. 347–365. Texas Tech Press, Lubbock.

Wible JR, Novacek MJ (1988). Cranial evidence for the monophyletic origin of bats. Am Mus Novit 2911:1–19.

Wimsatt WA, ed. (1970). Biology of Bats, Vol. 1. Academic Press, New York.

INDEX

accessory optic tract, 219, 220
acidosis, 73
aerodynamics, 24–29
African hammerhead bats, 253–54
aging, 137
altruism, 112, 259
ammonia, 92, 94
amygdala, 136
anastomoses, 49
anestrus, 241
Anoura geoffroyi, 218, 249, 269
anteroventral cochlear nucleus (AVCN), 172–73
antitragus, 160
Antrozoous pallidus, 242, 249
archipallium, 136
Artibeus jamaicensis
  embryonic diapause, 247
  feeding, 275
  harems, 250
  olfactory epithelium, 227
  pretectal nuclei, 221
  reproductive cycle, 243
  taste, 233
Artibeus lituratus, 228
arytenoid cartilages, 146
auditory cortex, 169, 180–81
auditory fovea, 170, 196–201
auditory system, 156–78
  functional anatomy of ear, 156–69
  pathways in central nervous system, 169–78
  See also echolocation; sound
AVCN. *See* anteroventral cochlear nucleus
azimuth, 185

basal ganglia, 136
basilar membrane, 164–68
Bat Conservation International, 285
bats
  auditory system, 156–78
  central nervous system, 117–37
  circulatory and respiratory systems, 43–60
  definition of, 4
  echolocation, 4, 140–205
  ecology, 262–85
  families, 5–8
  functional anatomy and locomotion, 9–40
  hibernation, 46, 70, 72–81
  homeothermy, 64–68
  and human activities, 282–85
  living quarters, 277–82
  migrations, 264–67
  olfaction, 224–32
  phylogeny, 287–88
  and plants, 267–77
  reproduction and development, 236–59
  taste, 233
  taxonomy, 4–5, 288–99
  teeth, 98–102
  torpor, 68–70
  vision, 21–23
  water balance, 82–94
best frequency, 168

binaural pathways, 173–74
binocular connections, 221–23
biotopes, 191, 192–205
birds, 15, 45, 55
blood, 53–55
  oxygen capacity, 54
  profiles, 55
  supply to head, 48
blood pressure, 45–46
blood vessels, 47–48, 50
brain
  cerebellum, 128–29
  diencephalon, 130–31
  encephalization, 117–18, 119, 133
  hindbrain, 124–27
  midbrain, 127–28
  telencephalon, 131–37
breathing. *See* respiration
brown adipose tissue, 78, 79
bulldog bats, 7
bumblebee bats, 6

calpain, 137
capillary networks, 49–52
Carioderma cor, 254
carnivorous bats, 39, 104, 263
Carollia perspicillata
  directional beaming, 155–56
  food selection, 273, 275
  harems, 250, 252
  menstruation, 239
  morphometric measures for nocturnal eyes, 212
caves, 277, 278
central acoustic tract, 175, 176
central gray, 128
central nervous system, 117–37
  and aging, 137
  auditory pathways, 169–78
  terminology, 120
  See also brain
cerebellum, 128–29
Cheiromeles parvidens, 100, 103
chemical communication, 229–30
chitin, 104
chorda tympani nerve, 233
choroid, 213, 215–16
choroid papillae, 213
chromosomes, 290–94
circadian rhythms, 78, 248–49
circannual rhythms, 249–50
cladistics, 289–90
clicks, 204
cochlea, 157, 163, 164, 166, 168
cochlear microphonic potential, 164–65
cochlear nucleus, 172
cold-blooded animals, 63
conduction, 63
convection, 63
corpus striatum, 137
cortex, 131–35, 180–81
cranial nerves, 126
Craseonycteridae, 6
cribriform plate, 224–25, 227, 229
cricoid cartilage, 146, 148
cricothyroid muscles, 147, 150, 152
cutaneous glands, 229

Dasypterus intermedius, 267
day roosts, 277–82
delay lines, 172, 182
dentition, 103
Desmodus rotundus
  echolocation, 170, 203, 205
  feeding, 109–10
  menstruation, 239
  nasal cavity, 225
  reproductive cycle, 243
diencephalon, 130–31
diestrus, 241
diet, 109–15, 263–64
digestion, 104, 111, 113–14
digestive organs, 99, 101
Dijkgraf, Sven, 140
directional beaming, 155–56
disc-winged bats, 7
diurnal clock, 78
Doppler effects, 165, 201
Doppler-shift compensation, 201–2
dorsal lateral geniculate, 220
duty cycle, 141

ear
inner, 162–69
middle, 160–62
outer, 156, 158–60
echo colors, 190, 192, 193
echolocation, 4, 140–205
calls, 144
directional beaming, 155–56
disadvantages, 141–42
discovery of, 140–41
evolution of, 142–43
general principles, 141
neural control of vocalization, 152–55
object discrimination, 187–92
in open spaces, 192–205
performance, 178–92
range determination, 179–84
See also sound
ecology, 262–85
diet, 263–64
geographic distribution, 262–64
Ectophylla alba, 280, 281
efferent pathways, 175–77
Eidolon helvum, 123, 246, 273
elevation, 185
Emballonuridae, 6
embryonic diapause, 247
encephalization, 117–18, 119, 133
endorphins, 81
energy balance, 102, 104–9
and basal metabolic rate, 71
in frugivorous bats, 115
in nectar-feeding bats, 114–15
in vampire bats, 112
Eonycteris spelaea, 268, 273–74
epiphysis, 249
epithalamus, 130
Epomophorus franqueti, 279
Epomops dobsoni, 239
Eptesicus fuscus
body temperature, 70
echolocation, 152, 170
hibernation, 75
in human settlements, 283
migration, 266
nursery colonies, 258
reproduction, 241
subglottic pressures, 148
vision, 216
vocal membrane, 150
water balance, 83, 89
Eptesicus pumilis, 161
erythrocytes. *See* red blood cells
estrus, 241
ethmoturbinal recess, 224
ethmoturbinate region, 224–26
evaporation, 63, 82–84, 90
evening bats, 7
See also specific species
excretion, 86–94
extrapyramidal motor system, 136–37
eye(s). *See* vision

facial nerves, 126
facial nucleus, 127
false vampires, 4, 6, 205
See also specific species
fasciculi proprii, 121
fastigial nucleus, 129
fatty acids, 73
female genital organs, 237–41
fertilization, 244
fisherman bats, 7, 263
flight
aerodynamics, 24–29
efficiency, 32
energetics, 29–32
evolution of wing, 12
horizontal, 31–32, 33, 35
hovering, 29, 30, 33–36
lift and thrust, 25–27
maneuverability, 32
muscles, 18–21
oxygen extraction during, 57
style and form of wings, 36–40
techniques, 32–40

flocculus, 129
flower-feeding bats, 112–15, 269, 271, 276
  See also nectar-feeding bats
flying foxes, 4, 5, 59
  altruistic behavior, 259
  brain, 127–28
  coevolution with plants, 268, 271, 273–74, 276
  digestion, 113
  echolocation, 144
  eye and vision, 211, 213–18, 220–23
  musky secretion, 230
  somatosensory field, 133–35
  spinal cord, 121, 123
  taxonomy, 295–99
  teeth, 101
  temperature control, 66–67
  See also specific species
follicle-stimulating hormone (FSH), 241, 244
food. *See* diet; digestion
foraging, 39, 143, 145, 191, 192–205
fossils, 287
fovea, 214
free-tailed bats, 8, 100, 233, 288, 291
frequency filter, 162
fruit-eating bats, 91–92, 113–15, 273–76
  See also nectar-feeding bats
FSH. *See* follicle-stimulating hormone
funiculus gracilis, 121
funnel-eared bats, 7
fur coloration, 279
Furipteridae. *See* smoky bats

gas exchange, 55–60
genital organs, 236–41
gills, 63
globular cells, 173
Glossophaga longirostris, 269
Glossophaga soricina, 112–13, 239, 250
glottis, 146, 147, 148
grasping, 22–24
gray matter, 120
Griffin, Donald, 140

habenula, 130
Haplonycteris fischeri, 247
harems, 250–53
head, 48
hearing. *See* ear; echolocation; sound
heart, 43–47, 74
heat, 63–68
hibernation, 46, 70, 72–81
  arousal from, 76–80, 109
  onset of, 74, 76
  physiological status during, 72–74
  regulation of, 80–81
hindbrain, 124–27
hippocampus, 136
Hipposideridae, 6, 185
Hipposideros speoris, 67, 170
homeothermy, 64–68
horizontal flight, 29, 31–32, 33, 35
horizontal localization, 185–87
horseshoe bats, 4, 6, 35
  Doppler-shift compensation, 201–2
  echolocation, 141–42, 185, 194–200
  hibernation, 74, 75
  innervation, 166
  nasolaryngeal tract, 151
  vocalization, 153, 154
  See also specific species
hovering flight, 30, 33–36
human activities, 282–85
human eye, 212, 216
hyperventilation, 57
hypothalamus, 130–31
Hypsignathus monstrosus, 253, 254, 273

Icaronycteris index, 287
impedance matching, 160
infants, 257–59
inferior colliculus, 169–72, 174–75, 183

inferior laryngeal nerve, 153
inhibition, 183
inner ear, 162–69
innervation, 165–66
insecticides, 283–84
insectivorous bats, 40, 88–90, 103–4
See also specific species
interpositus nucleus, 129
intestine, 101–2

Jacobson's organ. *See* vomeronasal organ
jaw, 98, 103, 270

karyotyping, 290–94
kidney, 86–93

lactation, 256–58
laminar flow, 27, 28, 29
larynx, 144, 145–46, 147
Lasiurus borealis, 241, 267
Lasiurus seminolus, 66, 267
latency, 183
lateral cricoarytenoid muscles, 146
lateral geniculate body, 219–21
Lavia frons, 219, 254, 255
law of diffusion, 55, 56
leaf dwellings, 279–81
legs, 19, 21–24
lek system, 253–54
Leydig cells, 236, 241, 247
LH. *See* luteinizing hormone
life span, 93
lift, 25–27
limbic system, 136
Loop of Henle, 86–87
luteinizing hormone (LH), 241

Macroderma gigas, 65, 133–34, 158, 214
Macroglossus minimus, 245
Macrotus californicus, 84, 106–7, 247
male genital organs, 236–37, 238, 296–98
male reproductive cycle, 247–48
mamillary complex, 131
mammals, 9, 10
blood profile, 55
cortex, 131–32
cranial nerves, 126
ear, 157
kidney, 86–88
legs, 19
milk composition, 257
olfactory system, 231
vision, 218, 219, 220
mating. *See* reproduction
mealworm-fed bats, 105
medial geniculate body, 169, 171, 175
medial superior olive (MSO), 173–74
medulla oblongata, 124
Megachiroptera. *See* flying foxes
Megaderma lyra, 170, 188–90, 203, 216, 218, 258, 282
Megadermatidae. *See* false vampires
melatonin, 249
menstruation, 239
mesencephalic nucleus of the trigeminal, 125
mesencephalon, 124
metabolc rate, 70, 71, 73, 105–6
metestrus, 241
Microchiroptera. *See* true bats
midbrain, 127–28
middle ear, 160–62
migrations, 264–67
local, 266
long-distance, 265
orientation during, 266–67
milk teeth, 99
Miniopterus schreibersi, 232, 243, 246, 257
minute volume, 57
molecular biology, 294–95
Molossidae. *See* free-tailed bats
Molossus ater, 174, 186, 239, 258
monaural pathways, 171–73
Mormoopidae. *See* mustached bats
Mormopterus planiceps, 238
mother-infant interactions, 257–59
moths, 204, 205
motor nuclei, 126–27

mouse-tailed bats, 6
MSO. *See* medial superior olive
muscles, 9, 15–21
mustached bats, 4, 7, 200
myelencephalon, 124
Myotis capaccinii, 178
Myotis dasycneme, 266
Myotis daubentoni, 250, 263, 266, 291
Myotis grisescens, 155, 266
Myotis lucifugus
  auditory cortex, 181
  energy expenditure in pregnant, 107
  eyes, 217
  flight membrane, 50
  lack of auditory fovea, 196
  male reproductive cycle, 242
  metabolic rate, 105
  migration, 266
  nursery colonies, 258
  paradoxical latency shift, 183
  taste, 233
  urine production, 88
  "venous heart", 53
  water balance/intake, 83, 84, 89
Myotis myotis
  brain, 128
  echolocation, 193
  hibernation, 76, 80
  in human settlements, 283
  olfaction, 228
  rearing of young, 256
  skeleton, 10
  vision, 216, 218
Myotis sodalis, 65, 212
Myotis velifer, 89–90, 91
Mystacina tuberculata, 281, 295
Mystacinidae, 7
Myzpodidae, 7

nasal chambers, 150
nasolaryngeal tract, 150–51
nasopalatine duct, 230
Natalidae, 7
nectar-feeding bats, 112–15, 269–77
neocortex, 131–32, 135
neopallium, 131–35
neural time windows, 183–84
neurons, 133, 165, 185–86, 203, 219
New World leaf-nosed bats, 4, 7
night roosts, 282
Noctilio albiventris, 243, 244
Noctilio leporinus, 263
Noctilionidae, 7
nocturnal eyes, 211, 212
Norberg, U., 24
nose. *See* olfaction
nuclei, 124
nucleus ambiguus, 153
nucleus marginalis, 124
nucleus of the solitary tract, 126, 233
nursery colonies, 256–57
Nyctalus noctula, 239, 250, 278
Nycteridae, 6
Nycticeius humoralis, 258

object discrimination, 187–92
  echo colors, 190, 192
  material and form, 187–88
  size, 187
  surface structure, 188–90
Old World leaf-nosed bats, 6
olfaction, 224–32
  chemical communication, 229–30
  functional anatomy of nose, 224–29
  morphometric data, 226–27, 228
  thresholds, 228–29
  vomeronasal organ, 230–32
olfactory bulb, 226
olfactory epithelium, 226
optics, 210–14
Oroxylum indicum, 268, 269, 271, 273–75
Otomops martienssi, 100, 103
outer ear, 156, 158–60
ovulation, 241, 244
oxygen, 54, 55–60

paleopallium, 136
papillary aperture, 211
parabrachial nucleus, 126
paradoxical latency shift, 183
paraflocculus, 129
passive hearing, 202–5
passive resonators, 150
pelvis, 11
penis, 237, 296–98
pesticides, 283–84
pheromones, 232
Phyllostomidae, 4, 7
Phyllostomus discolor, 258, 275
Phyllostomus hastatus, 212, 227, 228, 250, 251, 276
phylogeny, 287–88
pineal gland, 130
pinna, 157, 159–60, 185
Pipistrellus ceylonicus chrysothrix, 244
Pipistrellus hesperus, 240
Pipistrellus kuhli, 246
Pipistrellus nathusii, 266, 267
Pipistrellus pipistrellus, 239, 245, 264, 284
Pizonyx vivesi, 88–89, 90
plants, 267–77
Plecotus auritus, 34, 69, 80, 82, 159, 160, 256
Plecotus townsendii, 243, 244
point image, 221
pollination, 272
pons, 124
posterior funiculus, 121, 123
posteroventral cochlear nucleus (PVCN), 172–73
pregnancy, 255–56
premaxilla, 298
pretectal nuclei, 219, 221
prey, 143, 145, 194–95, 202–5, 227
principal nucleus of the trigeminal, 125
proestrus, 241
Pteronotus parnellii
  acoustic fova, 223
  auditory cortex, 179–81
  central acoustic tract, 176
  central nervous system, 122
  echolocation, 150, 165, 174, 179–80, 184, 200
  olivocochlear system, 177
  reproductive cycle, 244
  spinal cord, 123
  vocalization, 153
  vomeronasal organ, 232
Pteropodidae. *See* flying foxes
Pteropus giganteus, 212, 215, 217, 225, 253
Pteropus poliocephalus, 221
Pteropus rodricensis, 259
Pteropus rufus, 239
pterosaurs, 12, 15
PVCN. *See* posteroventral cochlear nucleus
Pye model of sound production, 152, 153
pyramidal tract, 135–36

radiation, 63
Rattus norvegicus, 212
red blood cells, 54
reproduction, 236–59
  asynchrony of male reproductive cycle, 247–48
  control of rhythms, 248–50
  cycles, 241–48
  delayed implantation, 246–47
  delayed ovulation and fertilization, 244
  embryonic diapause, 247
  harems, 250–53
  lek system, 253–54
  mating strategies and food availability, 254–55
  social facilitation, 253
  sperm storage, 245–46
resolution, sound, 142
respiration, 55–60
reticular formation, 124, 125, 127
retina, 214–15
Reynolds number, 27
Rhinolophidae. *See* horseshoe bats

Rhinolophus bocharicus, 227
Rhinolophus euryale, 160
Rhinolophus ferrumequinum
 auditory cortex, 196
 cribriform plate, 227
 echolocation, 155, 167, 202
 ethmoturbinate region, 225
 hair cells in cochlea, 163, 164
 hibernation, 75
 horizontal flight, 35
 larynx, 147
 middle ear, 161
 reproductive cycle, 251
 tuning curve, 168
Rhinolophus hildebrandti, 150–51
Rhinolophus rouxi
 auditory cortex, 181
 echolocation, 141, 195, 197, 200
 reproduction, 246
 retina, 214
Rhinopoma microphyllum, 248
Rhinopomatidae, 6
Rhogeessa tumida, 292–94
roosts, 277–82
Rousettus genus, 91, 144, 210, 223, 230, 295–96

Saccopteryx bilineata, 252
scent. *See* olfaction
sensory nerves, 126
sensory nuclei, 124–26
sheath-tailed bats, 6
shelters. *See* roosts
shivering, 68
short-tailed bats, 7
signal elements, 144, 145
signal frequency, 152
skeleton, 9, 10, 11
slit-faced bats, 6
smell. *See* olfaction
smoky bats, 7
somatic motor nerves, 126
sound
 envelope, 148
 frequency, 143, 167–68, 199
 intensity, 146, 148
 limited field and range, 141–42
 localization, 184–87
 modifications, 150–51
 passive hearing, 202–5
 path in horseshoe bat, 142
 powering emission, 148
 pressure, 143, 148–50
 production, 146–52
 ultrasound, 150
 See also auditory system
Spallanzani, Lazzaro, 140
sperm, 236, 245–46, 298–99
spherical bushy cells, 173
spinal nerves, 120
spinal trigeminal nucleus, 125
spine and spinal cord, 10–11, 120–24
stellate cells, 172
stereocilia, 164
stomach, 101–2
stroke volume, 45–46
subglottic pressure, 146, 148, 149
substantia gelantinosa, 122, 123
substantia nigra, 128
subthalamus, 130
sucker-footed bats, 7
superior colliculus, 127, 171, 219–23
superior laryngeal nerve, 152–53
superior olivary complex, 186
suprachiasmatic nucleus, 219
systematics. *See* taxonomy

Tadarida brasiliensis
 ammonia tolerance, 94
 cave colonies, 282, 283
 crawling, 22
 echolocation, 167, 197
 lactation, 256
 migration, 265
 nursery colonies, 257, 258
Tadarida fulminans, 241
Taphozous longimanus, 239, 240

taste, 233
taxonomy, 4–5, 288–99
teeth, 98–104
tegmentum, 124
telencephalon, 131–37
temperature, body, 66–68, 70, 73, 77
tent-making, 280
territory, 70
thermoneutral zone, 64–66
thrust, 25–27
thyroarytenoid muscles, 146, 147
thyroid cartilage, 146
thyroid hormone, 73
Thyroptera tricolor, 279, 280
Thyropteridae, 7
torpor, 68–70
tracheal chambers, 150
Trachops cirrhosus, 161, 164
tragus, 160, 185
traveling waves, 166, 168
trigeminal nerve, 126
true bats, 4
  brain, 127–28
  digestion, 113
  heart, 46, 47
  penis, 297–98
  spinal cord, 121–24
  taxonomy, 288–89, 296
  teeth, 101
  temperature control, 67
  vision, 210–11, 214, 217, 220–23
tuning curves, 168
turbinate bones, 224–25
twins, 239
Tylonycteris pachypus, 279–81
Tylonycteris robustula, 279–81

ultrasound, 150
urea, 93
uterus, 237–41, 291

vampire bats, 4
  altruism, 112
  digestion, 111
  energy balance, 112
  feeding, 109–11, 264
  harems, 252
  kidney function, 90–91, 92
  locomotion, 24
  nasal cavity, 225
  teeth, 110
  See also specific species
vasoactive intestinal peptide (VIP), 81
veins, 47–48
venous hearts, 52–53
ventral lateral geniculate, 220
ventrolateral thalamus, 130
vertical localization, 185
Vespertilio murinus, 239, 266
Vespertilionidae, 7
vestibular nuclei, 125
VIP. *See* vasoactive intestinal peptide
vision, 21–23
  accommodation, 211–12, 214
  binocular connections, 221–23
  eye size, 210
  functional anatomy of eye, 210–16
  lateral geniculate body, 219–21
  nocturnal eyes, 211, 212
  papillary aperture, 211
  pathways, 219–23
  performance of visual system, 216–19
  refractive power, 211, 212
vocal cords, 152
vocalization, neural control of, 152–55
vocal membranes, 150
vomeronasal organ, 230–32
vomeropalatine duct, 231
vortex theory, 24–25

warm-blooded animals, 63
water
  balance, 82–94
  excretion, 96–94
  intake, 84–86
white matter, 121
wing, 11–21
  circulation in membrane, 49–53

downstroke and upstroke, 18, 21, 148
during horizontal flight, 33
evolution of, 12
extension of, 12–17
folding of, 14–15
form of and flight style, 36–40
measurements, 38
membrane, 11–12
shape, 37
skeletal structure of, 11
See also flight
wingbeating insects, 194–202
wing brain, 124

CPSIA information can be obtained at www.ICGtesting.com
Printed in the USA
LVOW102301191012

303722LV00001B/9/P

9 780195 099515